AF780882

Fine Chemicals from Natural Sources with Potential Application in the Cosmetic/Pharmaceutical Industry

Fine Chemicals from Natural Sources with Potential Application in the Cosmetic/Pharmaceutical Industry

Editor

Agnieszka Feliczak-Guzik

MDPI

Basel • Beijing • Wuhan • Barcelona • Belgrade • Novi Sad • Cluj • Manchester

Editor
Agnieszka Feliczak-Guzik
Faculty of Chemistry
Adam Mickiewicz University
in Poznan
Poznan
Poland

Editorial Office
MDPI AG
Grosspeteranlage 5
4052 Basel, Switzerland

This is a reprint of articles from the Special Issue published online in the open access journal *Cosmetics* (ISSN 2079-9284) (available at: https://www.mdpi.com/journal/cosmetics/special_issues/9Y3MJB8648).

For citation purposes, cite each article independently as indicated on the article page online and as indicated below:

Lastname, A.A.; Lastname, B.B. Article Title. *Journal Name* **Year**, *Volume Number*, Page Range.

ISBN 978-3-7258-1585-2 (Hbk)
ISBN 978-3-7258-1586-9 (PDF)
doi.org/10.3390/books978-3-7258-1586-9

© 2024 by the authors. Articles in this book are Open Access and distributed under the Creative Commons Attribution (CC BY) license. The book as a whole is distributed by MDPI under the terms and conditions of the Creative Commons Attribution-NonCommercial-NoDerivs (CC BY-NC-ND) license.

Contents

About the Editor

Agnieszka Feliczak-Guzik

Dr. Agnieszka Feliczak-Guzik, Prof. Associate, obtained her PhD degree in 2012 and DSc (Habilitation) on the synthesis, characterization, and nanoporous properties of materials in 2019 from Adam Mickiewicz University (AMU, Poznan, Poland). She is currently an employee of the Faculty of Chemistry of AMU. Her scientific interests focus on the synthesis, characterization, and applications of porous materials, such as in the pharmaceutical and/or cosmetic industries.

cosmetics

MDPI

Editorial

Fine Chemicals from Natural Sources with Potential Application in the Cosmetic/Pharmaceutical Industry

Agnieszka Feliczak-Guzik

Faculty of Chemistry, Adam Mickiewicz University in Poznań, Uniwersytetu Poznańskiego 8, 61-614 Poznań, Poland; agaguzik@amu.edu.pl; Tel.: +48-618291747

There is no doubt that chemistry has provided countless products that have improved people's lives in almost every aspect. Unfortunately, some of these products are proving difficult to dispose of, are not biodegradable or are toxic. As a result, there is an increasing emphasis on reducing the number of synthetic products or additives and replacing them with natural products or additives [1]. Through the use of physical, chemical and biochemical processes, these raw materials can be transformed into intermediates that can then be used to produce high-quality chemicals, polymers, lubricants, solvents or surfactants. Significant effort has been also devoted to developing new processes to replace those that previously used fossil fuels.

The pharmaceutical, cosmetic or food industries are particularly interested in natural raw materials, as natural sources have long been a reservoir of fine chemicals. Natural products are an interesting and largely untapped source for developing potential new cosmetic ingredients. In some parts of the world, natural products still form the basis for the production of pharmaceuticals and cosmetics. In developed countries, the interest in pharmaceuticals/cosmetics based on plant-based raw materials has been increasing for many years [2]. As a result, there is also a growing global market demand for natural cosmetic ingredients such as plant extracts that can be used for depigmentation, anti-aging treatments, and other cosmeceutical applications [3–5].

Eleven manuscripts (eight articles and three reviews) were submitted for consideration to be published in the Special Issue "Fine Chemicals from Natural Sources with Potential Application in the Cosmetic/Pharmaceutical Industry", all of which were subjected to the *Cosmetics* review process. These contributions are briefly summarized below:

In the first contribution (1), the herb *Serratula coronata* was shown to be a valuable source of bioactive phytoecdysteroids that could be used for the development of skin care formulations. The results indicated that the cosmetics containing *S. coronata* extract were chemically and microbiologically stable, which contributed to their safety. Their effectiveness was due to the transdermal permeability of 20-hydroxyecdysone. In addition, the extract from *Serratula coronata* can support the treatment of various inflammatory skin diseases.

The authors of the second contribution (2) have taken advantage of the fact that environmental conditions provide Dead Sea Water (DSW) with a unique composition of ions in concentrations that produce comprehensive positive effects on skin health. The authors reviewed two potential modes of action of DSW, as well as the biological function of DSW and its associated complex in dermatology and skin care. The paper also describes the improvement in chronic skin diseases and the skin care efficacy of DSW and related complexes. In particular, these complexes can counteract skin aging in three different ways (promoting keratinocyte rejuvenation, photoprotection and raising cellular energy), indicating their great potential for application in the development of anti-aging cosmetics.

The third contribution (3) presents the possible, available and most commonly used methods for obtaining encapsulated fragrances and aromas, which can then be used in various industries. The paper points out the advantages and disadvantages of each of the

Citation: Feliczak-Guzik, A. Fine Chemicals from Natural Sources with Potential Application in the Cosmetic/Pharmaceutical Industry. *Cosmetics* **2024**, *11*, 67. https://doi.org/10.3390/cosmetics11030067

Received: 25 April 2024
Accepted: 29 April 2024
Published: 30 April 2024

Copyright: © 2024 by the author. Licensee MDPI, Basel, Switzerland. This article is an open access article distributed under the terms and conditions of the Creative Commons Attribution (CC BY) license (https://creativecommons.org/licenses/by/4.0/).

described encapsulation methods, in order to facilitate the selection of the appropriate technology for the production of encapsulated fragrances and aromas.

Paper (4) is focused on the development of a topical formulation with the potential to relieve arthritis complaints. The authors tested a combination of two active ingredients, icariin from *Epimedium* L. extract as a potential promoter of chondrogenesis and glucosamine sulfate as a precursor of cartilage tissue. Permeation studies confirmed the skin permeation potential of both substances; however, the in vitro release test did not accurately reflect the degree of skin permeation.

The authors of paper (5) focused on the fact that annual wine production is accompanied by the generation of large amounts of residues, which are often disposed of and not valorized. To date, various studies have been conducted on grape pomace, but less attention has been paid to other residues, such as wine lees and diatomaceous earth, which is used in wine filtration. Twenty-nine phenolic compounds, including twelve anthocyanins, were tentatively identified in the residues, with red grape pomace showing the greatest diversity of compounds. Diatomaceous earth showed the highest content of non-anthocyanin phenolic compounds, being particularly rich in flavan-3-ols and myricetin-O-hexoside, as well as two anthocyanins.

The authors of paper (6) demonstrated the potential of olive pomace as a source of an innovative ingredient (semi-solid paste) for use in facial mask formulations, which can be used as a sustainable source of both lipids and polar bioactive compounds. The lipid fraction of this ingredient is a source of antioxidants (such as vitamin E) and water-soluble compounds (such as phenols) that protect the skin from oxidative stress.

Paper (7) describes the properties of prepared oil-in-water (O/W) and water-in-oil (W/O) emulsions containing vitamins A, C and E at concentrations of 0.5, 1 and 2% by weight. Their pH and viscosity stability during storage were determined, as were the sunscreen and antioxidant properties of the obtained emulsions.

In paper (8), the authors confirmed the photoprotective properties of naringenin in human primary skin cells and demonstrated its ability to protect against pollution-induced skin damage by inhibiting MMP1, as well as CYP1A1. Combined with naringenin's ability to reduce pigmentation, it counteracts some of the major external features of skin aging caused by UVB radiation and pollution.

In paper (9), it was shown that bio-wax derived from hydroprocessing of crude palm oil shows great potential as a cosmetic ingredient. It can be used as an emollient, since it modifies the sensory properties of the developed formulations, and also as a thickening agent, since viscosity and lubricity showed a high dependence on the percentage of bio-wax used in the formulations studied.

The authors of paper (10) reviewed the antioxidant potential of catechins for use in cosmetic formulations and described the current status of clinical trials on catechins in cosmetics.

In paper (11), the overarching goal of the study was to evaluate the cytotoxic and antiproliferative activity of two herbal extracts of *Haberlea rhodopensis* Friv. and to compare the studied effects with those observed for the reference anticancer, non-selective compound doxorubicin. The authors observed a decrease in the inhibitory activity of both extracts compared to that of doxorubicin against all tested cell lines.

In conclusion, this Special Issue, "Fine Chemicals from Natural Sources with Potential Application in the Cosmetic/Pharmaceutical Industry", provides an up-to-date perspective on the use of natural products in the cosmetic/pharmaceutical industry.

Conflicts of Interest: The author declares no conflict of interest.

List of Contributions:

1. Kroma, A.; Feliczak-Guzik, A.; Pawlaczyk, M.; Osmałek, T.; Urbańska, M.; Micek, I.; Nawrot, J.; Gornowicz-Porowska, J. Analysis of Cosmetic Products Containing Serratula coronata Herb Extract. *Cosmetics* **2023**, *10*, 18.

2. Dai, D.; Ma, X.; Yan, X.; Bao, X. The Biological Role of Dead Sea Water in Skin Health: A Review. *Cosmetics* **2023**, *10*, 21.
3. Kłosowska, A.; Wawrzyńczak, A.; Feliczak-Guzik, A. Microencapsulation as a Route for Obtaining Encapsulated Flavors and Fragrances. *Cosmetics* **2023**, *10*, 26.
4. Pikosz, K.; Nowak, I.; Feliczak-Guzik, A. Potential of Icariin–Glucosamine Combination in the Treatment of Osteoarthritis by Topical Application: Development of Topical Formulation and In Vitro Permeation Study. *Cosmetics* **2023**, *10*, 36.
5. Duarte, C.N.; Taofiq, O.; Dias, M.I.; Heleno, S.A.; Santos-Buelga, C.; Barros, L.; Amaral, J.S. Chemical Characterization and Bioactive Properties of Wine Lees and Diatomaceous Earth towards the Valorization of Underexploited Residues as Potential Cosmeceuticals. *Cosmetics* **2023**, *10*, 58.
6. Rodrigues, R.; Lobo, J.C.; Ferreira, D.M.; Senderowicz, E.; Nunes, M.A.; Amaral, M.H.; Alves, R.C.; Oliveira, M.B.P.P. Chemical and Rheological Characterization of a Facial Mask Containing an Olive Pomace Fraction. *Cosmetics* **2023**, *10*, 64.
7. Bikiaris, N.D.; Koumentakou, I.; Hatzistamatiou, K.; Lykidou, S.; Barmpalexis, P.; Nikolaidis, N. Preparation and Investigation of the SPF and Antioxidant Properties of O/W and W/O Emulsions Containing Vitamins A, C and E for Cosmetic Applications. *Cosmetics* **2023**, *10*, 76.
8. Österlund, C.; Hrapovic, N.; Lafon-Kolb, V.; Amini, N.; Smiljanic, S.; Visdal-Johnsen, L. Protective Effects of Naringenin against UVB Irradiation and Air Pollution-Induced Skin Aging and Pigmentation. *Cosmetics* **2023**, *10*, 88.
9. Aguilar, L.; Hernández, J.; López-Giraldo, L.J.; Mercado, R. Effect of Incorporating a Biowax Derived from Hydroprocessing of Crude Palm Oil in a Facial Cream and a Blemish Balm Cream. *Cosmetics* **2023**, *10*, 123.
10. Mita, S.R.; Husni, P.; Putriana, N.A.; Maharani, R.; Hendrawan, R.P.; Dewi, D.A. A Recent Update on the Potential Use of Catechins in Cosmeceuticals. *Cosmetics* **2024**, *11*, 23.
11. Peeva, M.I.; Georgieva, M.G.; Balacheva, A.A.; Pavlov, A.; Tzvetkov, N.T. In Vitro Investigation of the Cytotoxic and Antiproliferative Effects of Haberlea rhodopensis Total Extract: A Comparative Study. *Cosmetics* **2024**, *11*, 46.

References

1. Nieto, G.; Martínez-Zamora, L.; Peñalver, R.; Marín-Iniesta, F.; Taboada-Rodríguez, A.; LópezGómez, A.; Martínez-Hernández, G.B. Applications of Plant Bioactive Compounds as Replacers of Synthetic Additives in the Food Industry. *Foods* **2024**, *13*, 47. [CrossRef] [PubMed]
2. Pauzi, N.A.M.; Cheema, M.S.; Ismail, A.; Ghazali, A.R.; Abdullah, R. Safety assessment of natural products in Malaysia: Current practices, challenges, and new strategies. *Rev. Environ. Health* **2021**, *37*, 169–179. [CrossRef] [PubMed]
3. Nadeeshani Dilhara Gamage, D.G.; Dharmadasa, R.M.; Chandana Abeysinghe, D.; Saman Wijesekara, R.G.; Prathapasinghe, G.A.; Someya, T. Global Perspective of Plant-Based Cosmetic Industry and Possible Contribution of Sri Lanka to the Development of Herbal Cosmetics. *Evid.-Based Complement. Altern. Med.* **2022**, *2022*, 9940548. [CrossRef] [PubMed]
4. Xie, M.; Jiang, Z.; Lin, X.; Wei, X. Application of plant extracts cosmetics in the field of anti-aging. *J. Dermatol. Sci. Cosmet. Technol.* **2024**, 100014. [CrossRef]
5. Michalak, M. Plant Extracts as Skin Care and Therapeutic Agents. *Int. J. Mol. Sci.* **2023**, *24*, 15444. [CrossRef]

Disclaimer/Publisher's Note: The statements, opinions and data contained in all publications are solely those of the individual author(s) and contributor(s) and not of MDPI and/or the editor(s). MDPI and/or the editor(s) disclaim responsibility for any injury to people or property resulting from any ideas, methods, instructions or products referred to in the content.

cosmetics

MDPI

Review

Microencapsulation as a Route for Obtaining Encapsulated Flavors and Fragrances

Agnieszka Kłosowska [1,2], Agata Wawrzyńczak [1] and Agnieszka Feliczak-Guzik [1,*]

[1] Faculty of Chemistry, Adam Mickiewicz University, Poznań, Uniwersytetu Poznańskiego 8, 61-614 Poznań, Poland
[2] Allsenses Poland, Święty Marcin 29/8, 61-806 Poznań, Poland
* Correspondence: agaguzik@amu.edu.pl

Abstract: Microencapsulation methods for active substances, such as fragrance compounds and aromas, have long been of interest to researchers. Fragrance compositions and aromas are added to cosmetics, household, and food products. This is often because the choice of a particular product is dictated by its fragrance. Fragrance compositions and aromas are, therefore, a very important part of the composition of these items. During production, when a fragrance composition or aroma is introduced into a system, unfavorable conditions often exist. High temperatures and strong mixing have a detrimental effect on some fragrance compounds. The environments of selected products, such as high- or low-pH surfactants, all affect the fragrance, often destructively. The simple storage of fragrances where they are exposed to light, oxygen, or heat also has an adverse effect. The solution to most of these problems may be the encapsulation process, namely surrounding small fragrance droplets with an inert coating that protects them from the external environment, whether during storage, transport or application, until they are in the right conditions to release the fragrance. The aim of this article was to present the possible, available and most commonly used methods for obtaining encapsulated fragrances and aromas, which can then be used in various industries. In addition, the advantages and disadvantages of each method were pointed out, so that the selection of the appropriate technology for the production of encapsulated fragrances and aromas will be simpler.

Keywords: microencapsulation; fragrances; flavors; encapsulation technology

Citation: Kłosowska, A.; Wawrzyńczak, A.; Feliczak-Guzik, A. Microencapsulation as a Route for Obtaining Encapsulated Flavors and Fragrances. *Cosmetics* **2023**, *10*, 26. https://doi.org/10.3390/cosmetics10010026

Academic Editor: Enzo Berardesca

Received: 14 December 2022
Revised: 13 January 2023
Accepted: 19 January 2023
Published: 29 January 2023

Copyright: © 2023 by the authors. Licensee MDPI, Basel, Switzerland. This article is an open access article distributed under the terms and conditions of the Creative Commons Attribution (CC BY) license (https://creativecommons.org/licenses/by/4.0/).

1. Introduction

Grain, bird eggs, or cells in membranes are examples of encapsulation found in nature, which protect the inside of the material, lengthen and facilitate the process of storage or transportation, and finally, protects the contents from the external environment. The first market application of encapsulation was reported in 1957 [1–3]. Since then, there has been a continuous development of knowledge and an increase in the use of encapsulation in many different industries, such as agriculture (pesticides), dietary supplements (vitamins and fish oil), food (flavorings, essential oils, lipids and dyes) and cosmetics (textiles and the fragrance industry) [4,5]. There are many different encapsulation techniques. Choosing the right technique will depend on several factors, including the size of the encapsulates, the chemical structure of the coating, its biodegradability, availability, and price, the final application, and most importantly, the core material that will be encapsulated. Fragrance compositions and flavors are mixtures of organic fragrance compounds, which can include naturally occurring compounds, such as essential oils or resins, that are synthetically produced, but have an equivalent in nature. They are created by several thousand chemical compounds, belonging to different types of organic compound classes, including alcohols, hydrocarbons, esters, aldehydes, ketones, lactones, terpenes, and others, as well as artificial compounds that are unidentifiable in nature, such as musk fragrances [6,7]. Usually, unstable chemical compounds, which have a high tendency to evaporate and are volatile, sensitive to light, heat, and the external environment, require additional protection [8,9]. To protect them, as well as enhance the organoleptic sensation in the product, they are applied in the encapsulated form.

The main purposes of encapsulation are as follows: (i) immobilization of the active material by encapsulating it, (ii) protection, including separating the core from the destructive influence of the environment, (iii) controlled release of the core material, so that exposure to the active material is prolonged, (iv) structure change, obtaining a solid from a liquid or gas, and (v) functionality.

Encapsulation of flavors and fragrances has a number of benefits, including the following [10]:

1. Extended shelf life;
2. Improved stability during processing and in the final product, with a change in the structure from liquid to solid; liquidity, dispersibility, and dosage accuracy in the final product are improved;
3. Gradual, controlled release of aroma compounds, prolonging exposure to odor or taste;
4. Masking of taste and odor;
5. Protection from external factors, separation of chemically unstable and highly volatile substances from environmental factors, protection from UV radiation, degradation reactions, from heat, oxidation, and dehydration;
6. Improved safety by reducing the flammability of volatile substances.

2. Morphological Characterization of Capsules

Encapsulation is a process in which a micron-sized particle of the main substance is coated and surrounded by a wall/carrier material that insulates and protects the contents from the external environment [11–13]. The resulting encapsulation product is called a capsule or encapsulate. The capsules can be divided into the following two parts: the outer, inert layer, usually called the shell, the coating, the carrier material, the wall material, the matrix or membrane, and the inner, active layer called the core, the active agent, the fill, the internal phase or the payload phase (Figure 1) [3,14]. The resulting microcapsule can be of various shapes and sizes, can be regular and irregular in shape, can be a small spherical sphere, a crystal, a jagged adsorbent particle, an emulsion, a suspension of solids or a suspension of smaller microcapsules [1]. Most of the resulting encapsulates are of the order of magnitude from a few micrometers to a few millimeters. Depending on the physical state and chemical type of the material (Table 1) to be encapsulated, as well as the subsequent use and the application of the encapsulation, different types of techniques are possible for their manufacture [4,10]. Table 1 shows the type of shell and core materials [11,15,16].

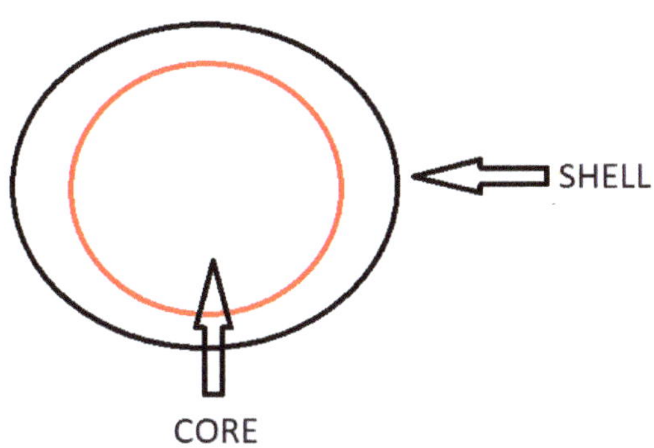

Figure 1. The general structure of the capsule, adapted from [10].

Table 1. Type of shell and core materials [11,15,16].

SHELL MATERIAL	CORE MATERIAL
• Gums: gum arabic, sodium alginate and carrageenan • Carbohydrates: starch, dextran and sucrose • Celluloses: carboxymethylcellulose and methylcellulose • Lipids: beeswax, stearic acid and phospholipids • Proteins: gelatin and albumin	Gas Liquid Solid Forms: solution, dispersion and emulsion
Composition of coating material • Inert polymer • Plasticizer • Coloring agent	Composition of core material • Flavor and fragrance • Drug or active constituent • Additives such as diluents • Stabilizers • Release rate enhancers

2.1. Effect of the Encapsulation Process on Capsule Size

The main factor that affects the stability and efficiency of the encapsulation process is the average particle size. Compared to small particles, large particles allow better protection, but exhibit low dispersion in a matrix such as food. On the other hand, a very small size can cause difficulties in encapsulation efficiency [9].

Based on size, the capsules can be divided into the following three main groups (Figure 2) [17–19]:

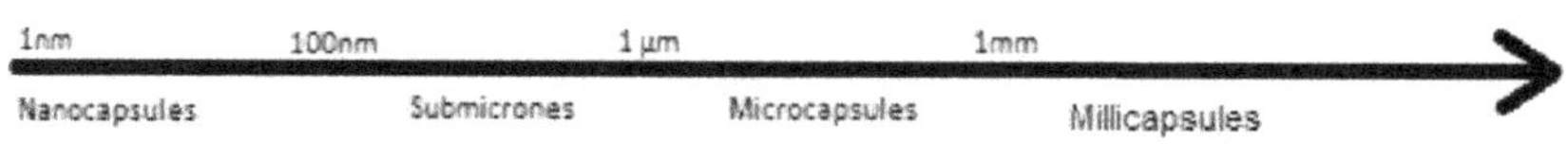

Figure 2. Division of the capsules by size [20].

1. Nanocapsules: <1 µm;
2. Microcapsules: 1 µm–1000 µm;
3. Millicapsules: >1 mm.

The resulting size of the microcapsules (microcapsules are defined as particles between 1 and 1000 µm in size, which contain an active agent—the core—coated with a natural or synthetic shell) depends on the method of obtaining them (the most commonly used methods are described in Section 4.1.), which may include the following: (i) coacervation (2–1200 µm); (ii) emulsion methods (0.5–1000 µm); supercritical liquid precipitation (about 1 µm); melt dispersion (1–50 µm); spray drying (5–5000 µm); coating (5–500 µm); polymerization (0.5–1100 µm); crosslinking (2–20 µm) [21].

2.2. Effect of the Encapsulation Process on Capsule Structure

The structure of the obtained microcapsule is influenced by the type of active ingredient and encapsulating material used in the encapsulation process, as well as the microencapsulation method used. The following structures can be distinguished (Figure 3) [1,4,11,13]:

1. Mononuclear, reservoir type—core/shell capsules and mononuclear encapsulates, in which a single shell is arranged around the core;
2. Polynuclear capsules—contain multiple cores surrounded by the shell;
3. Matrix encapsulation—the core is homogeneously distributed within the shell material. This is currently the most common type of encapsulation used in the pharmaceutical and food industries;
4. Multi-wall—a microcapsule made up of several coatings;
5. Coated matrix type—a combination of the matrix and mononuclear type.

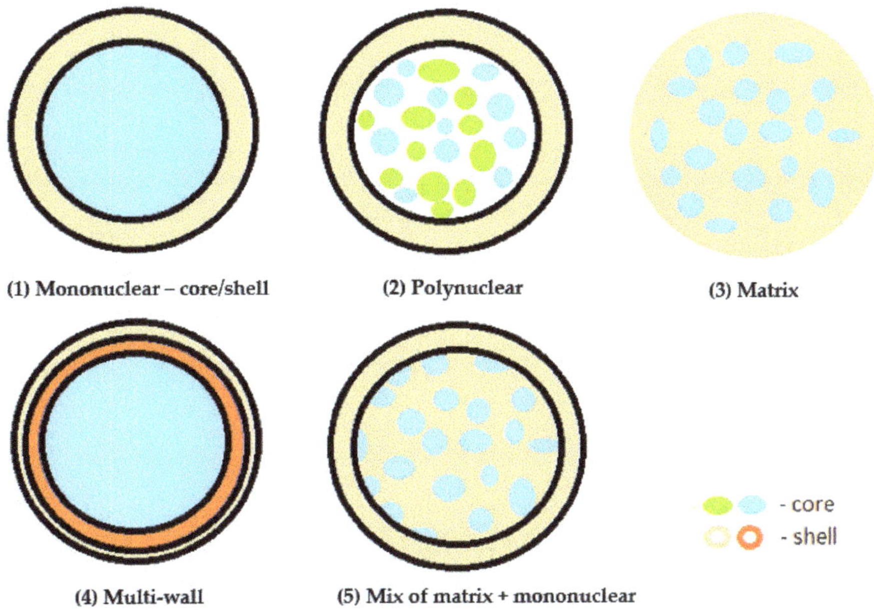

Figure 3. Encapsulate morphology, adapted from, adapted with permission from Ref. [1].

2.3. Selection of the Coating Substance

The determination of the appropriate coating material for the substance to be encapsulated is the most important step in the encapsulation process. The choice of coating material depends on the active ingredient and the desired properties of the final product [22,23]. The structure of the coating material determines the performance of the capsules. An ideal coating material should be characterized by the following: (i) the machinability during encapsulation should be easy; (ii) must display stable emulsifying properties towards active ingredient; (iii) it must not react with the core material during either encapsulation or long-term storage; (iv) it must be able to coat the active substance; and (vi) it must have the ability to be impermeable under processing and long-term storage conditions [2,24].

Due to the chemical properties of the coating, encapsulants are divided into the following categories [15,17]:

1. Polymeric, natural encapsulants, such as gum arabic, alginate, β-glucan, starch, plant protein and gelatin and synthetic encapsulants, e.g., polyesters (poly(lactide-co-glycolide) (PLGA);
2. Inorganic encapsulants, e.g., SiO_2, silica, which is a non-toxic, highly biocompatible, and mechanically stable substance that meets the requirements in pharmacy and biochemistry;
3. Polymers (inorganic).

2.4. Effect of Encapsulation on Prolonging the Aroma Experience

Encapsulation promotes the prolongation of the fragrance experience through the controlled release of fragrance. On the one hand, the volatile chemicals responsible for pleasant fragrances must be protected from evaporation and degradation reactions, ensuring a long shelf life; on the other hand, the key to excellent product performance is the consistent and long-lasting release of fragrance once it is deposited on fabrics or surfaces [25]. A direct consequence of the latter is that less fragrance material can be used in products, which has a positive impact on the environment [26].

Taking into an account the prolonged exposure to taste or odor, and release of aroma compounds under certain conditions, encapsulation can be divided into the following categories:

1. Impermeable sealed encapsulations;
2. Semi-permeable encapsulates;
3. Permeable open encapsulates. The coating on which the material is deposited can be salt or sugar and this process is cheap and sufficient in some cases, but unfavorable when considering the mixture of volatile compounds, as there is no barrier to oxidizing compounds.

3. Physicochemical Characterization of Microcapsules Obtained by Encapsulation Process

When very small capsules are considered, their characterization may concern the resulting powder (the entire mass of the product), rather than a single capsule. In this case, the sample is characterized by determining the (i) particle size distribution, (ii) flow properties, (iii) abrasion resistance, (iv) bulk density and porosity, (v) compressibility, (vi) dust index, and (vii) explosion classification [27]. In turn, the characterization of the resulting particles includes, among others, the size of the obtained capsules, density, abrasion resistance, and efficiency of encapsulation [27].

3.1. Size of the Capsules Obtained in the Encapsulation Process

The size of the capsules affects the various characteristics of the encapsulant, as its size indicates the amount of carrier material from which the encapsulant is constructed. Reactivity, hygroscopicity, and stability are the properties directly related to the size of the encapsulates. Dissolution rate is indirectly related to the size of the obtained capsules, even in combination with its shape and structure. An important characteristic affected by the size of the resulting encapsulates is sedimentation in the dispersion, and in many applications, it is very important to maintain a homogeneous distribution of encapsulates. The smaller the capsule, the greater the viscosity in the dispersion, and thus the slower the movement of the encapsulates in the final sample. In pharmacology, there is a division based on the size of the encapsulates in relation to the speed of delivery of the active material. Smaller capsules are designed for direct, immediate delivery of active ingredients, while larger capsules are designed for delayed delivery. The size of the encapsulates also affects the structure of the food product, and the addition of large particles is not recommended, as they will be felt during eating [16,28].

For spherical encapsulates (Figures 1 and 3), their size is determined from the length of the diameter, while for encapsulation products with irregular shapes, several additional parameters must be considered (Figure 4) [27].

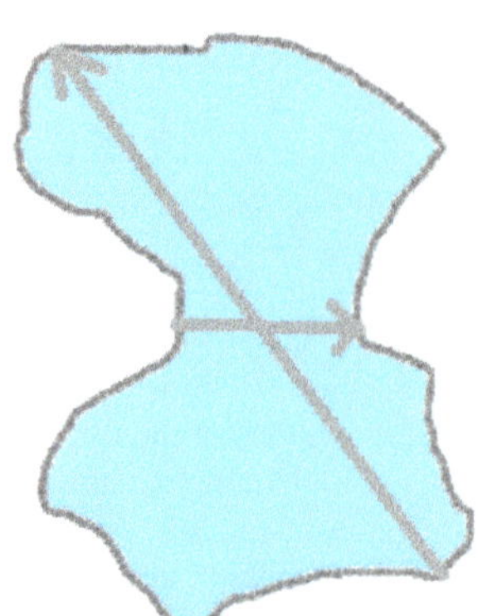

Characteristics of irregular particles:

(1) Weight;
(2) Maximum length;
(3) Minimum length;
(4) Surface area;
(5) Sedimentation velocity;
(6) Siving size;
(7) Projected area;
(8) Volume.

Figure 4. Irregular particle characteristics, adapted from [27].

The size of the encapsulates can be measured by techniques such as laser scattering [29] in the particle size range of 0.1 μm to 3 μm, where light from a coherent light source, such as a laser beam, scattered on particles creates a characteristic intensity distribution that depends on the scattering angle. The exact pattern of this distribution essentially depends on the size of the particle. In this method, the spherical shape of the particle is assumed. The particle size, shape, and morphology image can be obtained using SEM (scanning electron microscopy), where the resulting magnification image of the encapsulates is analyzed using image analysis software. In the case of encapsulates made of a coating of natural polymers, which are not conductive and, in effect, cause image distortion, they are covered with a conductive layer, such as a layer of gold [30]. SEM imaging is designed for encapsulations in the size range of 0.002 μm to 1 μm. In these techniques, precise knowledge of the shape of the encapsulant is needed, and the result is a digit [27]. In order to determine the size percentage of the encapsulates in a given range, mass evaluation techniques should be performed, such as centrifugal sedimentation, giving results in the size range of 0.01 μm to 10 μm, flow classification (gravity, centrifugal and cyclonic), in the range of 0.3 μm to 100 μm, sedimentation from 2 μm to 100 μm and sieving [31] in the range of 5 μm to 4000 μm. However, these techniques are designed for capsules of larger sizes. High-resolution imaging with electron microscopy or confocal laser scanning microscopy (CSLM) is useful for detailed examination of the microcapsule's morphology. CSLM can be combined with staining techniques to gain more insight into the characteristics and distribution of the hydrophobic core and hydrophilic shell [27].

The size distribution of the capsules is a determinant of proper encapsulation. The smaller the range of particle size distribution, the more efficient the process.

3.2. Density

When the presence and location of the pores in the capsules' shell material are taken under consideration, we can distinguish the following three types of density: (i) the true particle density (Equation (1)), which takes into account the open and closed pores in the encapsulating material, (ii) the apparent density (Equation (2)), which only takes into account the open pores of the encapsulating material, and is mostly determined using a pycnometer, where the gas passing through the encapsulant only recognizes the open pores (those that are outside the shell material) and (iii) the effective particle density (Equation (3)), which does not take into account the pores [27] and is also described as bulk density [32–34].

- True particle density

$$\text{True particle density} = \text{Mass of particle (g)}/\text{Volume excluding open and closed pores of particle material (ml)} \quad (1)$$

Equation (1). True particle density equation [27].

- Apparent particle density

$$\text{Apparent particle density} = \text{Mass of particle (g)}/\text{Volume excluding only open pores of particle material (ml)} \quad (2)$$

Equation (2). Apparent density determined using a picnometer [27,34].

- Effective particle density

$$\text{Effective particle density} = \text{Mass of particles (g)}/\text{Volume of particles (ml)} \quad (3)$$

Equation (3). Calculated from the relation of mass (g) and volume (mL) of a known quantity of the sample placed into a graduated cylinder (volume includes the open and closed pores of the particle material) [27,32,34].

- Porosity, €

$$€ = 1 - \frac{\text{Effective partice density}}{\text{Apparent particle density}} \quad (4)$$

Equation (4). Porosity calculations [33].

3.3. Abrasion Resistance

The determination of abrasion resistance is extremely important in the case of transportation, when encapsulates will rub against each other for hundreds of kilometers or when the next stage of production is the mixing of encapsulates, such as with tea. Knowing the abrasion resistance value, one can prevent possible production losses and the destruction of encapsulates. It is also important information for developing appropriate storage or further processing conditions [27].

3.4. Effectiveness and Efficiency of Encapsulation

This can be determined by GC-FID gas chromatography, preceded by extracting the encapsulated aroma compounds. Analysis of the total recovery is based on the chromatographic spectrum, or performed by GC-MS, when analyzing the recovery of individual aroma compounds.

4. Encapsulation Technologies

A number of methods are known for obtaining microcapsules, which include the following: (i) physicochemical methods (coacervation, emulsion methods, precipitation using supercritical fluid and melt dispersion), (ii) physicomechanical methods (spray drying, coextrusion, microfluidization methods, fluidized layer coating and coating in a drag drum), and (iii) chemical methods (polymerization methods and crosslinking method). The selection of the appropriate method for obtaining microcapsules is dictated by the properties of the active substance, the properties of the coating material, and the purpose of the final product. After the encapsulation process, the microcapsules are separated by decantation, filtration and centrifugation, and then washed with distilled water. At the final stage, they are subjected to drying at room temperature or in dryers (freeze-drying). Microcapsules can be used in the form of suspensions or powders. In addition, they can also be coated and used as a substrate for other drug forms [35,36].

As for the technologies used to produce encapsulated flavors and fragrances, the most commercially prevalent technologies are spray drying, extrusion, fluidized bed drying, spray chilling, spray cooling, and coacervation. In each of these technologies, the resulting capsules will vary in size, shape, type of release, or stability. All of these technologies have their own advantages and disadvantages [10].

In simple terms, the encapsulation process can be divided into four main stages, which are summarized in points I–IV.

I. Introduction of a core material, which will then be surrounded by a coating [21]. The active material can be either of the following:
    a. Liquid core:
        - Solution or melt;
        - Emulsion;
        - Suspension.
    b. Solid core (powder).

    Fragrance compositions and flavors are mostly liquids at room temperature.

II. Dispersion stage, where many different technologies are used to produce microcapsules, including the following [1,4,8,11,21]:
    a. Spraying;
    b. Dripping;
    c. Emulsification;
    d. Spray coating;
    e. Formation of suspension coating;
    f. Extrusion.

III. The proper process of encapsulation, which can be divided into the following three groups, in terms of the transformations that take place [13]:
   a. Physical methods:
      - Solidification;
      - Evaporation.
   b. Physical and chemical methods:
      - Gelation;
      - Coacervation.
   c. Chemical methods:
      - Polymerization;
      - Cross-linking.

IV. Scale-up and down processing [21].

4.1. The Most Common Methods Used to Encapsulate Flavors and Fragrances

Physicomechanical methods

4.1.1. Spray Drying

This is an extremely widely used process in the food industry, for example, in the production of milk powder, but also in the pharmaceutical and cosmetic industry [37]. Due to the widespread use of the process and the availability of apparatus, it has found application in encapsulation [12]. The general principle of the spray drying process is to dissolve and emulsify the core substance in an aqueous solution of the carrier material, and then atomize the mixture in a hot chamber, where smaller water molecules evaporate and a coating forms around the active particle. However, highly volatile fragrance compounds evaporate faster than water, so it is important to choose the right carrier to prevent the loss of volatile compounds, while allowing water to evaporate. When selecting a suitable carrier, the following parameters should be considered [1,4,10]:

- Good solubility in water;
- Good emulsifying properties;
- Low viscosity at high concentrations (<500 cps at >45% concentration);
- Low hygroscopicity;
- Taste and/or odor release under the right conditions;
- Low-cost and accessible material;
- Neutrality in taste;
- Stability.

The most commonly used carriers include hydrophilic polysaccharides, such as maltodextrins, chitosan, alginates, and various types of gums, and proteins, such as whey protein. The ratio of the carrier material to the main material is usually 4:1 [38].

The spray drying process can generally be divided into the following three stages (Figure 5) [21,39]:

1. Preparation of an emulsion or slurry from the main material and carrier. The emulsion is usually formed at high mixing speeds or using colloid mills, under high pressure. The product is then processed further by various mechanical means, such as high-pressure homogenization, microfluidization, and ultrasonic emulsification. These methods are used to stabilize the emulsion at least for a certain period of time. The viscosity of the emulsion affects the subsequent drying step and moderate values give the best encapsulation results. Emulsions with too high viscosity can clog the feeder nozzle, or settle on the walls of the chamber [27].
2. Atomization and dispersion of the emulsion. The emulsion is pumped into the drying chamber through an atomizer. Various techniques are used to atomize the emulsion and among the most common are high-pressure nozzles and centrifugal wheels. The atomizer separates the emulsions into small droplets (the size of the droplet depends

on the pore size of the atomizer) and sprays into the hot air in the chamber. The following three methods of atomizer air atomization are possible: co-current, counter-current and mixed. For fragrance compounds, co-current air is commonly used. Then, the moisture evaporates from the emulsion droplets, leading to the entrapment of the main material in the coating [38,40,41].

3. Particle collection. This includes the separation of the resulting encapsulates by a cyclone separator and collection in a receiver. The total duration of the encapsulation process is very short, ranging between a few milliseconds and a few seconds [38,40,41].

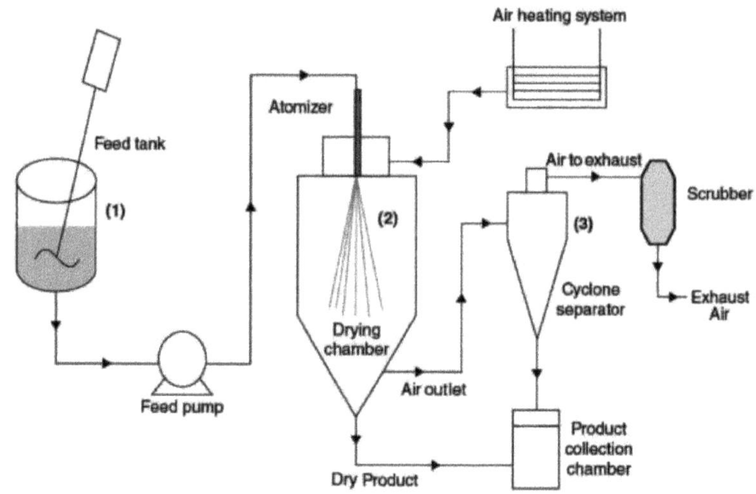

Figure 5. Spray drying technology. (1) Emulsification, (2) atomization and dispersion; (3) particle collection. Adapted from [42].

The finished product of spray drying can contain from 2% to 50% of the active substance. The advantages of spray drying are as follows: the availability of the apparatus, its low cost and good encapsulation results. Complementary to spray drying are product concentration and agglomeration methods. In both cases, the effect is to produce larger encapsulates. In the case of compaction, spray drying products are compressed under high pressure into lumps and crushed into small pieces from 0.7 mm to 3 mm (useful for tea applications). Compaction yields a product with a low porosity. Agglomeration involves fluidizing the spray drying product [41].

4.1.2. Extrusion

Extrusion is a relatively new encapsulation technique, compared to spray drying. Extrusion technology can be divided into the following five types: hot-melt extrusion, melt injection, centrifugal/coextrusion, electrostatic/electrospanning extrusion, and particle from gas saturated solution (PGSS) [43]. Melt injection was the first extrusion technology to be observed and is a vertical screwless extrusion technique, compared to melt extrusion, where, as is the case with most types of extruders, a twin-screw extruder is used [44]. On the other hand, the principle of operation is similar for all of them (Figure 6). The extrusion process involves mixing the active material with the carrier material by melting the carrier material and dispersing it with the core material, then the mixture is pressed through the die into a dehydration fluid, usually isopropyl alcohol, which hardens the coating, encapsulates the core material in the coating, and removes the residue of the active material on the coating. The resulting hard material is broken into small pieces, separated, and dried [45]. The product of extrusion is durable encapsulates that range in size from 250 μm to 2500 μm. The most commonly used carrier material is maltodextrin, simple sugars, starch and a combination of high- and low-molecular-weight carbohydrates, such as maltodextrin, with sucrose, lactose and maltose. The dosage of flavor or fragrance composition ranges from 7.5% to 40%, but in practice, the dosage amounts are much lower, in the range of 8–12% [46].

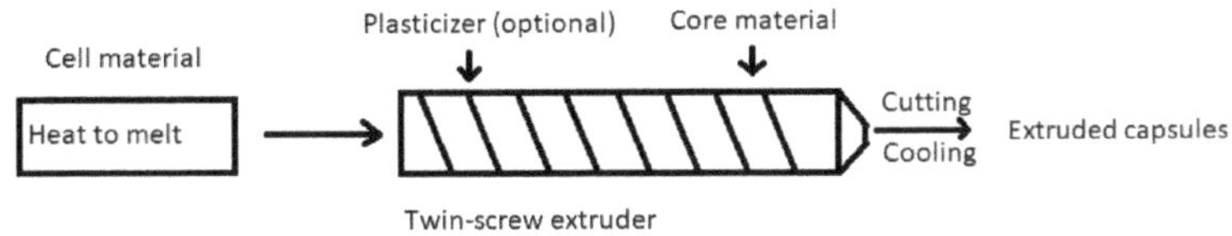

Figure 6. Extrusion process, adapted from [47].

4.1.3. Microfluidization Method

To obtain microcapsules by the aforementioned method, the Microfluidic® device is used, which consists of microchannels that allow the laminar flow of fluids and the production of double emulsions. By using this device, it is possible to obtain uniform sizes of microcapsules and repeatability of the process [48,49].

4.1.4. Pan Coating

Pan coating is used for the microencapsulation of solid particles with a diameter of more than 600 μm. In this apparatus, there are holes on the periphery and a disc that rotates in the direction opposite to the movement of the drum. A solution of the encapsulating material is fed into these holes, while the cores are fed into the center of the disc, from where they are transformed into a layer of the coating substance. If the mass of the core together with the shell reaches a certain value, where the centrifugal force overcomes the cohesion forces of the hole's membrane, the microcapsules are ejected outside the receiver. The solvent used is removed using warm air, while the shells are cured by physical or chemical methods. The advantage of this method is its high production efficiency, in addition to being fast, efficient, and relatively simple [21,50].

Physicochemical methods

4.1.5. Coacervation Method

Coacervation is a physicochemical method and is considered to be the method from which microencapsulation originated. There are two types of coacervation, simple and complex, depending on how the process is carried out [51]. In simple coacervation, there is only one type of polymer, with the addition of a highly hydrophilic agent in the colloidal solution. In complex coacervation, two or more polymers can be used. Complex coacervation relies on the ability of cationic and anionic polymers to interact with each other in water to form a polymer-rich liquid phase, called a complex coacervate. The most common cationic polymer used in this case is gelatin. A variety of natural and synthetic water-soluble polymers react with gelatin to form complex coacervates that are suitable for encapsulation, which occur in equilibrium with a clear liquid on the surface [17,25,52].

The coacervation process involves the separation of phases in a solution of colloids or polymers and the formation of two or more liquid phases as a result of the controlled modification of the solution environment, which includes changing the pH, ionic strength, temperature, solubility and addition of salt, or a countercharged polymer. The coacervation process produces a colloid-rich coacervate phase and an equilibrium phase [53]. The coacervation process (Figure 7) can be divided into the following four steps [54]:

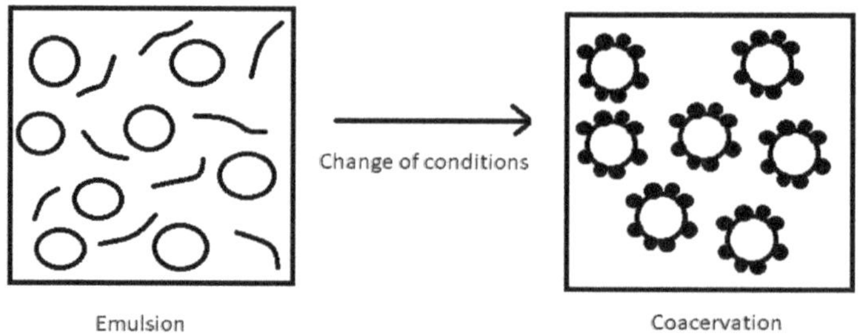

Figure 7. Coacervation scheme, adapted from [54].

1. Preparation of an aqueous solution of two or more polymers;
2. Emulsification of the aqueous phase of the polymers with the hydrophobic phase of the core. The active ingredient should not dissolve in water, as this may lead to losses, reducing the efficiency of the coacervation method (the maximum solubility of the core material in water is 20 mg/mL);
3. Change in the environmental conditions of the solution in order to proceed with coacervation and phase separation. A shell is formed around the core, and the phases are separated into a coacervate phase, with the core material surrounded by a carrier material, and an equilibrium phase (water);
4. Cooling of the system and addition of a cross-linking substance to harden the shell.

In simple coacervation, a single polymer is involved and coacervates are formed by a dehydration or water deficiency mechanism through the addition of salt, or desolvation fluid (coacervating agent or inducer). An example of simple coacervation is the separation of gelatin layers, following the addition of sodium sulfate (Na_2SO_4) or ethanol [53].

Complex coacervation is widely used in the pharmaceutical, food, agriculture, and textile industries. The process is extremely efficient, allowing the formation of 99% encapsulates. Complex coacervation primarily involves the use of two biopolymers with opposite charges, mainly proteins and polysaccharides, which can form a complex shell for the core material. The proteins used can be divided by origin into the following categories [53]:

- Animal: gelatin, whey, egg albumin and silk fibroin;
- Plant-based: soy protein, pea protein, wheat protein, lentil protein and chia protein.

Meanwhile, the most commonly used polysaccharides are alginate, chitosan, gum arabic, pectin, agar, carrageenan and carboxymethylcellulose.

Currently, coacervation is one of the most popular encapsulation methods used in the medicine and food industry.

4.1.6. Emulsion Methods

These methods are among the oldest and most widely used for obtaining microcapsules. They are used to encapsulate compounds that are both soluble and insoluble in water, and substances that are sensitive to external factors, such as temperature. Emulsion methods are based on the formation of single or double emulsions and immiscible liquids that contain an active substance and a polymer. As a result of the solidification of the polymer around the active substance when the solvent is removed, microcapsules are obtained [55–57].

The single-emulsion method involves preparing a solution of the encapsulating polymer in a volatile organic solvent (e.g., methylene chloride, chloroform, and ethyl acetate) that does not mix with water. Polymer solutions may additionally contain the addition of emulsifiers and stabilizers (e.g., Span 80 and Tween 80). The core material (solid form) is then dissolved or dispersed in the prepared polymer solution. The mixture thus obtained is emulsified with the aqueous phase with the addition of emulsifiers and/or stabilizers, resulting in an oil/water-type emulsion. Once the emulsion is obtained, the solvent is evaporated. During the evaporation of the solvent, the polymer settles around the cores and solidifies [58].

On the other hand, there is a two-step process, which usually involves the formation of water/oil/water or water/oil/oil emulsions. In the first stage, active substances in an aqueous medium are emulsified with a polymer solution in an organic solvent. In the next stage, the formed emulsion is added to the second aqueous phase, then a double emulsion of water/oil/water is formed or added to the other oil phase, leading to the water/oil/oil emulsion. These processes are accompanied by the addition of an emulsifier and/or stabilizers. Once the emulsion is formed, the volatile solvent is removed by evaporation [59].

Chemical methods

4.1.7. Polymerization

One of the chemical methods used to obtain encapsulates is the process of polymerization, including both interfacial and in situ polymerization. The former, interfacial polymerization, involves the formation of encapsulates by polymerizing monomers at the interface between the dispersing (water, glycerol, ethyl alcohol, chloroform and cyclohexane) and dispersed (animal fats, vegetable oils, and synthetic oils) phases [60,61]. The dispersion phase is mixed together with the suspended or dissolved active ingredient and the encapsulating monomer, until an o/w emulsion is obtained. The most commonly used monomers are multifunctional isocyanates and multifunctional organic acid chlorides. In addition to the above, methyl methacrylate, vinyl acetate, and a mixture of styrene and vinylbenzene or diamines are also used [50]. Monomers diffuse between the o/w phase, and the reaction results in a polymeric membrane (polyurea, polynylon, or polyurethane) that is insoluble in oil [55]. On the other hand, in situ polymerization, when compared to interfacial polymerization, occurs without the involvement of additional reactive agents. The polymer that is formed initially exhibits a low molecular weight, which increases with time. Subsequently, this polymer is deposited on the core material and forms a solid capsule shell [50].

The advantages and disadvantages of the above-described encapsulation methods are summarized in Table 2.

Table 2. Advantages and disadvantages of selected encapsulation methods [25,50,60].

Method	Advantages	Disadvantages
Spray drying	- Quick and easy process - High process efficiency - Microencapsulation of temperature-sensitive substances - Possibility to obtain sterile products	- Formation of polymorphic varieties of polymers that cause disintegration of microcapsules - Loss of material due to the adhesion of active substance particles to the drying chamber
Extrusion	- Fast and inexpensive process	- Formation of large-size capsules
Microfluidization method	- Small size of microcapsules - Uniformity of shape and size - Repeatability of process conditions	- Organic solvents and surfactants prevent microencapsulation of living cells
Pan coating	- Fast, effective and simple method	- Formation of large-size microcapsules
Coacervation	- Process is simple, fast and reproducible - Possibility of using low concentrations of surfactants - The process can be run in aqueous and anhydrous environments	- Agglomeration of droplets under the influence of solvent removal - Toxicity of the solvents used
Emulsion methods	- Ability to encapsulate hydrophilic and hydrophobic substances - High process efficiency	- Agglomeration of droplets under the influence of solvent removal - Toxicity of the solvents used
Polymerization	- Able to control the size of microcapsules by adding surfactants in low concentrations	- Reaction difficult to control - Requires significant washing to remove monomers, organic solvents and surfactants

5. Application of Encapsulates Containing Flavors and Fragrances

In recent years, scientific work on the application of encapsulation methods has focused primarily on obtaining encapsulates, which are used in the fields of drug delivery, pesticides, food preservation, and cosmetics [62]. Most of the patents that cover this topic area focus on the use of encapsulates in medical, hygiene, food, and cosmetic products [63–66].

Examples of their use are described below.

Microencapsulation is also used in the textile industry to improve the functional properties of fabrics, such as their antibacterial activity, odor reduction, mosquito repellency, UV protection and thermoregulation [67]. Fragrance can be imparted to the fabric directly by applying microencapsulation, using techniques such as impregnation, spraying, coating, or embossing [1]. This type of scenting is applied to products such as scarves, ties, and underwear, as well as home textiles, such as sofa throws, curtains, and aromatherapy pillows. "Scratch and sniff" T-shirts have also been produced [68]. Fragrance is also imparted by washing fabrics in scented washing powders or rinsing in scented softeners [69]. The fragrance microcapsules of the powder or softener adhere to the fabric, and the fragrance is released gradually under mechanical abrasion. Microcapsules are also used for the thermoregulatory function of fabrics. In this case, the encapsulates perform the function of absorbing or donating heat by melting and solidifying, depending on the temperature; these are called phase-change materials (PCMs), or smart materials [70]. Natural essential oils, which, in addition to their sensory properties, also possess antibacterial, antiviral, antifungal, anti-inflammatory, and insect repellent properties [8], are enjoying increasing consumer interest and are widely applied in the textile industry. Essential oils, trapped in microcapsules on clothing, can, after contact with skin, perform skin care, anti-aging, or odor control effects.

In the food industry, stabilized fragrance compounds are applied to tea, significantly extending the shelf life of the product, while maintaining the aroma profile. For instant coffee, the addition of microcapsules improves the product in terms of intensity. Encapsulated essential oils are also used as antifungal agents, or for fruit preservation.

In the cosmetics and household products industry, the possibilities for the use of encapsulates are numerous; products such as creams, shampoos, gels, soaps, surface washing liquids, and fragrances are hidden everywhere. Consumer demands are increasing and interest in natural fragrances is growing, and once the product is used, the scent is expected to remain for a long time. The use of encapsulated fragrance compounds can be a solution to high consumer expectations. On the other hand, the use of essential oils with antimicrobial properties can replace the use of preservatives, which will also be commercially attractive [71]. For aromatherapy purposes, encapsulates are added to the cellulose pulp to create, for example, scented wallpaper or scented wrapping paper.

Table 3 demonstrates examples of the use of the encapsulation process with respect to the aroma and fragrance compositions.

Table 3. Examples of the preparation of encapsulates with aromas and fragrances.

Core Material	Shell Material	Methodology	Goal	Morphology	Ref.
Vanilla oil	Chitosan/arabic gum	Complex coacervation	Controlled release and thermostability product for spice application in food industries	Spherical shape and smooth surface, 94.2% efficiency (VO/CS 2:1)	[72]
Limonene fragrance	Chitosan/cellulose	Freezing/thawing/stirring process	A fragrant component widely used in the flavor and fragrance industries; encapsulation prolonged the release of fragrances	Spherical shapes, with an average diameter of 2 mm, 51.3% efficiency	[73]
Fragrances: D-limonene, Claritone, Amarocit, Rose Oxide-High Cis, methyl salicylate, 1-octanal, 1-octanol, hydrocitronitrile, Majantol and ethyl 2-methylbutanoate	Bovine serum albumin and tetramethylrhodamine isothiocyanate-labeled BSA (TRITC-BSA)	Layer-by-layer (LbL)	Fragrance encapsulation. Controlling the release of fragrances; both TA and BSA are relatively cheap and available compounds	The encapsulation efficiency depends on the water solubility; the less water-soluble the ingredient, the smaller its losses upon LbL coating of emulsion in the filtration cell and the higher its relative content in released fragrance	[74]

Table 3. *Cont.*

Core Material	Shell Material	Methodology	Goal	Morphology	Ref.
Pink fruity fragrances and white floral fragrances	Protein, carbohydrate and lipid	Liquid–liquid dispersion	Fabric softener application and long-lasting property in textile applications	Spherical shape with an average size of 100–300 nm, efficiency of 69–75%	[69]
Tuna oil in water emulsion stabilized by lecithin-chitosan membrane, using an electrostatic layer-by-layer deposition process	Maltodextrin	Spray drying of two-layer emulsion	High oil-loaded microcapsules that may lead to a wide range of applications in food products	Spherical particles (except for oil:maltodextrin 1:1), efficiency of 89% for oil:maltodextrin 1:4 ratio and 56% for ratio of 1:1	[75]
Linoleic acid	Gum arabic or maltodextrin	Spray drying	Evaluation of influence of the encapsulation process on the stability of linoleic acid towards oxidation	Particles with an average size of 0.68 μm (gum arabic) or 1.68 μm (maltodextrin), efficiency of 75–99% (gum arabic) and 35–50% (maltodextrin)	[76]
Ascorbic acid	Maltodextrin	Extrusion	Vitamin encapsulation	Crystals with sharp edges, efficiency of 96% and load 19%	[77]
Orange terpenes	Maltodextrin and sucrose	Extrusion	Flavor encapsulation	Partly crystalline samples, about 1 mm particle size, efficiency of 34.5–67.3%; load 4.1%	[44]
Lemon oil	Sago and tapioca starch, gum arabic and stearic acid	Spray drying	Encapsulated agent for food industry	Efficiency of 49–59%	[78]
Orange oil in water emulsion	Lactose and caseinate	Spray drying	Application in different types of food or pharmaceutical products, where maximum protection for flavors or slow release are required	Particle size of 30.9 μm. Efficiency of 44.5%	[79]
Orange essential oil	Octenyl succinic anhydride, modified starch and maltodextrin	Vacuum spray drying	Application of vacuum spray drying	Efficiency of 99.89%	[80]
Gurum seed oil	Gum arabic, maltodextrin, pullulan and whey protein isolate	Spray drying	Evaluating the potential of combining maltodextrin with gum arabic and whey protein isolate	Efficiency of 97.38%	[81]
Citrus sinensil L. (essential oil)	Maltodextrin	Spray drying	Evaluating the factors affecting microencapsulation	Efficiency of 89.94%	[82]
Juniperus communis L. (essential oil)	Gum arabic, maltodextrin, sodium alginate and whey protein concentrate	Spray drying	Food flavoring agent and preservative	Efficiency of 82.66%	[83]
Rosemary (essential oil)	Maltodextrin and whey protein concentrate	Spray drying	The potential of combined emulsification and spray drying procedures to encapsulate polyphenolic components from rosemary	Efficiency of 27.09–42.93%	[84]
Cinnamon (essential oil)	Gum arabic, maltodextrin and whey protein concentrate	Spray drying	Effect of shell materials used on encapsulation efficiency	Efficiency of 13.8–50.1%	[85]
Syzygium Cumin Seed (essential oil)	Gum arabic	Spray drying	Antioxidant	Improvement of water vapor permeability; prolongation of oil oxidation	[86]

6. Conclusions

Although several encapsulated products have been developed and used in the cosmetic, pharmaceutical, and agribusiness industries, the number of studies on encapsulated fragrances and flavors is relatively small. The growing demand for modern cosmetic, pharmaceutical, agrochemical, or functional textiles and food products that contain flavor compounds in their composition contributes to the great interest in the issue of microencapsulation. "The Global Microencapsulation Report" forecasts a CAGR, namely a compound annual growth rate of 12.9% over 2022–2030. The benefits of encapsulation should overcome as many of the negative effects of processing flavor compounds as possible, such as the following:

- Additional costs;
- Increased complexity of the production process and/or supply chain;
- Unwanted consumer information (visual or tactile) about encapsulated foods;
- Stability of capsules during the processing and storage of the food product [12].

This paper presents the main technologies for obtaining encapsulates, in addition to their characteristics, and examples of applications. The challenge facing researchers will be the selection of a suitable method for encapsulate synthesis and its commercialization, since the final product is influenced by many factors, such as capsule size, carrier material, overall performance of the method, and the cost.

Author Contributions: Conceptualization, A.F.-G. and A.K.; writing—original draft preparation, A.K.; writing—review and editing, A.F.-G. and A.W.; visualization, A.K.; supervision, A.F.-G. and A.W. All authors have read and agreed to the published version of the manuscript.

Funding: This research received no external funding.

Institutional Review Board Statement: Not applicable.

Informed Consent Statement: Not applicable.

Data Availability Statement: Not applicable.

Conflicts of Interest: The authors declare no conflict of interest.

References

1. Saifullah, M.; Islam Shishir, M.R.; Ferdowsi, R.; Tanver Rahman, M.R.; Van Vuong, Q. Micro and nano encapsulation, retention and controlled release of flavor and armoma compounds: A critical review. *Trends Food Sci. Technol.* **2019**, *86*, 230–251. [CrossRef]
2. Desai, K.G.H.; Park, H.J. Recent Developments in Microencapsulation of Food Ingredients. *Dry. Technol.* **2005**, *23*, 1361–1394. [CrossRef]
3. Trojanowska, A.; Giamberini, M.; Tsibranska, I.; Nowak, M.; Marciniak, Ł.; Jastrzab, R.; Tylkowski, B. Microencapsulation in food chemistry. *JMSR* **2017**, *3*, 265–271.
4. Martins, E.; Poncelet, D.; Rodrigues, R.C.; Renard, D. Oil encapsulation techniques using alginate as encapsulating agent: Applications and drawbacks. *J. Microencapsul.* **2017**, *34*, 754–771. [CrossRef] [PubMed]
5. Gulumser, T. The role of microcapsules in masking bad odors of cotton fabrics. *Industria Textila.* **2017**, *68*, 275–282. [CrossRef]
6. Winkler, M.; Kopf, G.; Hauptvogel, C.; Neu, T. Fate of artificial musk fragrances associated with suspended particulate matter (SPM) from the River Elbe (Germany) in comparison to other organic contaminants. *Chemosphere* **1998**, *37*, 1139–1156. [CrossRef]
7. Horst, S.; Johannes, P. *Common Fragrance and Flavor Materials: Preparation, Properties and Uses*; Wiley-VCH: Weinhem, NY, USA, 2016.
8. Perinelli, D.R.; Palmieri, G.F.; Cespi, M.; Bonacucina, G. Encapsulation of flavours and fragrances into polymeric capsules and cyclodextrins inclusion complexes: An update. *Molecules* **2020**, *25*, 5878–5911. [CrossRef] [PubMed]
9. Reis, D.R.; Ambrosi, A.; Di Luccio, M. Encapsulated essential oils: A perspective in food preservation. *Future Food* **2022**, *5*, 100126. [CrossRef]
10. Mishra, M.K. *Handbook of Encapsulation and Controlled Release*; CRC Press: Boca Raton, FL, USA, 2016.
11. Zhu, G.Y.; Xiao, Z.B.; Zhou, R.J.; Yi, F.P. Fragrance and flavor microencapsulation technology. *Adv. Mater. Res.* **2012**, *535–537*, 440–445. [CrossRef]
12. Nedovic, V.; Kalusevic, A.; Manojlovic, V.; Levic, S.; Bugarski, B. An overview of encapsulation technologies for food applications. *Procedia Food Sci.* **2011**, *1*, 1806–1815. [CrossRef]
13. Mamusa, M.; Resta, C.; Sofroniou, C.; Baglioni, P. Encapsulation of volatile compounds in liquid media: Fragrances, flavors, and essential oils in commercial formulations. *Adv. Colloid Interface Sci.* **2021**, *298*, 102544. [CrossRef] [PubMed]
14. Gibbs, F.; Kermasha, S.; Inteaz Al, B. Encapsulation in the food industry: A review. *Int. J. Food Sci. Nutr.* **1999**, *50*, 213–224. [PubMed]
15. Wandrey, C.; Bartkowiak, A.; Harding, S.E. *Materials for Encapsulation. Encapsulation Technologies for Active Food Ingredients and food Processing*; Springer: New York, NY, USA, 2009; pp. 31–100.
16. Kaushik, P.; Dowling, K.; Barrow, C.J.; Adhikari, B. Microencapsulation of omega-3 fatty acids: A review of microencapsulation and characterization methods. *JFF* **2015**, *19*, 868–881. [CrossRef]
17. He, L.; Hu, J.; Deng, W. Preparation and application of flavor and fragrance capsules. *Polym. Chem.* **2018**, *9*, 4926–4946. [CrossRef]
18. Daneshniya, M.; Nezhad, H.J.; Maleki, M.H.; Jalali, V.; Behrouzian, M. A Review of encapsulation of bioactive peptides with antimicrobial and antioxidant activity. *Int. J. Acad. Eng. Res.* **2020**, *4*, 7.
19. Temiz, U.; Öztürk, E. Encapsulation methods and use in animal nutrition. *Selcuk J. Agr. Food Sci.* **2018**, *32*, 624–631. [CrossRef]
20. Poncelet, D.; Oxley, J. Introduction on Microencapsulation. Available online: https://www.youtube.com/watch?v=MNMbki8W1W8 (accessed on 16 October 2020).
21. Ghosh, S.K. *Functional Coatings and Microencapsulation: A General Perspective*; Ghosh, S.K., Ed.; Functional coatings by polymer microencapsulation; John Wiley & Sons: New York, NY, USA, 2006; pp. 12–25.
22. Zuidam, N.J.; Shimoni, E. Overview of microencapsulates for use in food products or processes and methods to make them. In *Encapsulation Technologies for Active Food Ingredients and Food Processing*; Springer: New York, NY, USA, 2010; pp. 3–29.
23. Gökmen, S.; Palamutoğlu, R.; Sarıçoban, C. Encapsulation applications in food industry. *J. Food Technol.* **2012**, *7*, 36–50.
24. Tarhan, Ö.; Gökmen, V.; Harsa, Ş. Applications of nanotechnology in the field of food science and technology. *Nutr. J.* **2010**, *35*, 219–225.
25. Bruyninck, K.; Dusselier, M. Sustainable chemistry considerations for the encapsulation of volatile compounds in laundry-type applications. *ACS Sustain. Chem. Eng.* **2019**, *7*, 8041–8054. [CrossRef]
26. Salvador Cesa, F.; Turra, A.; Baruque-Ramos, J. Synthetic fibers as microplastics in the marine environment: A review from textile perspective with a focus on domestic washings. *Sci. Total Environ.* **2017**, *598*, 1116–1129. [CrossRef]

27. Weissbordt, J. Microcapsule Practical Characterization. Available online: https://www.youtube.com/watch?v=oS2iXHGwMog. (accessed on 21 February 2022).
28. Champagne, C.P.; Fustier, P. Microencapsulation for the improved delivery of bioactive compounds into foods. *Curr. Opin. Biotechnol.* **2007**, *18*, 184–190. [CrossRef]
29. *Standard ISO 13320:2020*; Technical Committee: ISO/TC 24/SC 4 Particle characterization. ICS: 19.120 Particle size analysis. Sieving. January 2020.
30. Rosenberg, M.; Kopelman, I.J.; Talmon, Y. A Scanning electron microscopy study of microencapsulation. *J. Food Sci.* **2006**, *50*, 139–144. [CrossRef]
31. Procedure DIN 66165-1.
32. Getachew, A.T.; Chun, B.S. Optimization of coffee oil flavor encapsulation using response surface methodology. *Food Sci. Technol.* **2016**, *70*, 126–134. [CrossRef]
33. Ferrari, C.C.; Germer, S.P.M.; Alvim, I.D.; Vissotto, F.Z.; de Aguirre, J.M. Influence of carrier agents on the physicochemical properties of blackberry powder produced by spray drying. *Int. J. Food Sci.* **2012**, *47*, 1237–1245. [CrossRef]
34. Comunian, T.A.; da Silva, A.G.; Bezerra, E.O.; Moraes, I.C.F.; Hubinger, M.D. Encapsulation of pomegranate seed oil by emulsification followed by spray drying: Evaluation of different biopolymers and their effect on particle proper. *Food Bioproc. Tech.* **2019**, *13*, 53–66. [CrossRef]
35. Lee, Y.; Ra Ji, Y.; Lee, S.; Choi, M.J.; Cho, Y. Microencapsulation of probiotic *Lactobacillus acidophilus* KBl409 by extrusion technology to enhance survival under simulated intestinal and freeze-drying conditions. *J. Microbiol. Biotechnol.* **2019**, *29*, 721–730. [CrossRef]
36. Pellicer, J.A.; Fortea, M.I.; Trabal, J.; Rodríguez-López, M.I.; Gabaldón, J.A.; Núñez-Delicado, E. Stability of microencapsulated strawberry flavour by spray drying, freeze drying and fluid bed. *Powder Technol.* **2019**, *347*, 179–185. [CrossRef]
37. Altay, Ö.; Köprüalan, Ö.; İlter, I.; Koç, M.; Ertekin, F.K.; Jafari, S.M. Spray drying encapsulation of essential oils; process efficiency, formulation strategies, and applications. *Crit. Rev. Food Sci. Nutr.* **2022**, 1–20. [CrossRef]
38. Costa, S.S.; Machado, B.A.S.; Martin, A.R.; Bagnara, F.; Ragadalli, S.A.; Costa Alves, A.R. Drying by spray drying in the food industry: Micro-encapsulation, process parameters and main carriers used. *Afr. J. Food Sci.* **2015**, *9*, 460–470.
39. Mohammed, N.K.; Tan, C.P.; Manap, Y.A.; Muhialdin, B.J.; Hussin, A.S.M. Spray drying for the encapsulation of oils—A review. *Molecules* **2020**, *25*, 3873–3889. [CrossRef]
40. Patel, R.P.; Patel, M.P.; Suthar, A.M. Spray drying technology: An overview. *Afr. J. Food Sci.* **2009**, *2*, 44–47. [CrossRef]
41. Jafari, S.M.; Assadpoor, E.; He, Y.; Bhandari, B. Encapsulation efficiency of food flavours and oils during spray drying. *Dry. Technol.* **2008**, *26*, 816–835. [CrossRef]
42. Pignatelllo, R.; Musumeci, T. *Biomaterials: Physics and Chemistry—New Edition Edited*; IntechOpen: London, UK, 2018.
43. Bamidele, O.P.; Emmambux, M.N. Encapsulation of bioactive compounds by "extrusion" technologies: A review. *Crit. Rev. Food Sci. Nutr.* **2020**, *61*, 1–19. [CrossRef]
44. Tackenberg, M.W.; Krauss, R.; Schuchmann, H.P.; Kleinebudde, P. Encapsulation of orange terpenes investigating a plasticisation extrusion process. *J. Microencapsul.* **2015**, *32*, 408–417. [CrossRef] [PubMed]
45. Risch, S.J. *Encapsulation: Overview of Uses and Techniques*; ACS Symposium Series: Washington, DC, USA, 1995; pp. 2–7.
46. Porzio, M. Advances in flavor encapsulation. *IFT* **2012**, 6.
47. Reineccius, G.A. *Edible Films and Coatings for Food Applications*; Springer: New York, NY, USA, 2009; pp. 269–294.
48. Rabanel, J.M.; Banquy, X.; Zouaoui, H.; Mokhtar, M. Technology in microencapsulation methods for cell therapy. *Biotechnol. Prog.* **2009**, *25*, 946–963. [CrossRef]
49. Dendukuri, D.; Doyle, P.S. The synthesis and assembly of polymeric microparticles using Microfluidics. *Adv. Mater.* **2009**, *21*, 4071–4086. [CrossRef]
50. Jyothi, N.V.; Prasanna, P.M.; Sakarkar, S.N.; Prabha, K.S.; Ramaiah, P.S.; Srawan, G.H. Microencapsulation techniques, factors influencing encapsulation efficiency. *J. Microencapsul.* **2010**, *27*, 187–197. [CrossRef]
51. Przybysławska, M.; Winnicka, K. Technologies of microencapsulation. *Farm Pol.* **2012**, 283–289.
52. Sagiri, S.S.; Anis, A.; Pal, K. Review on encapsulation of vegetable oils: Strategies, preparation methods, and applications. *Polym. Plast. Technol. Eng.* **2016**, *55*, 291–311. [CrossRef]
53. Timilsena, Y.P.; Akanbi, T.O.; Khalid, N.; Adhikari, B.; Barrow, C.J. Complex coacervation: Principles, mechanisms and applications in microencapsulation. *Int. J. Biol. Macromol.* **2019**, *121*, 1276–1286. [CrossRef]
54. Eghbal, N.; Choudhary, R. Complex coacervation: Encapsulation and controlled release of active agents in food system. *LWT.* **2018**, *90*, 254–264. [CrossRef]
55. Bansode, S.S.; Banarjee, S.K.; Gaikwad, D.D.; Jadhav, S.L.; Thorat, R.M. Microencapsulation: A review. *Int. J. Pharm. Sci. Rev. Res.* **2010**, *1*, 38–43.
56. Sonawane, S.H.; Bhanvase, B.A.; Sivakumar, M.; Potdar, S.B. Current overview of encapsulation. In *Encapsulation of Active Molecules and Their Delivery System*; [s.l.]; Elsevier Inc.: Amsterdam, The Netherlands, 2020; pp. 1–8.
57. Sedaghat Doost, A.; Nikbakht Nasrabadi, M.; Kassozi, V.; Nakisozi, H.; Van der Meeren, P. Recent advances in food colloidal delivery systems for essential oils and their main components. *Trends Food Sci.Technol.* **2020**, *99*, 474–486. [CrossRef]
58. Lu, W.; Kelly, A.L.; Miao, S. Emulsion-based encapsulation and delivery systems for polyphenols. *Trends Food Sci. Technol.* **2016**, *47*, 1–9. [CrossRef]

59. Stasse, M.; Laurichess, E.; Ribaut, T.; Anthony, O.; Héroguez, V.; Schmitt, V. Formulation of concentrated oil-in-water-in-oil double emulsions for fragrance encapsulation. *Colloids Surf. A Physicochem. Eng. Asp.* **2020**, *592*, 124564. [CrossRef]
60. Sachen, K.N.; Singh, B.; Rama, R.K. Controlled drug delivery through microencapsulation. *Malays. J. Pathol.* **2006**, *4*, 65–81.
61. Park, Y.; Yeo, Y. Microencapsulation technology. *Encycl. Pharm. Technol.* **2007**, *3*, 2315–2325.
62. Vishwakarma, G.S.; Gautam, N.; Babu, J.N.; Mittal, S.; Jaitak, V. Polymeric encapsulates of essential oils and their constituents: A review of preparation techniques, characterization, and sustainable release mechanisms. *Poly. Rev.* **2016**, *56*, 668–701. [CrossRef]
63. European Patent Office, United States Patent and Trademark Office, Cooperative Patent Classification, Scheme and Definitions 2020, (n.d.).
64. Geffroy, C.; Schreiber, S.S.; Goodall, M.J.; Fadel, A.; Harrison, I.M. Encapsulation of Perfumes. US2015/0044262 A1, 12 February 2015.
65. Brahms, J.; Lei, Y.; Wieland, J.; Xu, L.; Popplewell, L.M. Reloadable Microcapsules. WO2017192648A1, 9 November 2017.
66. Morrison, E. Ibuprofen Nanoparticle Carriers En-capsulated with Hermatic Surfactant Films. US20170296496A1, 23 August 2007.
67. Singh, N.; Sheikh, J. Microencapsulation and its application in production of functional textiles. *IJFTR* **2020**, *45*, 495–509.
68. Nelson, G. Application of microencapsulation in textiles. *Int. J. Pharm.* **2002**, *242*, 55–62. [CrossRef] [PubMed]
69. Pithanthanakul, U.; Vatanyoopaisarn, S.; Thumthanaruk, B.; Puttanlek, C.; Uttapap, D.; Kietthanakorn, B.; Rungsardthong, V. Encapsulation of fragrances in zein nanoparticles and use as fabric softener for textile application. *Flavour Fragr. J.* **2021**, *36*, 365–373. [CrossRef]
70. Cárdenas-Ramírez, C.; Jaramillo, F.; Gómez, M. Systematic review of encapsulation and shape-stabilization of phase change materials. *J Energy Storage.* **2020**, *30*, 101495–101516. [CrossRef]
71. Azevedo, S.G.; de Andrade, C.P.; D'Ambros, N.C.d.S.; Pérez, M.T.M.; Manzato, L. Biotechnological Applications of Nanoencapsulated Essential Oils: A Review. *Polymers* **2022**, *14*, 5495.
72. Yang, Z.; Peng, Z.; Li, J.; Li, S.; Kong, L.; Li, P.; Wang, Q. Development and evaluation of novel flavour microcapsules containing vanilla oil using complex coacervation approach. *Food Chem.* **2014**, *145*, 272–277. [CrossRef] [PubMed]
73. Lopes, S.; Afonso, C.; Fernandes, I.; Barreiro, M.F.; Costa, P.; Rodrigues, A.E. Chitosan-cellulose particles as delivery vehicles for limonene fragrance. *Ind. Crop. Prod.* **2019**, *139*, 111407. [CrossRef]
74. Sadovoy, A.V.; Lomova, M.V.; Antipina, M.N.; Braun, N.A.; Sukhorukov, G.B.; Kiryukhin, M.V. Layer-by-layer assembled multilayer shells for encapsulation and release of fragrance. *ACS Appl. Mater. Interfaces* **2013**, *5*, 8948–8954. [CrossRef]
75. Kwamman, Y.; Klinkesorn, U. Influence of oil load and maltodextrin concentration on properties of tuna oil microcapsules encapsulated in two-layer membrane. *Dry. Technol.* **2014**, *33*, 854–864. [CrossRef]
76. Minemoto, Y.; Hakamata, K.; Adachi, S.; Matsuno, R. Oxidation of linoleic acid encapsulated with gum arabic or maltodextrin by spray-drying. *J. Microencapsul.* **2002**, *19*, 181–189. [CrossRef]
77. Chang, D.; Abbas, S.; Hayat, K.; Xia, S.; Zhang, X.; Xie, M.; Kim, J.M. Original article: Encapsulation of ascorbic acid in amorphous maltodextrin employing extrusion as affected by matrix/core ratio and water content. *Int. J. Food Sci.* **2010**, *45*, 1895–1901. [CrossRef]
78. Varavinit, S.; Chaokasem, N.; Shobsngob, S. Studies of flavor encapsulation by agents produced from modified Sago and Tapioca starches. *Starch—Stärke* **2001**, *53*, 281–287. [CrossRef]
79. Edris, A.; Bergnståhl, B. Encapsulation of orange oil in a spray dried double emulsion. *Nahrung. Food.* **2001**, *45*, 133–137. [CrossRef] [PubMed]
80. de Melo Ramos, F.; Silveira Júnior, V.; Prata, A.S. Assessing the vacuum spray drying effects on the properties of orange essential oil microparticles. *Food Bioproc Technol.* **2019**, *12*, 1917–1927. [CrossRef]
81. Karrar, E.; Ali Mahdi, A.; Sheth, S.; Mohamed Ahmed, I.A.; Faisal Manzoor, M.; Wei, W.; Wang, X. Effect of maltodextrin combination with gum arabic and whey protein isolate on the microencapsulation of gurum seed oil using a spray-drying method. *Int. J. Biol. Macromol.* **2021**, *171*, 208–216. [CrossRef]
82. Nguyen, P.T.N.; Nguyen, H.T.A.; Hoang, Q.B.; Nguyen, T.D.P.; Nguyen, T.V.; Cang Mai, H. Influence of spray drying parameters on the physicochemical characteristics of microencapsulated orange (*Citrus sinensis* L.) essential oil. *Mater. Today: Proc.* **2022**, *60*, 2026–2033. [CrossRef]
83. Bajac, J.; Nikolovski, B.; Lončarević, I.; Petrović, J.; Bajac, B.; Đurović, S.; Petrović, L. Microencapsulation of juniper berry essential oil (*Juniperus Communis* L.) by spray drying: Microcapsule characterization and release kinetics of the oil. *Food Hydrocoll.* **2022**, *125*, 107430. [CrossRef]
84. Bušić, A.; Komes, D.; Belščak-Cvitanović, A.; Vojvodić Cebin, A.; Špoljarić, I.; Mršić, G.; Miao, S. The potential of combined emulsification and spray drying techniques for encapsulation of polyphenols from rosemary (Rosmarinus officinalis L.) Leaves. *FTB* **2018**, *56*, 494–505. [CrossRef] [PubMed]

85. Felix, P.H.C.; Birchal, V.S.; Botrel, D.A.; Marques, G.R.; Borges, S.V. Physicochemical and thermal stability of microcapsules of cinnamon essential oil by spray drying. *J. Food Process. Preserv.* **2017**, *41*, 12919–12928. [CrossRef]
86. Abdin, M.; El-Beltagy, A.E.; El-sayed, M.E.; Naeem, M.A. Production and characterization of sodium alginate/gum arabic based films enriched with Syzygium cumini seeds extracts for food application. *J. Food Process. Preserv.* **2022**, *30*, 1615–1626. [CrossRef]

Disclaimer/Publisher's Note: The statements, opinions and data contained in all publications are solely those of the individual author(s) and contributor(s) and not of MDPI and/or the editor(s). MDPI and/or the editor(s) disclaim responsibility for any injury to people or property resulting from any ideas, methods, instructions or products referred to in the content.

cosmetics

MDPI

Article

Chemical Characterization and Bioactive Properties of Wine Lees and Diatomaceous Earth towards the Valorization of Underexploited Residues as Potential Cosmeceuticals

Cristina N. Duarte [1,2,3], Oludemi Taofiq [1,2,4], Maria Inês Dias [1,2], Sandrina A. Heleno [1,2], Celestino Santos-Buelga [3], Lillian Barros [1,2] and Joana S. Amaral [1,2,*]

1 Centro de Investigação de Montanha (CIMO), Instituto Politécnico de Bragança, Campus de Santa Apolónia, 5300-253 Bragança, Portugal
2 Laboratório Associado Para a Sustentabilidade e Tecnologia em Regiões de Montanha (SusTEC), Instituto Politécnico de Bragança, Campus de Santa Apolónia, 5300-253 Bragança, Portugal
3 Grupo de Investigación en Polifenoles (GIP-USAL), Facultad de Farmacia, Universidad de Salamanca, E-37007 Salamanca, Spain
4 Nutrition and Bromatology Group, Department of Analytical Chemistry and Food Science, Faculty of Science, Universidade de Vigo, E-32004 Ourense, Spain
* Correspondence: jamaral@ipb.pt; Tel.: +351-273303138

Abstract: Annually, wine production is responsible for generating large quantities of residues, which are frequently disposed of and not valorized. So far, different studies have been conducted on grape pomace, yet less attention has been paid to other residues, such as wine lees and diatomaceous earth used in wine filtration. In this context, this study aimed to evaluate and compare the phenolic profile of these underexploited winemaking residues and assess their biological potential based on their antioxidant, antimicrobial, cytotoxic, and anti-aging activities (inhibition of tyrosinase and collagenase). Twenty-nine phenolic compounds, including twelve anthocyanins, were tentatively identified in the residues, with red grape pomace showing the highest diversity of compounds. The diatomaceous earth presented the highest content of non-anthocyanin phenolic compounds, being particularly rich in flavan-3-ols and myricetin-O-hexoside, and also presenting two anthocyanins. This sample also showed a high antioxidant activity, evidencing the best result in the reducing power assay. The red wine lees extract, despite showing a low content of phenolic compounds and less antioxidant activity, presented the highest inhibition capacity of bacteria growth. The extracts did not exhibit cytotoxicity against keratinocyte (up to 400 μg/mL) and fibroblast (up to 100 μg/mL) skin cell lines. However, the capacity of inhibiting tyrosinase and collagenase was low for the lees and diatomaceous earth, contrary to the grape pomace, seeds, and skins extracts that showed promising results, evidencing its potential as a cosmeceutical. Overall, this study highlights for the first time the potential of diatomaceous earth, an underexploited winemaking waste, in the obtention of added-value extracts and/or ingredients for cosmetic industry.

Keywords: winemaking residues; antioxidant activity; skin enzymes; keratinocytes; fibroblasts

Citation: Duarte, C.N.; Taofiq, O.; Dias, M.I.; Heleno, S.A.; Santos-Buelga, C.; Barros, L.; Amaral, J.S. Chemical Characterization and Bioactive Properties of Wine Lees and Diatomaceous Earth towards the Valorization of Underexploited Residues as Potential Cosmeceuticals. *Cosmetics* **2023**, *10*, 58. https://doi.org/10.3390/cosmetics10020058

Academic Editor: Agnieszka Feliczak-Guzik

Received: 4 March 2023
Revised: 24 March 2023
Accepted: 27 March 2023
Published: 29 March 2023

Copyright: © 2023 by the authors. Licensee MDPI, Basel, Switzerland. This article is an open access article distributed under the terms and conditions of the Creative Commons Attribution (CC BY) license (https://creativecommons.org/licenses/by/4.0/).

1. Introduction

Polyphenols are secondary metabolites produced by plants that play important roles in plant systems, such as providing them with protection against different types of attacks, including from parasites, insect pests, ultraviolet (UV) radiation, and other environmental threats [1]. Polyphenols may also have beneficial effects on health as they show different biological activities, including antioxidant, anti-inflammatory, antimutagenic, anticarcinogenic, and antiproliferative properties [2]. One of the most interesting characteristics of polyphenols is their antioxidant activity, owing to their reducing properties and their capacity to neutralize free radicals, such as singlet oxygen, which have been related to the ability to avoid or delay the oxidation of biomolecules (mainly lipids) in biological systems [3,4]. Moreover, the phenolic composition is an important quality parameter in certain foods and drinks, such as wine, contributing to various organoleptic attributes, such as bitterness, astringency, color, flavor, and oxidative stability [5].

Over recent years, different by-products from the food industry have been suggested as potential sources of polyphenols, since a high proportion of interesting bioactive compounds remains in their composition. The winemaking industry is amongst the ones that generate a large number of by-products, mainly grape pomace (consisting of grape skins, seeds, and stems), but also wine lees and other undervalued residues, such as diatomaceous earth used in filtrations. At the end of the winemaking process, residues containing yeasts, bacteria, and organic matter are deposited at the bottom of the wine tanks, corresponding to the lees. To avoid undesirable flavors and aromas, the wine should not remain in contact with these deposits, thus being separated from them [6]. Moreover, to guarantee the desired cleanness and stabilization of the wine, besides transfers, the winemaking industry carries out further operations such as filtration and tartaric stabilization. When properly treated, diatomaceous earth (silica deposits consisting of the accumulation of fossilized shells from microscopic marine algae) yields a material of high porosity non-deformable structure and of a large specific surface. This material is frequently used in wine filtration to eliminate suspended particles, rendering a clear and bright aspect to the wine [7], being an additional residue generated by this industry. While the pomace seeds can be used to produce grapeseed oil and the pomace can further undergo a distillation process or be used as a fertilizer or as animal feed, the stems, lees, and diatomaceous earth are generally discarded and perceived as an environmental problem [8]. However, recent studies suggest that the wine lees can be exploited for their richness in polyphenols, since these compounds can be adsorbed in their colloidal state to the yeast's cell walls as well as being dissolved in the liquid fraction of the wine lees [9,10]. Nevertheless, so far, only a few studies have advanced beyond the evaluation of the total phenolic compounds and carried out the identification of individual phenolics in wine lees. Most of these works concerned red wine lees [11–15], and only limited information is available on white wine lees [16,17]. Likewise, data on the diatomaceous earth from winemaking residues are very scarce. Recently, the feasibility of applying molecularly imprinted polymers to extracts prepared from winemaking diatomaceous earth residues to obtain quercetin-enriched fractions was demonstrated [18]. However, the study mainly focused on some selected compounds without a detailed phytochemical characterization of this by-product being carried out.

In the framework of the United Nations Sustainable Development Goals (SDGs), particularly SDG12 "Ensure sustainable consumption and production patterns", and within a circular economy perspective, there is now a considerable interest in changing the paradigm of wine industry wastes by turning them into by-products that other industries can further use. In this sense, grape pomaces from the wine and juice industries have been widely studied as a source of polyphenols with potential nutraceutical properties to be exploited by the food industry [19–21]. Furthermore, the use of bioactive compounds extracted from these sources is increasingly common in the cosmetic field, mostly based on their antioxidant capacity. In the human body, oxidative stress occurs when the balance between the generation of reactive oxygen species (ROS) and the antioxidant defense systems is compromised. In the skin, environmental stress, such as ultraviolet light, exacerbates ROS production, resulting in cell damage and degradation of the extracellular matrix proteins such as collagen and elastin, ultimately leading to skin aging. The main changes that take place at the dermal connective tissue level, which are essentially translated into the loss of mature collagen and alterations in the elastic network [22], allow uneven pigmentation of the epidermis to occur with aging due to changes in tyrosinase activity in the melanocytes [23]. Distinct phenolic compounds present in grapes, such as anthocyanins, not only possess renowned antioxidant properties, but can also directly inhibit enzymes involved in the skin aging process—namely, tyrosinase and collagenase [24,25]. For this reason, winemaking by-products can be potential sources of innovative functional ingredients for cosmetics, and more studies are required concerning the so far less studied residues.

This work aimed at evaluating several winemaking by-products, with particular focus on underexploited residues for which there is a scarcity of data and knowledge (wine lees and diatomaceous earth), and aimed to compare those with the more widely studied grape pomace, skins, and seeds. The extracts obtained from the residues have been evaluated for their phenolic compounds profile, their antioxidant properties, and their antimicrobial and anti-fungicidal activity. Moreover, to evaluate their potential as cosmeceutical ingredients, the inhibition of tyrosinase and collagenase, as enzymes involved in the skin aging process, as well as the cytotoxicity in fibroblasts (HFF-1) and keratinocytes (HaCaT) cell lines, have also been assessed.

2. Materials and Methods

2.1. Winemaking Industry by-Products

Winemaking by-products were supplied by Caves Campelo S.A. (Portugal) during the harvest season in September 2021. The seeds and skins were manually separated from the red grape pomace. All the other samples were sent individually. The samples of whole pomace, seeds, skins, and stems from red grapes were arranged on separated trays and dried in an oven (Scientific, Series 9000, Scientific Engineering, Johannesburg, South Africa) at 40 °C for 3 days, and then ground into powder in a mill (IKA, A11 basic) to pass a sieve of 20 mesh. The red lees were sent on a plastic bottle and the separation of phases was visible. Thus, they were placed on falcon tubes and centrifuged at 7871× g for 5 min. The liquid phase was then pipetted into a flask and separated from the solid phase. After eliminating the ethanol by rotoevaporation (Julabo, TW12), the liquid phase was lyophilized (Telstar, LyoQuest—55 Plus). The separated solid phase from red lees as well as the white lees were also submitted to lyophilization. All the samples were stored at −20 °C inside a container with silica, and were protected from light.

2.2. Extraction of Compounds

Phenolic compounds were extracted using ethanol/water 80:20 (v/v). A total of 5 grams of powder of each sample was extracted with 100 mL of solvent by sonication (SONO SWISS, sw1) for 5 min followed by stirring (Lbx instruments, S03 series) for 1 h at ambient temperature and protected from light. The samples were centrifuged (Eppendorf, Centrifuge 5810 R) at 7871× g for 5 min, and the pellet was re-extracted. The supernatants were collected and joined and filtered, and the ethanol (Fisher Scientific, Loughborough, UK) evaporated under vacuum on a rotary evaporator (Julabo, TW12, Julabo Labortechnik, Seelbach, Germany) at 40 °C. Finally, the extracts were lyophilized, and the obtained dry residues were stored at −20 °C in flasks kept inside a container with silica for further use or weighted and redissolved in ethanol/water 80:20 (v/v) to a final concentration of 10 mg/mL.

2.3. Total Phenolic Content

The total phenolic compounds content of the hydroethanolic extracts prepared was determined by spectrophotometry in a microplate reader (Epoch 2, Biotek, BioTek Instruments, Inc., Winooski, VT, USA) using the Folin–Ciocalteu reagent. To obtain the calibration curve, an ethanolic solution of gallic acid was used as a standard, within the range of 0.005–0.5 mg/mL. For analysis, different concentrations of sample extracts were prepared from a stock solution (5 mg/mL) prepared from the lyophilizate. In a test tube, 250 µL of extract from each sample or standard were added with 1.25 mL of Folin-Ciocalteu reagent (1:10 v/v in water) and 1 mL of sodium carbonate (75 g/L). The tubes were vortexed for 15 s and placed to rest in the dark for 30 min at 40 °C for colour development. After this, the tubes were centrifuged at 12,000× g rpm for 2 min and the solution transferred to a 96-well microplate, and the absorbance read at 765 nm [26]. Ethanol was used as blank control. The results were expressed as mg gallic acid equivalents per gram of dry extract (mg GAE/g extract).

2.4. Characterization of Phenolic Compounds by HPLC-DAD-ESI/MS^n

Individual phenolic compounds were separated, identified, and quantified by high-performance liquid chromatography coupled to double diode array and tandem mass spectrophotometry detection (HPLC-DAD-ESI-MS/MS). Dry extracts (40 mg) were redissolved in 2 mL of ethanol/water 80:20 (v/v) and filtered through 0.22 μm disposable filter disks. The analysis was carried out in a Dionex Ultimate 3000 UPLC (Thermo Scientific, San Jose, CA, USA) system equipped with a degasser, a quaternary pump, an auto sampler (kept at 5 °C), an automated thermostatically controlled column compartment, and a diode array detector (DAD) coupled to a mass spectrometer with an electrospray ionization (ESI) source.

Chromatographic separation was achieved with a Waters Spherisorb S3 ODS-2C18 (3 μm, 4.6 mm×150 mm, Waters, Mil-ford, MA, USA) column maintained at 35 °C. The solvents used were: (A) 0.1% formic acid in water, and (B) acetonitrile. The elution gradient established was 15% B (5 min), 15% B to 20% B (5 min), 20–25% B (10 min), 25–35% B (10 min), and 35–50% B (10 min), followed by the re-equilibration of the column in a flow rate of 0.5 mL/min. Double online detection was carried out in the DAD using 280, 320, 370, and 520 nm as preferred wavelengths and in a mass spectrometer (MS) connected to HPLC system via the DAD cell outlet. MS detection was performed using a Orbitrap Exploris 120 mass spectrometer (ThermoFinnigan, San Jose, CA, USA) equipped with an ESI source operating in negative or positive mode for non-anthocyanin and anthocyanins, respectively. Nitrogen served as the sheath gas (50 psi), and the system was operated with a spray voltage of 5 kV, a source temperature of 325 °C, and a capillary voltage of −20 V. The tube lens offset was kept at a voltage of −70 V. The full MS and MS2 scans covered the mass range from m/z 100–1200. The collision energy used was 35 (arbitrary units). Data acquisition was carried out with Xcalibur® data system (ThermoFinnigan, San Jose, CA, USA).

The phenolic compounds were identified by comparing their retention times, UV-Vis, and mass spectra with those obtained from standard compounds, when available. Otherwise, compounds were tentatively identified comparing the obtained information with available data reported in the literature. For quantitative analysis, a calibration curve for each available phenolic standard was constructed based on the areas of the peaks recorded at 280 nm, except for flavonols (370 nm) and for anthocyanin compounds (520 nm). For the identified phenolic compounds for which a commercial standard was not available, the quantification was performed through the calibration curve of the most similar available standard. The results were expressed as mg/g of extract (dry weight).

2.5. Antioxidant Activity

Antioxidant activity assays were performed as previously described by Iyda et al. (2019) [26]. The extracts described in 2.1 were re-dissolved in ethanol/water (80:20, v/v) to obtain stock solutions (with various concentrations), which were further diluted to obtain a range of working concentrations to evaluate the antioxidant activity. DPPH radical-scavenging activity was evaluated by using an Epoch 2 microplate reader (BioTek Instruments, Inc., Winooski, VT, USA), and was calculated as a percentage of DPPH discoloration using the formula: $[(A_{DPPH} - AS)/A_{DPPH}] \times 100$, where AS is the absorbance of the sample solution at 515 nm, and A_{DPPH} is the absorbance of the DPPH solution. Reducing power was evaluated by the capacity to convert Fe^{3+} into Fe^{2+} by measuring the absorbance at 690 nm, as described in the literature [27], using the same microplate reader. The lipid peroxidation inhibition was evaluated using porcine brain tissue due to its richness in lipids, and the ability of the assay in generating free radicals produced by oxidative stress. Lipid oxidation originates different products, the most common being malonaldehyde (MDA), which is determined based on the reaction with thiobarbituric acid

(TBA), generating a complex of pink coloration, the absorbance of which was measured at 532 nm. The results were expressed as EC_{50} values, corresponding to the sample concentration providing 50% of antioxidant activity or 0.5 of absorbance in the reducing power assay. Trolox was used as positive control.

2.6. Antibacterial Activity

The bacterial strains were clinical isolates obtained from patients hospitalized in various departments at the North-Eastern Local Health Unit (Bragança, Portugal) and Hospital Center of Trás-os-Montes and Alto Douro (Vila Real, Portugal), isolated in a previous work [28]. Five Gram-negative bacteria, namely, *Escherichia coli* (isolated from urine, VRU12881), *Klebsiella pneumoniae* (isolated from urine, VRI17214), *Proteus mirabilis* (isolated from wound exudate, VRI78844), *Pseudomonas aeruginosa* (isolated from expectoration, VRU14123), and *Morganella morganii* (isolated from urine, VRU14272), and three Gram-positive bacteria, namely, *Enterococcus faecalis* (isolated from urine, VRU14041), *Listeria monocytogenes* (isolated from cerebrospinal fluid, VRU17684), and methicillin-resistant *Staphylococcus aureus* (MRSA) (isolated from expectoration, VRI17654), were all tested. These microorganisms were selected based on their antibiotic resistance profile previously described [29]. All microorganisms were incubated at 37 °C in appropriate fresh medium during the 24 h before analysis in order to maintain the exponential growth phase. The minimum inhibitory concentration (MIC) determination was conducted based on the colorimetric broth microdilution assay [30]. The samples were first dissolved in 5% (*v*/*v*) dimethyl sulfoxide (DMSO) and 95% autoclaved distilled water to give a final concentration of 20 mg/mL for the stock solution. Afterwards, the samples were successively diluted with tryptic soy broth (TSB) culture medium to obtain the concentration ranges of 10 to 0.03125 mg/mL. 10 µL of inoculum (standardized at 1.5×10^6 Colony Forming Unit (CFU)/mL) was then added to each well containing 90 µL of sample dilutions, reaching a final concentration of 1.5×10^5 CFU/mL. Two negative controls were prepared, one with TSB and another with the extract only. Two positive controls were also prepared, one with TSB and each inoculum and another with TSB, antibiotics, and bacteria. Ampicillin was used for all the tested bacteria, and imipenem was used against all the strains, except for *E. faecalis* and MRSA, in which vancomycin was used. The microplates were incubated at 37 °C for 24 h and the MIC was determined following the addition (40 µL) of 0.2 mg/mL *p*-iodonitrotetrazolium chloride (INT) and incubation at 37 °C for 30 min. The MIC was defined as the lowest concentration that inhibits the visible bacterial growth determined by color change from yellow to pink if the microorganisms are viable. For the minimum bactericidal concentration (MBC) determination, 10 µL from each well that showed no change in color was plated on Blood agar (7% sheep blood) solid medium and incubated at 37 °C for 24 h. The lowest concentration that yielded no growth was established as the MBC. The MIC and MBC results are expressed in mg/mL.

2.7. Antifungal Activity

The antifungal activity was determined against *Aspergillus fumigatus* (ATCC 204305) and *Aspergillus brasiliensis* (ATCC 16404) according to the methodology described in the literature [31]. The micromycetes were maintained on malt agar and the cultures stored at 4 °C. Before the assay, they were further placed in a new medium and incubated at 25 °C for 72 h. To investigate the antifungal activity, the fungal spores were washed from the surface of agar plates with sterile 0.85% saline containing 0.1% Tween 80 (*v/v*). The spore suspension was adjusted with sterile saline to a concentration of approximately 1.0×10^5 in a final volume of 100 µL per well. The extracts were first dissolved in 5% (*v/v*) dimethyl sulfoxide (DMSO) and 95% of autoclaved distilled water to give a final concentration of 20 mg/mL for the stock solution. Afterwards, 90 µL of this concentration was added in the first well (96-well microplate) in duplicate with 100 µL of Malt Extract Broth (MEB). The remaining wells were added with 90 µL of MEB, and then the samples were successively diluted to obtain a concentration range from 10 to 0.03125 mg/mL. The MIC was established as the lowest concentration without visible growth (at the binocular microscope). The minimum fungicidal concentration (MFC) was determined by the sub-cultivation of 2 μL of each well that presented no visible growth into microplates containing 100 μL of MEB per well, and then further incubation 72 h at 26 °C. The lowest concentration with no visible growth was defined as MFC, indicating a 99.5% killing of the original inoculum. Commercial fungicide ketoconazole (Frilabo, Porto, Portugal) was used as positive control. The MIC and MFC were expressed in mg/mL.

2.8. Cytotoxicity Assay in Skin Cell Lines

In vitro spontaneously transformed keratinocytes from histologically normal skin (HaCaT) and fibroblast from human foreskin (HFF-1) cells were cultured in Dulbecco's modified Eagle's medium containing 10% fetal bovine serum (FBS), 100 U/mL penicillin, and 100 μg/mL streptomycin. The cryopreserved culture of HaCaT was purchased from Cell Line Service (Germany), while HFF-1 was purchased from ATCC. Cells were routinely sub-cultured in 75 cm^2 flasks in humidified 5% CO_2 atmosphere at 37 °C [32]. After cultured cells reached the appropriate confluence, the culture medium was removed from the culture plate, washed twice with Hank's Balanced Salt Solution (HBSS), and then trypsinised using 0.25% trypsin/0.53 mM EDTA, and cells were suspended in a fresh medium and seeded into 96-well plates at 2×10^4 cells/well for 24 h. The cytotoxic effect of the extract was investigated by using Sulforhodamine B (SRB). Briefly, different concentrations of extracts were added to each well, and the control wells were exposed to Triton X-100 at a final concentration of 1% (*w*/*v*). After the 48 h incubation period, an ice-cold solution of trichloroacetic acid (TCA) (10%, *w*/*v*) was added to each well, followed by incubation for 1 h at 4 °C. Afterwards, the microplates were washed with water and dried at room temperature. A solution of SRB (0.057%, *w*/*v*) was then added to each well. Wells were washed three times with an acetic acid solution (1%, *v*/*v*) and left to dry at room temperature. Finally, the SRB was solubilised with Tris (10 mM, 200 μL), the absorbance was measured at 540 nm, and the results were expressed as the percentage of cell viability.

2.9. Tyrosinase Inhibitory Activity

The tyrosinase enzyme inhibition activity was evaluated using L-3,4-dihydroxy-phenylalanine (L-DOPA) as a substrate with a concentration of 5 mM and 3-methyl-2-benzothiazolinone-hydrazonehydrochloride (MBTH) with a concentration of 20.7 mM as a chromogenic stabilizing agent in 96-well microplates, following the procedure previously described [33,34]. Briefly, test samples (10 μL at 10 mg/mL), phosphate buffer (pH 7.1, 0.1 M), L-DOPA (60 μL), and MBTH (87 μL) were mixed and pre-incubated at 25 °C for 10 min. Subsequently, 6 μL of mushroom tyrosinase enzyme (142 Units of enzyme/mL) were added to each well and the plates were incubated for 30 min at 25 °C. The blank was made with a phosphate buffer (pH 7.1, 0.1 M) and butylresorcinol was used as positive control. The formation of the dopaquinone-MBTH complex was evaluated at 505 nm, using a microplate spectrophotometer (SPECTROstar Nano Multi-Detection Microplate Reader; BMG Labtech, Ortenberg, Germany). The percentage of tyrosinase enzyme inhibition was calculated using the following equation:

$$\%\ \text{tyrosinase inhibition} = ((\text{Absorbance Blank} - \text{Absorbance Sample})/(\text{Absorbance Blank})) \times 100$$

2.10. Collagenase Activity Colorimetric Assay

The collagenase inhibition assay was performed using the colorimetric assay kit from Sigma-Aldrich (MAK293) and according to manufacturer's instructions. The eight samples of the extracts from the winemaking residues were prepared with ultrapure water to reach a concentration of 5 mg/mL. A volume of 10 μL of each sample (50 μg of extract) was placed in a 96 well plate in addition with 80 μL of collagenase assay buffer and 10 μL of the provided enzyme (collagenase 0.35 U/mL). For positive control, 10 μL of enzyme only (0.35 U/mL) was used. For inhibitor control, a volume of 10 μL of provided collagenase (0.35 U/mL) and 2 μL of inhibitor (1,10-phenanthroline) was placed into desired wells.

The volume of positive control and inhibitor control wells were adjusted to 100 μL with collagenase assay buffer. The plate was incubated at 37 °C for 10 min. A mixture containing 60 μL of collagenase assay buffer and 40 μL of collagenase substrate FALGPA (N-(3-[2-furyl]acryloyl)-L-leucylglycyl-L-prolyl-L-alanine), was prepared and 100 μL of it was added into the desired wells. Subsequently, the absorbance was read at 345 nm in a microplate reader (Epoch 2, BioTek Instruments, Inc., USA) for 1 h. The percentage of collagenase enzyme inhibition was calculated using the following equation:

$$\% \text{ collagenase inhibition} = ((\text{Activity enzyme} - \text{Activity inhibitor})/(\text{Activity enzyme})) \times 100$$

with the enzyme activity being calculated as indicated by the manufacturer—-namely, by taking the absorbance (A1 and A2) at two time points (T1 and T2) in the linear range with at least two readings in between and at least 1 min apart. Therefore, T1 was considered the first absorbance reading and T2 was established at a 10 min reading.

2.11. Statistical Analysis

For the eight samples evaluated, all the assays were carried out in triplicate and the results were expressed as mean values and standard deviation (SD). Statistical analysis was performed using one way analysis of variance (ANOVA) followed by Tukey's HSD Test with $p = 0.05$. When necessary, a Student's t-test was used to determine the significant difference between less than three different samples, with $p = 0.05$. These analyses were carried out using IBM SPSS Statistics for Windows, Version 23.0. (IBM Corp., Armonk, NY, USA).

3. Results and Discussion

3.1. Chemical Characterization

3.1.1. Total Phenolic Compounds

Several works reported on the phenolic composition of different winemaking residues, either regarding their total content in phenolic compounds or by identifying and quantifying the respective individual compounds. However, most focused on grape pomace, particularly from red grape cultivars [35–37]. In this work, besides red wine pomace, other less studied residues were also evaluated—-namely, wine lees and diatomaceous earth from wine filtration. The wine lees from red wine production were supplied by the industry separated into two phases (solid and liquid phases), which were individually assessed. As far as we know, this is the first report on the chemical characterization and bioactive properties of extracts obtained from winemaking diatomaceous earth residue.

Considering the potential interest as a screening assay for the industry, allowing us to estimate which batches of residues can be more promising to be exploited, all extracts were evaluated for its total phenolic content (TPC) based on the reaction with Folin–Ciocalteu. The results for TPC are presented in Figure 1, showing that the highest content was observed for the extracts of grape seeds from red wine pomace, diatomaceous earth, and red wine pomace with values of 146.6 ± 0.6, 111 ± 2, and 126.9 ± 0.9 mg GAE/g extract, respectively. Interestingly, the results suggest that diatomaceous earth is a very promising residue, since it presented a TPC content even higher than the red wine pomace. On the contrary, the extracts from wine lees showed the lowest contents, particularly the one from white wine production (19.7 ± 0.4 mg GAE/g extract).

The obtained results of TPC are consistent with values previously reported for red wine grape pomace by Maicas et al. (between 55.5 to 153.8 mg GAE/g) [38] and by Matos et al. (83.9 ± 2.0 mg GAE/g extract) [39]. Higher values (254 ± 24 mg GAE/g dry extract) were obtained in other studies [40] when evaluating wine lees from red wine production, extracted with a similar solvent as in the current study here (ethanol:water 75:25). However, this can be related to the fact that the lees sample was obtained after the wine had been aged for 12 months in American oak barrels, while the red wine lees herein analyzed were obtained after the fermentation phase (7 to 10 days) of a red wine produced in stainless steel tanks.

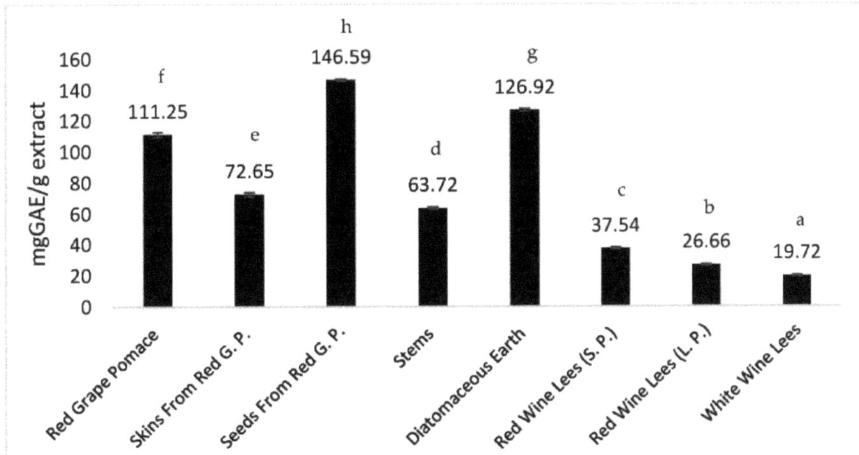

Figure 1. Total phenolic content (TPC) (mg GAE/g dry extract) of hydroethanolic winemaking extracts. (Red Wine Lees (S. P.)—Red Wine Lees Solid Phase; Red Wine Lees L. P.—Red Wine Lees Liquid Phase. Values indicate the means, and error bars represent the standard deviation. In each extract, different letters mean significant differences ($p < 0.05$)).

3.1.2. Phenolic Compounds Profile

Twenty-nine phenolic compounds (non-anthocyanins and anthocyanins) were identified or tentatively identified in the ethanol/water (80:20, v/v) extracts prepared from the winemaking residues studied in this work—-namely, whole pomace, seeds, skins, stems from red wine production, red and white wine lees, and diatomaceous earth used in red wine production (Table 1). Seventeen non-anthocyanin phenolic compounds were tentatively identified: one phenolic acid (*p*-coumaric acid hexoside), six flavan-3-ols ((epi)catechin derivatives), four flavonols (*O*-glycosylated myricetin, and quercetin derivatives), three dihydroflavonols (taxifolin hexosides), two resveratrol derivatives (resveratrol tetramer and trans-resveratrol), and ethyl-gallate. In addition, twelve anthocyanins were also identified, including *O*-glycosylated and acylated derivatives of delphinidin, cyanidin, malvidin, petunidin, and peonidin.

Most of the detected compounds have been previously described in grape pomace residues from winemaking and other winery byproducts [35,41–44]. Peaks **4** and **9** ([M-H]$^-$ at m/z 289) were identified as (+)-catechin and (-)-epicatechin, respectively, by comparing their retention time and UV spectra with the available standard compounds. Taxifolin and myricetin aglycones were previously reported in red wine samples from China [45], but, in the present work, these two compounds were found to be linked to sugar moieties. Taxifolin derivatives, peaks **6**, **7**, and **8**, all with a deprotonated ion [M-H]$^-$ at m/z 465, presented a unique MS^2 fragment at m/z 303 (taxifolin aglycone), corresponding to the loss of 162 u (hexosyl moiety) and thus being tentatively assigned as taxifolin-*O*-hexosides. Peak **11** ([M-H]$^-$ at m/z 479), with an MS2 fragment at m/z 317 (-162 u; myricetin aglycone), was tentatively identified as myricetin-*O*-hexoside. Peak **15** ([M-H]$^-$ at m/z 521) also released an MS^2 fragment at m/z 317 from the loss of an acetyl-hexoside unit (42 u + 162 u) that allowed its identification as myricetin-*O*-acetyl-hexoside. Peak 12 presented a deprotonated [M-H]$^-$ ion at m/z 197 and MS^2 fragment ions at m/z 169 ([M-H-28]$^-$; gallic acid) and at m/z 125 ([M-H-28-44]$^-$) that corresponded to the successive loss of ethyl (28 u) and CO_2 residues, thus tentatively identified as ethyl gallate. This compound has been previously reported in wine and grape pomace samples [45–47]. In this case, its quantification was not carried out due to a lack of similar standard compounds in our research lab that could be used to construct the calibration curve. Peaks **16** ([M-H]$^-$ at m/z 905) and **17** ([M-H]$^-$ at m/z 227) were tentatively identified as stilbene type-compounds—-namely, a resveratrol tetramer (cis) and trans-resveratrol, respectively. This tentative identification was performed comparing the obtained results with mass fragmentation data reported in the literature for those compounds [48,49].

Table 1. Retention time (Rt), wavelength at the maximum absorption (λmáx), mass spectra (*m/z*), and tentative identification of the non-anthocyanin and anthocyanin phenolic compounds found in winemaking residues from the year 2021.

Peak	**Rt (min)**	**λmáx (nm)**	**[M-H]⁻ (*m/z*)**	**MS^2 (*m/z*)**	**Tentative Identification**
Non-anthocyanin phenolic compounds					
1	4.58	270	865	739(8), 577(45), 575(8), 425(20), 407(30), 289(11), 287(25)	*β*-type procyanidin trimer
2	4.83	279	577	451(100), 575(39), 425(5), 407(5), 289(5), 287(12)	*β*-type procyanidin dimer
3	5.05	278	1153	865(3), 863(14), 577(7),575(5), 289(14), 287(5)	*β*-type procyanidin tetramer
4	5.45	279	289	245(100), 205(45), 179(13)	(+)-Catechin
5	5.63	279	577	451(100), 575(39), 425(5), 407(5), 289(5), 287(12)	*β*-type procyanidin dimer
6	6.17	298	465	303(100)	Taxifolin-*O*-hexoside
7	6.37	298	465	303(100)	Taxifolin-*O*-hexoside
8	6.7	298	465	303(100)	Taxifolin-*O*-hexoside
9	6.91	282	289	245(100), 205(25), 179(12), 203(13), 231(5), 271(3), 161(3)	(-)-Epicatechin
10	7.16	277	325	163, 145, 119	*p*-Coumaric acid hexoside
11	9.8	274	479	317(100)	Myricetin-*O*-hexoside
12	12.33	277	197	169(100), 124(25)	Ethyl gallate
13	13.78	353	477	301(100)	Quercetin-*O*-hexuronoside
14	14.37	343	463	301(100)	Quercetin-*O*-hexoside
15	18.54	358	521	317(100)	Myricetin-*O*-acetyl-hexoside
16	29.64	284	905	811, 717, 357, 451, 611, 887	Resveratrol tetramer (*cis*)
17	35.24	284	227	186, 159, 143	*trans*-Resveratrol
Peak	Rt (min)	λmax (nm)	[H]⁺ (*m/z*)	MS^2 (*m/z*)	Tentative Identification
Anthocyanin phenolic compounds					
18	27.23	523	465	303(100)	Delphidin-3-*O*-glucoside
19	28.64	523	449	287(100)	Cyanidin-3-*O*-glucoside
20	29.61	525	479	317(100)	Petunidin-3-*O*-glucoside
21	31.76	520	463	301(100)	Peonidin-3-*O*-glucoside
22	32.92	526	493	331(100)	Malvidin-3-*O*-glucoside
23	36.78	517	507	303(100)	Dephinidin-3-*O*-acetylglucoside
24	40.13	529	521	317(100)	Petunidin-3-*O*-acetylglucoside
25	43.06	529	535	331(100)	Malvidin-3-*O*-acetylglucoside
26	44.81	531	655	331(100)	Malvidin-3-*O*-caffeoylglucoside
27	45.57	531	625	317(100)	Petunidin-3-*O*-*p*-coumaroylglucoside
28	47.42	526	609	301(100)	Peonidin-3-*O*-*p*-coumaroylglucoside
29	47.78	531	639	331(100)	Malvidin-3-*O*-*p*-coumaroylglucoside

Anthocyanins were identified based on their chromatographic and mass information, in agreement with the data reported in the literature [35,50–54]. The presence of acylated anthocyanins in the winemaking residues is particularly interesting, as they can have greater color stability when exposed to light compared with non-acylated anthocyanins [48,49], thereby making them more suitable to be used as food colorants.

Table 2 presents the quantification (mg/g dry extract) of the non-anthocyanin and anthocyanin phenolic compounds found in all the studied samples. The diatomaceous earth and the red wine pomace presented the highest amounts of non-anthocyanin phenolic compounds, which is in good agreement with the data presented in Figure 1. Moreover, the one of diatomaceous earth was also the richer extract in (-)-epicatechin, showing an amount significantly higher of this flavan-3-ol than the remaining extracts. Only the stems and the seeds from red wine pomace revealed the presence of stilbenes, namely, resveratrol tetramer and trans-resveratrol, respectively, and yet only in trace amounts below the level of quantification. For the white wine lees, only two phenolic compounds were identified, although it was the sample that presented the highest amount of (+)-catechin.

Table 2. Quantification of the non-anthocyanin and anthocyanin phenolic compounds found in the hydroethanolic extracts of winemaking residues from the year 2021 (mean ± standard deviation).

Peak	Red Grape Pomace	Skins	Seeds	Stems	Diatomaceous Earth	Red Wine Lees		White Wine Lees
						(Solid Phase)	(Liquid Phase)	
Non-anthocyanin phenolic compounds (mg/g extract)								
1	0.0686 ± 0.0065 ^{a}	nd	0.0483 ± 0.0028 ^{b}	0.0653 ± 0.0014 ^{a}	nd	nd	nd	nd
2	0.0413 ± 0.0053 ^{e}	nd	0.0637 ± 0.0101 ^{d}	0.167 ± 0.002 ^{a}	0.1554 ± 0.0147 ^{b}	0.0832 ± 0.001 ^{c}	0.1713 ± 0.0163 ^{a}	nd
3	0.0234 ± 0.0047	nd	nd	nd	nd	nd	nd	nd
4	0.0551 ± 0.0066 ^{e}	nd	0.101 ± 0.0191 ^{d}	0.139 ± 0.0047 ^{b}	nd	0.098 ± 0.0293 ^{d}	0.122 ± 0.00 ^{c}	0.1510 ± 0.0236 ^{a}
5	0.0266 ± 0.0015 ^{c}	nd	0.087 ± 0.0048 ^{b}	nd	0.1208 ± 0.0292 ^{a}	nd	nd	nd
6	0.0424 ± 0.0004 ^{d}	0.070 ± 0.0034 ^{a}	nd	nd	nd	0.0536 ± 0.0052 ^{c}	0.057 ± 0.00 ^{b}	nd
7	0.0286 ± 0.0019 ^{d}	nd	0.0335 ± 0.0021 ^{c}	nd	0.0807 ± 0.0171 ^{a}	nd	nd	0.0597 ± 0.0042 ^{b}
8	0.0302 ± 0.0007 ^{c}	nd	0.0525 ± 0.0102 ^{b}	0.0664 ± 0.0091 ^{a}	nd	nd	nd	nd
9	0.0591 ± 0.0024 ^{e}	0.099 ± 0.0141 ^{b}	0.0337 ± 0.0020 ^{f}	0.0753 ± 0.0046 ^{d}	0.254 ± 0.009 ^{a}	0.0729 ± 0.0007 ^{d}	0.0851 ± 0.0138 ^{c}	nd
10	0.0033 ± 0.0002	nd	tr	nd	nd	nd	nd	nd
11	nd	nd	nd	nd	0.532 ± 0.004	nd	nd	nd
12	nd	nd	nd	nd	0.201 ± 0.009	nd	nd	nd
13	0.114 ± 0.0004 ^{e}	0.1140 ± 0.0036 ^{b}	0.0939 ± 0.0001 ^{e}	0.1226 ± 0.0036 ^{a}	0.101 ± 0.0005 ^{d}	0.1051 ± 0.0010 ^{c}	0.0944 ± 0.0001 ^{e}	nd
14	0.0973 ± 0.0014 ^{g}	0.0992 ± 0.0012 ^{d}	0.1065 ± 0.0011 ^{b}	0.097 ± 0.0004 ^{c}	0.108 ± 0.0023 ^{a}	0.0964 ± 0.0011 ^{e}	0.0938 ± 0.0002 ^{f}	nd
15	0.5016 ± 0.0003 ^{a}	nd	nd	nd	nd	nd	nd	nd
16	nd	nd	nd	tr	nd	nd	nd	nd
17	nd	nd	tr	nd	nd	nd	nd	nd
TF3O	0.27 ± 0.0034 ^{a}	0.0998 ± 0.0141 ^{h}	0.334 ± 0.0252 ^{b}	0.443 ± 0.0254 ^{d}	0.730 ± 0.0146 ^{c}	0.254 ± 0.0283 ^{f}	0.3782 ± 0.0025 ^{e}	0.1510 ± 0.0236 ^{g}
TOF	0.81 ± 0.0006 ^{e}	0.2830 ± 0.001 ^{c}	0.286 ± 0.0092 ^{d}	0.2859 ± 0.0058 ^{b}	0.822 ± 0.0159 ^{a}	0.2551 ± 0.0032 ^{d}	0.2460 ± 0.0001 ^{f}	0.0597 ± 0.0042 ^{g}
TPC	1.092 ± 0.004 ^{e}	0.3829 ± 0.015 ^{c}	0.6202 ± 0.0345 ^{d}	0.7322 ± 0.0195 ^{b}	1.553 ± 0.0305 ^{a}	0.509 ± 0.0315 ^{d}	0.6242 ± 0.0025 ^{f}	0.2107 ± 0.0194 ^{g}
Anthocyanin phenolic compounds (mg/g extract)								
18	7.568 ± 0.186 ^{a}	4.134 ± 0.071 ^{b}	2.25 ± 0.003 ^{d}	1.161 ± 0.001 ^{e}	nd	1.075 ± 0.004 ^{f}	2.82 ± 0.043 ^{c}	nd
19	2.886 ± 0.101 ^{a}	1.562 ± 0.021 ^{d}	2.24 ± 0.001 ^{c}	1.166 ± 0.014 ^{e}	nd	1.067 ± 0.005 ^{f}	2.673 ± 0.038 ^{b}	nd
20	7.481 ± 0.174 ^{a}	4.35 ± 0.064 ^{b}	2.009 ± 0.003 ^{d}	1.074 ± 0.004 ^{e}	nd	0.985 ± 0.004 ^{f}	2.446 ± 0.024 ^{c}	nd
21	4.273 ± 0.130 ^{a}	2.182 ± 0.016 ^{c}	1.997 ± 0.001 ^{d}	1.144 ± 0.021 ^{e}	nd	0.964 ± 0.007 ^{f}	2.485 ± 0.051 ^{b}	nd
22	27.142 ± 0.389 ^{a}	16.78 ± 0.469 ^{b}	2.090 ± 0.004 ^{d}	1.614 ± 0.012 ^{e}	nd	1.065 ± 0.004 ^{f}	3.362 ± 0.0359 ^{c}	nd
23	2.743 ± 0.208 ^{a}	1.843 ± 0.04 ^{d}	2.239±0.001 ^{c}	1.071 ± 0.013 ^{f}	nd	1.111 ± 0.003 ^{e}	2.65 ± 0.05 ^{b}	nd
24	3.254 ± 0.043 ^{a}	1.931 ± 0.01 ^{d}	1.993 ± 0.003 ^{c}	0.982 ± 0.006 ^{e}	nd	0.975 ± 0.01 ^{e}	2.379 ± 0.038 ^{b}	nd
25	5.426 ± 0.246 ^{a}	2.779 ± 0.046 ^{b}	2.006 ± 0.0004 ^{d}	1.086 ± 0.009 ^{e}	nd	1.048 ± 0.008 ^{e}	2.619 ± 0.004 ^{c}	nd
26	4.146 ± 0.00 ^{a}	2.262 ± 0.076 ^{c}	2.002 ± 0.003 ^{d}	0.997 ± 0.007 ^{e}	nd	0.983 ± 0.013 ^{e}	2.337 ± 0.036 ^{b}	nd
27	2.861 ± 0.169 ^{a}	2.152 ± 0.038 ^{c}	1.999 ± 0.0057 ^{d}	1.014 ± 0.003 ^{e}	nd	1.031 ± 0.008 ^{e}	2.345 ± 0.047 ^{b}	nd
28	3.157 ± 0.054 ^{b}	1.545 ± 0.104 ^{e}	1.995 ± 0.0005 ^{d}	1.479 ± 0.005 ^{f}	3.401 ± 0.023 ^{a}	0.997 ± 0.009 ^{g}	2.272 ± 0.048 ^{c}	nd
29	9.901 ± 0.06 ^{a}	6.505 ± 0.04 ^{b}	2.033 ± 0.0002 ^{e}	1.274 ± 0.009 ^{f}	3.441 ± 0.012 ^{c}	1.024 ± 0.004 ^{g}	2.371 ± 0.007 ^{d}	nd
TA	80.837 ± 0.861 ^{a}	48.023 ± 0.583 ^{b}	24.852 ± 0.009 ^{d}	14.063 ± 0.013 ^{f}	6.842 ± 0.035 ^{g}	12.324 ± 0.026 ^{e}	30.76 ± 0.368 ^{c}	nd

TF3O—Total flavan-3-ols; TOF—Total Other flavonoids; TPC—Total phenolic compounds; TA—Total anthocyanin. nq—non quantifiable, nd—non detectable, tr—trace amount. Standard calibration curves used for quantification: cyanidin-3-*O*-glucoside ($y = 134{,}578x - 3 \times 10^6$, $R^2 = 0.9986$, LOD = 0.94 μg/mL; LOQ = 3.13 μg/mL, peaks 19, 20, and 24); (-)—epicatechin ($y = 13{,}304x - 7786.3$; $R^2 = 0.9998$; LOD = 0.07 μg/mL; LOQ = 0.23 μg/mL, peaks 1, 2, 3, 4, 5, and 9); quercetin-3-*O*-glucoside ($y = 28{,}555x + 3032.3$, $R^2 = 0.9996$, LOD = 0.02 μg/mL; LOQ = 0.07 μg/mL, peaks 11 and 16); *p*-coumaric acid ($y = 76{,}029x + 102{,}258$, LOD = 0.71 μg/mL; LOQ = 2.38 μg/mL peak 10); peonidin-3-*O*-glucoside ($y = 151{,}438x - 3 \times 10^6$, $R^2 = 0.9965$, LOD = 0.13 μg/mL; LOQ = 0.40 μg/mL, peaks 21, 22, 23, 25, 26, 27, 28, 29, and 30); quercetin-3-*O*-glucoside ($y = 28{,}555x + 3032.3$, $R^2 = 0.9996$, LOD = 0.02 μg/mL; LOQ = 0.07 μg/mL, peaks 13 and 14); taxifolin ($y = 39{,}133x - 13{,}647$, $R^2 = 0.02$; LOD = 0.07 μg/mL; LOQ = 2.02 μg/mL, peaks 6, 7, and 8) and resveratrol ($y = 54{,}835x - 29{,}986$, $R^2 = 0.9949$; LOD = 0.03 μg/mL; LOQ = 0.11 μg/mL, peaks 17 and 18). ANOVA analysis—In each row different letters mean significant differences ($p < 0.05$).

In general, the non-anthocyanin and anthocyanin phenolic composition was in accordance with what previously reported in red wine pomaces [35,54]. Interestingly, peonidin 3-*O*-*p*-coumarylglucoside and malvidin 3-*O*-*p*-coumarylglucoside were also detected in the diatomaceous earth extract, which was also the sample that presented the highest and third highest level of these two compounds, respectively. It should be noticed that the samples supplied by the Caves Campelo S.A. company were obtained as part of their normal routines and seasonal production. Therefore, the sample of diatomaceous earth supplied corresponded to the final discarded residue, which at industry level corresponds to around four tons that were used to filtrate several batches of wine (obtained from different cultivars, geographical origins, etc., with the requisite of all filtrated wines being red or all being white). For this reason, a direct comparison of the composition of the pomace and that of the diatomaceous earth is not possible with the obtained results. Nevertheless, for the first time, these results highlight this residue as being a potential source of phenolic compounds that otherwise would just be discarded as a waste. The high content of polyphenols verified for the diatomaceous earth sample suggest that the compounds become adsorbed in the silicon atoms during wine filtration, which can afterwards be desorbed with an adequate solvent, allowing extracts to be obtained with a high phenolic concentration.

3.2. Bioactive Properties

3.2.1. Antioxidant Activity

Table 3 shows the results of the antioxidant activity of the extracts obtained from the winemaking residues re-dissolved in ethanol/water (80:20) at different concentrations. Overall, the extracts of grape seeds, red grape pomace, and diatomaceous earth gave the lowest EC_{50} values (meaning the highest antioxidant activity) in all of the assays performed—namely, DPPH scavenging activity, reducing power, and TBARS inhibition. For the TBARS assay, the sample of skins from red grape pomace also performed very well, showing the best activity with an EC_{50} value of 0.018 mg/mL. The wine lees from red and white wine revealed the highest values (i.e., lowest antioxidant activity), yet still presented an interesting activity since all showed EC_{50} lower than 1 mg/mL. In general, the results were in good agreement with those obtained for the phenolic content, since the wine lees were also the samples evidencing the lowest contents of phenolic compounds (Figure 1, Table 2). Compared to the herein obtained results, Maluf et al. [54] reported a lower EC_{50} (6.9 μg/mL) in the DPPH assay for a grape pomace extract. However, the extract was obtained with a different solvent (75% acetone/water) and from a different grape species (*Vitis lambrusca*), thus resulting in different concentrations of phenolic compounds (69.830 mg EAG/g). Recently, the antioxidant activity of red wine lees extracted with 50% methanol/water was investigated by Lopez-Fernandez et al. [55] that reported an EC_{50} of 0.0129 mg/mL in DPPH assay, but their sample also presented a higher content of total phenolic compounds (148.03 ± 0.48 mg GAE/g). Greater EC_{50} than ours was reported in the literature [30] for the skins obtained from red grape pomace (0.563 mg/mL), while a lower value was observed for the seeds (0.023 mg/mL).

Table 3. Antioxidant activity of extracts obtained from winemaking residues.

Hydroethanolic Extracts	Antioxidant Activity EC_{50} Values (mg/mL)		
	DPPH Scavenging Activity	**Reducing Power**	**TBARS Inhibition**
Red Grape Pomace	0.123 ± 0.007	0.17 ± 0.004	0.025 ± 0.0173
Skins From Red Grape Pomace	0.217 ± 0.005	0.24 ± 0.004	0.018 ± 0.0045
Seeds From Red Grape Pomace	0.081 ± 0.005	0.11 ± 0.007	0.021 ± 0.00289
Stems	0.21 ± 0.005	0.41 ± 0.004	0.11 ± 0.00682
Diatomaceous Earth	0.17 ± 0.01	0.11 ± 0.004	0.195 ± 0.0321

Table 3. *Cont.*

Hydroethanolic Extracts	Antioxidant Activity EC_{50} Values (mg/mL)		
	DPPH Scavenging Activity	**Reducing Power**	**TBARS Inhibition**
Red Wine Lees (solid phase)	0.506 ± 0.007	0.68 ± 0.002	0.609 ± 0.00458
Red Wine Lees (liquid phase)	0.578 ± 0.005	0.78 ± 0.001	0.346 ± 0.0185
White Wine Lees	0.93 ± 0.08	0.83 ± 0.09	0.376 ± 0.0224

EC_{50} values correspond to the sample concentration achieving 50% of antioxidant activity or 0.5 of absorbance in reducing power assay. Trolox EC50 values: 0.0536 mg/mL (DDPH), 0.0451 mg/mL (reducing power), and 0.00251 mg/mL (TBARS inhibition).

3.2.2. Antimicrobial Assays

The results of the antibacterial and antifungal activities are presented in Tables 4 and 5, respectively. None of the extracts showed bactericidal activity at the tested concentrations. However, for all the bacteria, bacteriostatic activity was observed for at least one of the assayed samples. In general, the extracts were more effective against Gram-positive than Gram-negative bacteria, with the lowest MIC values being obtained for the red wine lees extracts against MRSA (Methicillin-resistant *Staphylococcus aureus*). Stronger activity against Gram-positive bacteria was also reported [56] when testing extracts obtained from peels, seeds, and stems of two red grape cultivars. Interestingly, the lowest MIC values were generally observed for the extracts of the red wine lees, which together with the extract from white wine lees, were able to inhibit the growth of all tested bacteria. On the other hand, the grape stems presented the weakest activity, since they showed no inhibition of Gram-negative bacteria, despite its activity against Gram-positive strains, particularly *L. monocytogenes* (MIC = 5 mg/mL).

Table 4. Antimicrobial activity of hydroethanolic extracts (mg/mL) obtained from winemaking residues from the harvest of 2021.

	Red Grape Pomace		**Skins**		**Seeds**		**Stems**		**Diatomaceous Earth**		**Red Wine Lees (Solid Phase)**	
	MIC	**MBC**	**MIC**	**MBC**	**MIC**	**MBC**	**MIC**	**MBC**	**MIC**	**MBC**	**MIC**	**MBC**
Gram-negative bacteria												
Escherichia coli	>10	>10	>10	>10	10	>10	>10	>10	10	>10	10	>10
Klebsiella pneumoniae	>10	>10	>10	>10	>10	>10	>10	>10	10	>10	5	>10
Morganella morganii	5	>10	5	>10	>10	>10	>10	>10	10	>10	5	>10
Proteus mirabilis	10	>10	10	>10	10	>10	>10	>10	10	>10	10	>10
Pseudomonas aeruginosa	>10	>10	>10	>10	>10	>10	>10	>10	>10	>10	10	>10
Gram-positive bacteria												
Enterococcus faecalis	10	>10	10	>10	10	>10	10	>10	10	>10	10	>10
Listeria monocytogenes	10	>10	10	>10	5	>10	5	>10	10	>10	5	>10
MRSA	10	>10	10	>10	10	>10	10	>10	5	>10	2.5	>10

	Red Wine Lees (Liquid Phase)		White Wine Lees		Ampicillin (20 mg/mL)		Imipenem (1 mg/mL)		Vancomycin (1 mg/mL)	
	MIC	MBC	MIC	MBC	MIC	MBC	MIC	MBC	MIC	MBC
Gram-negative bacteria										
Escherichia coli	5	>10	10	>10	<0.15	<0.15	<0.0078	<0.0078	n.t.	n.t.
Klebsiella pneumoniae	10	>10	10	>10	10	>10	<0.0078	<0.0078	n.t.	n.t.
Morganella morganii	5	>10	10	>10	>10	>10	<0.0078	<0.0078	n.t.	n.t.
Proteus mirabilis	10	>10	10	>10	<015	<0.15	<0.0078	<0.0078	n.t.	n.t.
Pseudomonas aeruginosa	10	>10	10	>10	>10	>10	0.5	1	n.t.	n.t.
Gram-positive bacteria										
Enterococcus faecalis	5	>10	10	>10	<0.15	<0.15	n.t.	n.t.	<0.0078	<0.0078
Listeria monocytogenes	5	>10	10	>10	<0.15	<0.15	<0.0078	<0.0078	n.t.	n.t.
MRSA	2.5.	>10	5	>10	<0.15	<0.15	n.t.	n.t.	0.25	0.5

MIC values are defined as the lowest concentration of an antimicrobial agent that prevents the visible growth of bacteria. MBC values corresponds to the lowest concentration of an anti-bactericidal agent required to kill the bacteria. MRSA—Methicillin resistant *Staphylococcus aureus*. n.t.—not tested.

Table 5. Antifungal activity of hydroethanolic extracts (mg/mL) obtained from winemaking residues from the harvest of 2021.

	Aspergillus brasiliensis		*Aspergillus fumigatus*	
	MIC	**MIC**	**MFC**	**MFC**
Red Grape Pomace	10	10	>10	>10
Skins from Red Grape Pomace	5	10	>10	>10
Seeds from Red Grape Pomace	>10	10	>10	>10
Stems	10	10	>10	>10
Diatomaceous earth	>10	>10	>10	>10
Red Wine Lees Solid Phase	10	10	>10	>10
Red Wine Lees Liquid Phase	>10	10	>10	>10
White Wine Lees	10	10	>10	>10
Ketoconazole	0.06	0.5	0.125	1

MIC values are defined as the lowest concentration of an antifungal agent that prevents the visible growth of fungus. MFC values correspond to the lowest concentration of an antifungal agent required to kill the fungus.

The results obtained for the antifungal activity were similar to the antibacterial ones, since none of the extracts demonstrated fungicidal potential and most extracts showed only a fungistatic activity at the highest tested concentration (10 mg/mL).

3.2.3. Cytotoxicity Assay in Skin Cell Lines

Due to the possible interest of the studied residues as a source of multifunctional bioactive molecules for cosmetic formulation development, the cell viability effect of the prepared extracts in HaCaT (keratinocytes) and HFF-1 (fibroblast) skin cell lines was also evaluated. As shown in Figure 2, Triton-X was utilised as a positive control to induce a decrease in cell viability. In contrast, cell culture media (DMEM) was used as the negative control, maintaining 100% cell viability. Keratinocytes are a biologically relevant target for skin irritants because they are the first layer of skin cells that come in contact with topically applied compounds. HaCaT cells also present several morphological and functional features typical of normal epidermal keratinocytes, making them a good model for skin toxicity tests [35]. By comprising a mixture of different compounds, the hydroethanolic extracts from winemaking residues may potentiate the development of reactions (allergic or irritant) when applied for dermatologic purposes, thus imposing the need to perform a cytotoxicity evaluation in skin cells and access the minimum concentration, reducing the viability of the tested cell lines. Regarding HaCaT cells, after exposure to the eight extracts (Figure 2), up to 100% cell viability was maintained at 50 μg/mL, while the highest tested concentration (400 μg/mL) was found to significantly inhibit cell viability, particularly for the red wine lees and diatomaceous earth extracts. Concerning HFF-1 cells, the cell viability was maintained up to 100% at 50 μg/mL and more than 100% viability was also observed for all the tested extracts at the highest concentration (400 μg/mL), with one exception (sample C, white wine lees). Therefore, it can be concluded that, in general, all the extracts did not show evidence of affecting the viability of HFF-1 cells. Similar results were observed in previous studies [39] that reported a percentage of cell viability around 100% when HaCaT cells were incubated with 2 mg/mL of Port wine lees extract, and 0.5 mg/mL of red wine lees extract, and when HFF cells were treated with 1 mg/mL and 0.5 mg/mL of the same extracts, respectively.

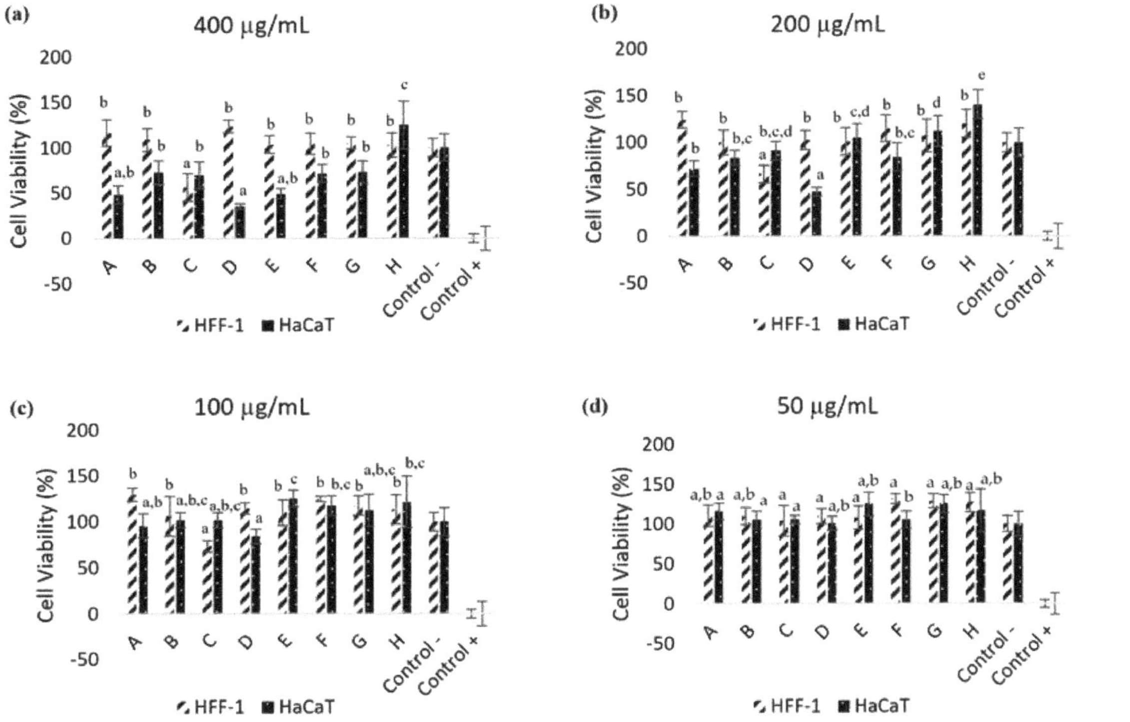

Figure 2. Cell viability effects of hydroethanolic extracts from winemaking residues on HFF-1 and HaCAt cells at different concentrations: (**a**) 400 μg/mL; (**b**) 200 μg/mL; (**c**) 100 μg/mL; and (**d**) 50 μg/mL. Values are expressed as means ± SD. In each extract, different letters mean significant differences between the concentrations ($p < 0.05$). A—Red Wine Lees (Solid Phase); B—Red Grape Pomace; C—White Wine Lees; D—Diatomaceous Earth; E—Red Wine Lees (Liquid Phase); F—Seeds; G—Skins; H—Stems.

3.3. Anti-Tyrosinase and Anti-Ageing Activities

3.3.1. Tyrosinase Inhibition

With aging, pigmentation disorders tend to appear on the skin, which has attracted the attention of cosmetic industries and led to the search for compounds with anti-hyperpigmentation potential. These pigmented lesions are caused by alterations resulting in melanin accumulation. Therefore, inhibiting melanin production is the most explored approach in this field. Because tyrosinase is the enzyme limiting the synthesis of melanin, it is a promising target for the development of skin-whitening cosmetic products. Due to their aromatic structural characteristics, phenolic compounds may have some similarities with tyrosine, the tyrosinase substrate that initiates melanin synthesis. Therefore, phenolic compounds have been described as potential tyrosine analogues that can act as competitive inhibitors of the enzyme [57].

The results obtained for the tyrosinase inhibition assay are presented in Table 6 as the percentage of enzyme inhibition. The extract that showed the most promising result was that prepared from the seeds (45.3%). On the other hand, the red lees showed negligible activity, while the white wine lees and the diatomaceous earth did not show any inhibitory activity against tyrosinase enzyme. The inhibitory action of grape pomace extract on tyrosinase enzyme was also reported [39], which demonstrated that the microwave-treated lees extract presented the best inhibitory effect (50% inhibition at 0.14 mg/mL). Red and white grape stems (1 mg/mL) were also reported to inhibit tyrosinase enzyme, with % inhibition ranging from 41.47% and 53.83% for all grape stem extracts [58]. Phenolic compounds such as those belonging to chalcones, simple phenols, and hydroxystilbenes families have been referred in the literature as effective tyrosinase inhibitors [59]. However, in this work, it was not always the extracts with the highest content in non-anthocyanin phenolic and anthocyanin phenolic compounds that translated into better tyrosinase inhibition, such as in the case of diatomaceous earth. This indicates that other biomolecules present in the extract might be the most important contributors to this effect. Butylresorcinol is a well-known inhibitor of tyrosinase enzyme that has presented better effectiveness when compared to kojic acid, arbutin, and hydroquinone [60]. In the present work, butylresorcinol presented 100% inhibition of tyrosinase enzyme at 36.52 ± 2.85 μg/mL. Overall, these findings suggest that these winemaking by-products can be suitable raw materials for recovering biomolecules with potent anti-tyrosinase activity.

Table 6. Percentage of tyrosinase and collagenase inhibition towards extracts of winemaking residues.

Extracts	% Collagenase Inhibition	% Tyrosinase Inhibition
Red Wine Lees (Solid Phase)	NA	1.22 ± 0.03 [a,b]
Red Wine Pomace	89 ± 2	28.2 ± 0.1 [c,d]
White Wine Lees	NA	NA
Diatomaceous Earth	18 ± 3	NA
Red Wine Lees (Liquid Phase)	NA	0.61 ± 0.02 [a,b]
Seeds	87 ± 8	45.3 ± 0.1 [e]
Skins	75 ± 7	34.55 ± 0.05 [b,c]
Stems	82 ± 3	29.93 ± 0.02 [d,e]

NA—No Activity. The results identified with different letters in the same column are statistically different (p-value ≤ 0.05). Positive controls: tyrosinase assay (butylresorcinol) presented 100% inhibition at 36.52 ± 2.85 μg/mL; collagenase assay (1,10-phenanthroline) presented 100% inhibition at the conditions suggested by the kit's manufacturer.

3.3.2. Collagenase Inhibition

Collagenase is an enzyme in the matrix metalloproteinase family that breaks down collagen, assisting in the degradation of the extracellular matrix, a key step in the skin aging process. Collagenase has been used to treat Dupuytren's contracture, a disease characterized by the thickening of connective tissue. In this study, a collagenase activity assay kit was used, providing a quick and easy way to measure the inhibition of collagenase activity by the winemaking residues based on the use of a synthetic peptide (FALGPA) that mimics collagen's structure. The obtained results are shown in Table 6. The red grape pomace, the skins, the seeds, and the stems were the extracts that evidenced a higher percentage of collagenase inhibition. Hydroethanolic extract (600 μg/mL) from grape seed, abundant in phenolic compounds, also presented 100% inhibition of collagenase enzyme in a previous study [15,61]. The grape pomace extract with abundant anthocyanin phenolic compounds, such as delphidin-3-*O*-glucoside, malvidin-3-*O*-glucoside, and malvidin-3-*O*-acetylglucoside, presented the best collagenase inhibitory effect. In this assay, the diatomaceous earth also presented some activity, but it showed a much weaker performance compared to the pomace extracts. Both the white and red wine lees did not exhibit any activity. However, a previous study proved that both conventional and microwave-pretreated red wine lees extract presented 50% inhibition of MMP-1 activity [39]. MMP-1 is a collagenase enzyme and its inhibition is vital in preventing extracellular matrix degradation and restoring aging skin. These results showed that these wine by-products could be explored as raw materials to scale up the recovery of structurally diverse biomolecules with a broad spectrum of biological activities relevant to skin care, in contrast to the current disposal approach.

4. Conclusions

Grape pomace has been widely studied concerning its chemical composition and bioactivity, since there is considerable interest from the food, cosmetic, and pharmaceutical industries to turn this by-product rich in bioactive compounds into innovative ingredients. However, other residues from the winemaking process, namely, the wine lees and the diatomaceous earth used for wine filtration, are less studied and are underexploited products. In this study, both of these residues demonstrated that they are potential sources of phenolic compounds, with diatomaceous earth being rich in flavan-3-ols and presenting the highest amount of epicatechin among all of the extracts. This extract also showed the highest amount of non-anthocyanin phenolic compounds, mainly due to its relatively high content in myricetin-*O*-hexoside. As expected, the red grape pomace exhibited the highest amounts of anthocyanins, mainly due to the presence of malvidin derivatives. Interestingly, a few anthocyanins were also found in the diatomaceous earth.

Together with the extracts from the seeds and red pomace, the diatomaceous earth extract also revealed a promising antioxidant activity, since it presented the best results in the reducing power assay. The lees extract, particularly the red ones, showed the highest capacity to inhibit bacterial growth, presenting a bacteriostatic effect against all tested bacteria. In general, the absence of toxicity in HFF-1 cells was observed at the highest tested concentration of extracts. However, some toxicity effects were observed in the HaCaT lines for the highest concentrations of the lees and diatomaceous earth extracts. These extracts also presented the least inhibitory effect against tyrosinase and collagenase enzymes. Therefore, while the extracts from grape pomace, seeds, skins, and stems presented results that suggest their interest as potential ingredients in cosmetic formulations (due to their antioxidant properties and their inhibition of enzymes associated with skin pigmentation and loss of firmness), the extracts of diatomaceous earth and wine lees did not perform as well, since their results suggest a possible cytotoxicity in keratinocytes at high concentrations. Even so, these undervalued residues showed a high potential for producing extracts with high antioxidant properties that can be further exploited by other industries, or for the isolation of added-value compounds, such as anthocyanins, that can be used as natural colorants. For their potential use as cosmetic ingredients, further studies are needed that consider the purification of the extracts in view of decreasing/eliminating the potential cytotoxic effect on keratinocytes.

This work reveals that the wine lees and diatomaceous earth, which are generally discarded and considered by the wine industry as waste, can be valorized as potential sources of bioactive molecules.

Author Contributions: Conceptualization, J.S.A., L.B. and C.S.-B.; methodology, C.N.D., O.T. and M.I.D.; formal analysis, S.A.H., L.B., C.S.-B. and J.S.A.; writing—original draft preparation, C.N.D., O.T. and M.I.D.; writing—review and editing, J.S.A., C.S.-B. and L.B.; supervision, J.S.A., L.B. and C.S.-B.; funding acquisition, J.S.A. and L.B. All authors have read and agreed to the published version of the manuscript.

Funding: This work was funded by "BacchusTech–Integrated Approach for the Valorization of Winemaking Residues" (POCI-01-0247-FEDER-069583), supported by the Competitiveness and Internationalization Operational Program (COMPETE 2020), under the PORTUGAL 2020 Partnership Agreement, through the European Regional Development Fund (ERDF). The authors are grateful to the Foundation for Science and Technology (FCT, Portugal) for financial support through national funds FCT/MCTES (PIDDAC) to CIMO (UIDB/00690/2020 and UIDP/00690/2020) and SusTEC (LA/P/0007/2021), and national funding by FCT, P.I., through the institutional scientific employment program-contract for L. Barros and M.I. Dias, and to COST Action 20133 FULLRECO4US, supported by COST (European Cooperation in Science and Technology). This work is also supported by MICINN for the Juan de la Cierva Formation contract of T. Oludemi (FJC2019-042549-I). The GIP-USAL is indebted to Junta de Castilla y León (Spain) for financial support through the Project SA093P20.

Institutional Review Board Statement: Not applicable.

Informed Consent Statement: Not applicable.

Data Availability Statement: Data available on request.

Acknowledgments: To Óscar Lima (Caves Campelo S.A.) for supplying the winemaking residues used in this work.

Conflicts of Interest: The authors declare no conflict of interest.

References

1. Troncozo, M.I.; Liesevic, M.; Beskoski, V.; Andelkovic, B.; Ballati, P.; Saparrat, M.C.N. Fungal Transformation and Reduction of Phytotoxicity of Grape Pomace Waste. *Chemosphere* **2019**, *237*, 124458. [CrossRef]
2. Gomes, C.; Sadoyan, G.; Dias, R.C.S.; Costa, M.R.P.F.N. Development of Moleculary Imprinted Polymers to Target Polyphenols Present in Plant Extracts. *Processes* **2017**, *5*, 72. [CrossRef]
3. Anastasiadi, M.; Pratsinis, H.; Kletsas, D.; Skaltsounis, A.L.; Haroutounian, S.A. Bioactive non-coloured Polyphenols Content of Grapes, Wine and Vinification by-products: Evaluation of the Antioxidant Activities of their Extracts. *Food Res. Int.* **2010**, *43*, 805–813. [CrossRef]
4. Gerogiannaki-Christopoulou, M.; Kyriakidis, N.V.; Athanasopoulos, P.E. Effect of Grape Variety (*Vitis vinifera* L.) and Grape Pomace Fermentation Conditions on some Volatile Compounds of the Produced Grape Pomace Distillate. *J. Int. Des Scien. De La Vigne Et Du Vin* **2004**, *38*, 225–230. [CrossRef]
5. Fontana, A.R.; Bottini, R. High-throughput Method Based on Quick, Easy, Cheap, Effective, Rugged and Safe Followed by Liquid Chromatography-multi-wavelength Detection for the Quantification of Multiclass Polyphenols in Wines. *J. Chrom. A* **2014**, *1342*, 44–53. [CrossRef]
6. Tao, Y.; Wu, D.; Qing-An, Z.; Da-Wen, S. Ultrasound-assisted Extraction of Phenolics from Wine Lees: Modeling, Optimization and Stability of Extracts During Storage. *Ultra. Sonochem.* **2014**, *21*, 706–715. [CrossRef]
7. Abe, L.T.; Vieira da Mota, R.; Lajolo, F.M.; Genovese, M.I. Phenolic Compounds and Antioxidant Activity of *Vitis labrusca* and *Vitis vinifera* cultivars. *Food Scien. Tech.* **2007**, *27*, 394–400. [CrossRef]
8. Da Silva, F.R.T. Caracterização e Valorização Energética de Resíduos da Poda da Vinha. Master's Thesis, Instituto Superior de Engenharia do Porto, Porto, Portugal, 2017. Available online: https://recipp.ipp.pt/bitstream/10400.22/11684/1/DM_FilipaSilva_2017_MES.pdf (accessed on 14 December 2022). (In Portuguese).
9. Guchu, E.; Ebeler, S.E.; Lee, J.; Mitchell, A.E. Monitoring Selected Monomeric Polyphenol Composition in pre- and post-fermentation products of *Vitis vinifera* L. cv. *Airén and cv. Grenache noir. LWT—Food Scien. Tech.* **2015**, *60*, 552–562. [CrossRef]
10. Ricci, A.; Mejia, J.A.A.; Versari, A.; Chiarello, E.; Bordoni, A.; Parpinello, G.P. Microencapsulation of Polyphenolic Compounds Recovered from Red Wine Lees: Process Optimization and Nutraceutical Study. *Food Bioprod. Process.* **2022**, *132*, 1–12. [CrossRef]
11. Delgado de la Torre, M.P.; Priego-Capote, F.; Luque de Castro, M.D. Characterization and Comparison of Wine Lees by Liquid Chromatography—Mass Spectrometry in High-Resolution Mode. *J. Agric. Food Chem.* **2015**, *63*, 1116–1125. [CrossRef]
12. Dujmic, F.; Ganic, K.K.; Curic, D.; Karlovic, S.; Bosiljkov, T.; Jezek, D.; Vidrih, R.; Hribar, J.; Zlatic, E.; Prusina, T.; et al. Non-Thermal Ultrasonic Extraction of Polyphenolic Compounds from Red Wine Lees. *Foods* **2020**, *9*, 472. [CrossRef]
13. Giacobbo, A.; Dias, B.B.; Onorevoli, B.; Bernardes, A.M.; Norberta de Pinho, M.; Camarão, E.B.; Rodrigues, E.; Jacques, R.A. Wine Lees from the 1st and 2nd Rackings: Valuable By-Products. *J. Food Sci. Tech.* **2019**, *56*, 1559–1566. [CrossRef] [PubMed]
14. Pérez-Serradilla, J.A.; Luque de Castro, M.D. Microwave-assisted Extraction of Phenolic Compounds from Wine Lees and Spray-Drying of the Extract. *Food Chem.* **2011**, *124*, 1652–1659. [CrossRef]
15. Wu, J.J.; Lin, J.C.; Wang, C.H.; Jong, T.T.; Yang, H.L.; Hsu, S.L.; Chang, C.J. Extraction of antioxidative compounds from wine lees using supercritical fluids and associated anti-tyrosinase activity. *J. Supercrit. Fluids* **2009**, *50*, 33–41. [CrossRef]
16. Zhijing, Y.; Shavandi, A.; Harrison, R.; Bekhit, A.E. A Characterization of phenolic compounds in wine lees. *Antioxidants* **2018**, *7*, 48. [CrossRef] [PubMed]
17. Barcia, M.T.; Pertuzatti, P.B.; Gómez-Alonso, S.; Godoy, H.T.; Hermosín-Gutiérrez, I. Phenolic Composition of Grape and Winemaking By-Products of Brazilian Hybrid Cultivars BRS Violeta and BRS Lorena. *Food Chem.* **2014**, *159*, 95–105. [CrossRef] [PubMed]
18. Bzainia, A.; Dias, R.C.S.; Costa, M.R.P.F.N. Enrichment of Quercetin from Winemaking Residual Diatomaceous Earth via a Tailor-Made Imprinted Adsorbent. *Molecules* **2022**, *27*, 6406. [CrossRef]
19. Antonic, B.; Jancikova, S.; Dordevic, D.; Tremlova, B. Grape Pomace Valorization: A Systematic Review and Meta-Analysis. *Foods* **2020**, *9*, 1627. [CrossRef]
20. Bianchi, F.; Lomuscio, E.; Rizzi, C.; Simonato, B. Predicted Shelf-Life, Thermodynamic Study and Antioxidant Capacity of Breadsticks Fortified with Grape Pomace Powders. *Foods* **2021**, *10*, 2815. [CrossRef]
21. Tolve, R.; Simonato, B.; Rainero, G.; Bianchi, F.; Rizzi, C.; Cervini, M.; Giuberti, G. Wheat Bread Fortification by Grape Pomace: Nutritional, Technological, Antioxidant, and Sensory Properties. *Foods* **2021**, *10*, 75. [CrossRef]
22. Thakur, R.; Batheja, P.; Kaushik, D.; Michniak, B. Structural and Biochemical Changes in Aging Skin and Their Impact on Skin Permeability Barrier. In *Skin Aging Handbook*; Andrew, W., Ed.; Elsevier: Amsterdam, The Netherlands, 2009; pp. 55–90, ISBN 9780815519799.
23. Gilchrest, B.A.; Blog, F.B.; Szabo, G. Efects of Aging and Chronic Sun Exposure on Melanocytes in Human Skin. *J. Investig. Dermat.* **1979**, *73*, 141–143. [CrossRef]

24. Wittenauer, J.; Mäckle, S.; Sußmann, D.; Schweiggert-Weisz, U.; Carle, R. Inhibitory Effects of Polyphenols from Grape Pomace Extract on Collagenase and Elastase Activity. *Fitoterapia* **2015**, *101*, 179–187. [CrossRef] [PubMed]
25. Lin, Y.S.; Chen, H.J.; Huang, J.P.; Lee, P.C.; Tsai, C.R.; Hsu, T.F.; Huang, W.Y. Kinetics of Tyrosinase Inhibitory Activity Using *Vitis vinifera* Leaf Extracts. *BioMed Res. Int.* **2017**, *2017*, 5232680. [CrossRef] [PubMed]
26. Iyda, J.H.; Fernandes, A.; Ferreira, F.D.; Alves, M.J.; Pires, T.C.S.P.; Barros, L.; Amaral, J.S.; Ferreira, I.C.F.R. Chemical composition and bioative properties of the wild edible plant *Rapahnus raphanistrum* L. *Food Res. Int.* **2019**, *121*, 714–722. [CrossRef] [PubMed]
27. Barros, L.; Pereira, E.; Calhelha, R.C.; Dueñas, M.; Carvalho, A.M.; Santos-Buelga, C.; Ferreira, I.C.F.R. Bioactivity and Chemical Characterization in Hydrophilic and Lipophilic Compounds of *Chenopodium ambrosioides* L. *J. Funct. Foods* **2013**, *5*, 1732–1740. [CrossRef]
28. Alves, M.J.; Ferreira, I.C.F.R.; Lourenço, I.; Costa, E.; Martins, A.; Pintado, M. Wild Mushroom Extracts as Inhibitors of Bacterial Biofilm Formation. *Pathogens* **2014**, *3*, 667–679. [CrossRef]
29. Dias, M.I.; Barros, L.; Morales, P.; Cámara, M.; Alves, M.J.; Oliveira, M.B.P.P.; Santos-Buelga, C.; Ferreira, I.C.F.R. Wild *Fragaria vesca* L. fruits: A Rich Source of Bioactive Phytochemicals. *Food Funct.* **2016**, *11*, 4471–4722. [CrossRef]
30. Pires, T.C.S.P.; Dias, M.I.; Barros, L.; Calhelha, R.C.; Alves, M.J.; Santos-Buelga, C.; Ferreira, I.C.F.R. Phenolic Compounds Profile, Nutritional Compounds and Bioactive Properties of *Lycium barbarum* L.: A comparative study with Stems and Fruits. *Indust. Crops Prod.* **2018**, *122*, 574–581. [CrossRef]
31. Heleno, S.A.; Stojkovic, D.; Barros, L.; Glamoclija, J.; Sokovic, M.; Martins, A.; Queiroz, M.J.R.P.; Ferreira, I.C.F.R. A Comparative Study of Chemical Composition, Antioxidant, and Antimicrobial Properties of *Morchella esculenta* (L.) Pers. From Portugal and Serbia. *Food Res. Int.* **2013**, *51*, 236–243. [CrossRef]
32. Taofiq, O.; Rodrigues, F.; Barros, L.; Barreiro, M.F.; Ferreira, I.C.F.R.; Oliveira, M.B.P.P. Mushroom Ethanolic Extracts as Cosmeceuticals Ingredients: Safety and ex vivo Skin Permeation Studies. *Food Chem. Tox.* **2019**, *127*, 228–236. [CrossRef]
33. Winder, A.J. A Stopped Spectrophotometric Assay for the Dopa Oxidase Activity of Tyrosinase. *J. Bio. Bioph. Methods* **1994**, *28*, 173–183. [CrossRef]
34. Taofiq, O.; Heleno, S.H.; Calhelha, R.C.; Alves, M.J.; Barros, L.; González-Paramás, A.M.; Barreiro, M.F.; Ferreira, I.C.F.R. The potential of *Ganoderma lucidum* extracts as bioactive ingredientes in topical formulations, beyond its nutritional benefits. *Food Chem. Toxicol.* **2017**, *108*, 139–147. [CrossRef] [PubMed]
35. Peixoto, C.M.; Dias, M.I.; Alves, M.J.; Calhelha, R.C.; Barros, L.; Pinho, S.; Ferreira, I.C.F.R. Grape Pomace as a Source of Phenolic Compounds and Diverse Bioactive Properties. *Food Chem.* **2018**, *253*, 132–138. [CrossRef] [PubMed]
36. Santos-Buelga, C.; Francia-Aricha, E.M.; Escribano-Bailón, M.T. Comparative flavan-3-ol Compostition of Seeds from Different Grape Varieties. *Food Chem.* **1995**, *51*, 197–201. [CrossRef]
37. Montealegre, R.R.; Peces, R.R.; Vozmediano, J.L.C.; Gascueña, J.M.; Romero, E.G. Phenolic Compounds in Skins and Seeds of Ten Grape *Vitis vinifera* Varieties Grown in a Warm Climate. *J. Food Comp. Anal.* **2006**, *19*, 687–693. [CrossRef]
38. Maicas, S.; Mateo, J. Sustainability of Wine Production. *Sustainability* **2020**, *12*, 559. [CrossRef]
39. Matos, M.S.; Romero-Diaz, R.; Álvarez, A.; Bronze, M.R.; Rodriguez-Rojo, S.; Mato, R.B.; Cocero, M.J.; Matias, A.A. Polyphenol-Rich Extract Obtained from Winemaking Waste Streams as Natural Ingredients with Cosmeceutical Potential. *Antioxidants* **2019**, *8*, 355. [CrossRef]
40. Romero-Díez, R.; Rodríguez-Rojo, S.; Cocero, M.J.; Duarte, C.M.M.; Matias, A.A.; Bronze, M.R. Phenolic Characterization of Aging Wine Lees: Correlation with Antioxidant Activities. *Food Chem.* **2018**, *259*, 188–195. [CrossRef]
41. Gil-Sánchez, I.; Ayuda-Durán, B.; González-Manzano, S.; Santos-Buelga, C.; Cueva, C.; Martín-Cabrejas, M.A.; Sanz-Buenhombre, M.; Guadamarra, A.; Moreno-Arribas, M.V.; Bartolomé, B. Chemical characterization and in vitro colonic fermentation of grape pomace extracts. *J. Sci. Food Agric.* **2017**, *97*, 3433–3444. [CrossRef]
42. Jara-Palacios, M.J.; Hernanz, D.; González-Manzano, S.; Santos-Buelga, C.; Escudero-Gilete, M.L.; Heredia, F.J. Detailed phenolic composition of white grape by-products by RRLC/MS and measurement of the antioxidant activity. *Talanta* **2014**, *125*, 51–57. [CrossRef]
43. Jara-Palacios, M.J.; González-Manzano, S.; Escudero-Gilete, M.L.; Hernanz, D.; Dueñas, M.; González-Paramás, A.M.; Heredia, F.J.; Santos-Buelga, C. Study of Zalema grape pomace: Phenolic composition and biological effects in *Caenorhabditis elegans*. *J. Agric. Food Chem.* **2013**, *61*, 5114–5121. [CrossRef]
44. González-Paramás, A.M.; Esteban-Ruano, S.; Santos-Buelga, C.; Pascual-Teresa, S.; Rivas-Gonzalo, J.C. Flavanol content and antioxidant activity in winery byproducts. *J. Agric. Food Chem.* **2004**, *52*, 234–238. [CrossRef] [PubMed]
45. Sun, J.; Liang, F.; Bin, Y.; Li, P.; Duan, C. Screening Non-colored Phenolics in Red Wines using Liquid Chromatography/Ultraviolet and Mass Spectrometry/Mass Spectrometry Libraries. *Molecules* **2007**, *12*, 679–693. [CrossRef] [PubMed]
46. Milincic, D.D.; Stanisavljevic, N.S.; Kostic, A.Z.; Bajic, S.S.; Kojic, M.O.; Gasic, U.M.; Barac, M.B.; Stanojevic, S.P.; Tesic, Z.L.; Pesic, M.B. Phenolic compounds and biopotential of grape pomace extracts from Prokupac red grape variety. *LWT* **2021**, *138*, e110739. [CrossRef]
47. Lingua, M.S.; Fabani, M.P.; Wunderlin, D.A.; Baroni, M.V. In vivo antioxidant activity of grape, pomace and wine from three red varieties grown in Argentina: Its relationship to phenolic profile. *J. Funct. Foods* **2016**, *20*, 332–345. [CrossRef]
48. Montero, L.; Sáez, V.; Von Baer, D.; Cifuentes, A.; Herrero, M. Profiling of *Vitis vinifera* L. canes (poly) phenolic compounds using comprehensive two-dimensional liquid chromatography. *J. Chromatogr. A* **2018**, *1536*, 205–215. [CrossRef] [PubMed]

49. Careri, M.; Corradini, C.; Elviri, L.; Nicoletti, I.; Zagnoni, I. Direct HPLC analysis of quercetin and trans-resveratrol in red wine, grape, and winemaking byproducts. *J. Agric. Food Chem.* **2003**, *51*, 5226–5231. [CrossRef]
50. Alcalde-Eon, C.; Escribano-Bailón, M.T.; Santos-Buelga, C.; Rivas-Gonzalo, J.C. Changes in the detailed pigment composition of red wine during maturity and ageing. A comprehensive study. *An. Chim. Acta* **2006**, *563*, 238–2549. [CrossRef]
51. Hebrero, E.; Santos-Buelga, C.; Rivas-Gonzalo, J.C. High Performance Liquid Chromatography Diode Array Spectroscopy Identification of Anthocyanins of Vitis vinifera variety Tempranillo. *Am. J. Enol. Vitic.* **1988**, *39*, 3. [CrossRef]
52. García-Beneytez, E.; Cabello, F.; Revilla, E. Analysis of grape and wine anthocyanins by HPLC-MS. *J. Agric. Food Chem.* **2003**, *51*, 5622–5629. [CrossRef]
53. García-Marino, M.; Hernández-Hierro, J.M.; Santos-Buelga, C.; Rivas-Gonzalo, J.C.; Escribano-Bailón, M.T. Multivariate analysis of the polyphenol composition of Tempranillo and Graciano red wines. *Talanta* **2011**, *85*, 2060–2066. [CrossRef] [PubMed]
54. Maluf, D.F.; Gonçalves, M.M.; D'Angelo, R.W.O.; Girassol, A.B.; Tulio, A.P.; Pupo, Y.M.; Farago, P.V. Cytoprotection of Antioxidant Biocompounds from Grape Pomace: Further Exfoliant Phytoactive Ingredients for Cosmetic Products. *Cosmetics* **2018**, *5*, 46. [CrossRef]
55. López-Férnandez-Sobrino, R.; Margalef, M.; Torres-Fuentes, C.; Ávila-Román, J.; Aragonès, G.; Muguerza, B.; Bravo, F.I. Enzyme-Assisted Extraction to Obtain Phenolic-Enriched Wine Lees with Enhanced Bioactivity in Hypertensive Rats. *Antioxidants* **2021**, *10*, 517. [CrossRef]
56. Silva, V.; Igrejas, G.; Falco, V.; Santos, T.P.; Torres, C.; Oliveira, A.M.P.; Pereira, J.E.; Amaral, J.S.; Poeta, P. Chemical composition, antioxidante and antimicrobial activity of phenolic compounds extracted from wine industry by-products. *Food Cont.* **2018**, *92*, 516–552. [CrossRef]
57. Castejón, N.; Thorarinsdottir, K.A.; Einarsdottir, R.; Kristbergsson, K.; Mateinsdóttir, G. Exploring the Potential of Icelandic Seaweeds Extracts Produced by Aqueous Electric Fields-Assisted Extraction for Cosmetic Applications. *Mar. Drugs* **2021**, *19*, 662. [CrossRef] [PubMed]
58. Leal, C.; Gouvinhas, I.; Santos, R.A.; Rosa, E.; Silva, A.M.; Saavedra, M.J.; Barros, A.I.R.N.A. Potential application of grape (*Vitis vinifera* L.) stem extracts in the cosmetic and pharmaceutical industries: Valorization of a by-product. *Indust. Crops Prod.* **2020**, *154*, 112675. [CrossRef]
59. Panzella, L.; Napolitano, A. Natural and Bioinspired Phenolic Compounds as Tyrosinase Inhibitors for the Treatment of Skin Hyperpigmentation: Recent Advances. *Cosmetics* **2019**, *6*, 57. [CrossRef]
60. Kolbe, L.; Mann, T.; Gerwat, W.; Batzer, J.; Ahlheit, S.; Scherner, C.; Wenck, H.; Stab, F. 4-n-butylresorcinol, a highly effective tyrosinase inhibitor for the topical treatment of hyperpigmentation. *JEADV* **2013**, *27* (Suppl. S1), 19–23. [CrossRef]
61. Michailidis, D.; Angelis, A.; Nikolaou, P.E.; Mitakou, S.; Skaltsounis, A.L. Explitation of *Vitis vinifera, Foeniculum vulgare, Cannabis sativa* and *Punica granatum* By-Product Seeds as Dermo-Cosmetic Agents. *Molecules* **2021**, *26*, 731. [CrossRef]

Disclaimer/Publisher's Note: The statements, opinions and data contained in all publications are solely those of the individual author(s) and contributor(s) and not of MDPI and/or the editor(s). MDPI and/or the editor(s) disclaim responsibility for any injury to people or property resulting from any ideas, methods, instructions or products referred to in the content.

cosmetics

MDPI

Article

Preparation and Investigation of the SPF and Antioxidant Properties of O/W and W/O Emulsions Containing Vitamins A, C and E for Cosmetic Applications

Nikolaos D. Bikiaris [1], Ioanna Koumentakou [1], Katerina Hatzistamatiou [1], Smaro Lykidou [1], Panagiotis Barmpalexis [2] and Nikolaos Nikolaidis [1,*]

[1] Laboratory of Chemistry and Technology of Polymers and Dyes, Department of Chemistry, Aristotle University of Thessaloniki, 54124 Thessaloniki, Greece; katerinahatzistamatiou@gmail.com (K.H.)
[2] Department of Pharmaceutical Technology, School of Pharmacy, Aristotle University of Thessaloniki, 54124 Thessaloniki, Greece; pbarmp@pharm.auth.gr
* Correspondence: nfnikola@chem.auth.gr

Abstract: In the current work, Oil in Water (O/W) and Water in Oil (W/O) emulsions containing Vitamins A, C and E in 0.5, 1 and 2% wt concentrations were prepared. The pH and viscosity stability over storage, as well as the sunscreen and antioxidant properties of the obtained emulsions, were investigated. The results obtained showed that vitamins slightly increased the pH of the blank emulsions; however, their pH values were within the acceptable values (pH = 4–6). Nevertheless, all emulsions presented excellent pH stability during storage for up to 90 days. Similar results were observed by rheological measurements as the prepared emulsions did not exhibit viscosity instabilities deriving during storage. Moreover, emulsions containing Vitamin A exhibited higher UV protection than the other emulsions, as the W/O emulsion containing 2% wt Vitamin A presented the highest SPF value at 22.6.

Keywords: emulsions; oil in water (O/W); water in oil (W/O); vitamin A; vitamin E; vitamin C; UV protection; antioxidant properties

Citation: Bikiaris, N.D.; Koumentakou, I.; Hatzistamatiou, K.; Lykidou, S.; Barmpalexis, P.; Nikolaidis, N. Preparation and Investigation of the SPF and Antioxidant Properties of O/W and W/O Emulsions Containing Vitamins A, C and E for Cosmetic Applications. *Cosmetics* **2023**, *10*, 76. https://doi.org/10.3390/cosmetics10030076

Academic Editor: Agnieszka Feliczak-Guzik

Received: 10 April 2023
Revised: 2 May 2023
Accepted: 4 May 2023
Published: 10 May 2023

Copyright: © 2023 by the authors. Licensee MDPI, Basel, Switzerland. This article is an open access article distributed under the terms and conditions of the Creative Commons Attribution (CC BY) license (https://creativecommons.org/licenses/by/4.0/).

1. Introduction

Skin is the primary defensive barrier of the human organism against pathogens and chemical threats. It comprises three main layers: the stratum corneum (SC), the dermis (DE) and the hypodermis [1–3]. Ultraviolet (UV) irradiation, naturally abundant in the environment through sun exposure, subsidizes a variety of skin maladies, including degenerative aging, inflammation and cancer. Moreover, UV exposure is responsible for the generation of reactive oxygen species (ROS) that may have detrimental effects on the skin's health. In addition, a great number of active substances have been applied to the surface of the skin for therapeutic and pharmaceutical purposes through cosmetic products, such as body, face and hand emulsions [4–6].

Emulsions are complex formulations used for optimizing various end-use properties of dermatological substances, such as the stability and delivery of active compounds [7]. The basic formulation of a standard emulsion contains at least water as the continuous phase, oil as the dispersed phase, and emulsifiers ensuring the stability of the dispersion of oil droplets [8]. In the past, oil in water (O/W) and water in oil (W/O) emulsions have been extensively used and studied for the dermal delivery [9] of numerous active ingredients such as vitamins [10,11], retinoids [12], drugs [13], proteins and peptides [14].

Vitamins, one the most-known antioxidants used in the cosmetics industry, are naturally occurring in human skin, where they prevent lipid peroxidation at the cell membrane level, stabilizing especially those membranes with high content of polyunsaturated fatty acids [15]. Vitamin A (Vit A) is a generic term for a family of organic lipid-soluble retinal that is essential for the performance of multiple biological functions, such as normal growth and development [16]. The most effective forms of Vit A are retinal and retinol, which contain an aldehyde or an alcohol end group, respectively [17]. Vitamin C is a major antioxidant in the aqueous cell compartment with the capability to neutralize oxidative stress generated by various factors, with the most common source being UV exposure [18,19]. In certain skin disorders, the use of ascorbic acid (Vit C) can create an acidic environment that contributes to wound healing by regulating the wound infection and barrier function, increasing the antimicrobial activity and enhancing epithelization and angiogenesis [20]. Vitamin E (Vit E) can be found in high concentrations in the deepest layers of the stratum corneum (SC), forming the primary defense of the skin to the oxidative stress induced by UV exposure. Topical delivery of Vitamin E protects the skin against UV-caused cutaneous harm, and carcinogenic and mutagenic activity of chemical agents [21,22].

Yoshida, et al. [23], proposed the preparation of O/W/O emulsions for the stabilization of Vitamin A. Maia, et al. [24] investigated the influence of sodium metabisulfite and glutathione antioxidants on the stability of Vitamin C in O/W emulsions. It was found that the storage condition of 40.0 $\pm$ 0.5 °C accelerated the degradation rate of Vitamin C, despite the presence of the antioxidants. Fraj, et al. [25] studied the in vitro release of Vitamins C and E from microcapsules stabilized by gelatin sodium caseinate system in double W/O/W emulsions. Moyano and Segall [26] evaluated the stability of Vitamin A palmitate in the presence of Vitamin E and other antioxidants in lipophilic/hydrophilic medium (O/W emulsions) at pH 3.0, 5.0 and 7.0.

In the current work, a series of O/W and W/O emulsions containing Vitamins A, C and E have been prepared, and their sunscreen activity and their antioxidant properties are studied for the first time.

2. Materials and Methods

2.1. Materials

Vit A (Retinyl palmitate), Vit C (L-Ascorbic acid) and Vit E (Alpha-Tocopherol) were received by Tzimas Cosmetics (Thessaloniki, Greece). For the preparation of O/W and W/O emulsions, olive oil, shea butter, glycerin, cetearyl alcohol, cetyl alcohol, sodium citrate, beeswax, xanthan gum, polysorbate 60 (Tween 60), stearic acid, ethylhexylglycerin and phenoxyethanol were purchased by Novita Group (Thessaloniki, Greece). All other materials and reagents used in this study were of analytical grade of purity.

2.2. Preparation of the O/W and W/O Emulsions

The preparation of the O/W and W/O emulsions was conducted according to a technique described in previous works [27,28]. Briefly, Table 1 summarizes the two phases and the utilized substances for the emulsions' fabrication. In summary, the water phase was completely dissolved in a water bath at 70 °C. The oil phase was heated separately in a water bath at 70 °C until all the substances were dissolved and were then added dropwise to the water phase under constant mechanical stirring (600 RPM) using a 2020 RZR (Heidolph, Schwabach, Germany) mechanical stirrer until complete homogenization. The prepared emulsions were left under mechanical stirring (400 RPM) for 1 h; vitamins were added at 45 °C in various weight ratios (0.5, 1 and 2% *w*/*w*) as illustrated in Table 2, and mechanical stirring continued for 1 more hour. Afterwards, ethylhexylglycerin and phenoxyethanol were added as preservatives at room temperature, and mechanical stirring was continued for 30 min. Regarding the O/W formulations, the water phase and the oil phase of the final emulsions were 25% *w*/*w* and 75% *w*/*w*, respectively. On the other hand, W/O emulsions contained 75% *w*/*w* of the oil phase and 25% *w*/*w* of the water phase. Lastly, two blank emulsions (O/W and W/O) without the addition of any vitamins were also prepared as control samples. The final volume/weight ratio of the produced emulsions was 150 mL/200 g. The fabricated formulations were stored in sterilized glass sample jars (250 g) at room temperature.

Table 1. Water and oil phases for the preparation of the O/W and W/O emulsions.

	Water Phase	
Ingredients	**O/W Emulsions (75%)**	**W/O Emulssions (25%)**
Water	70 gr	23.5 gr
Glycerin	3.5 gr	1.15 gr
Citric acid	0.5 gr	0.175 gr
Xanthan gum	1 gr	0.175 gr
	Oil phase	
Ingredients	**O/W emulsions (25%)**	**W/O emulsions (75%)**
Olive oil	11 gr	43.75 gr
Cetyl alcohol	2 gr	3.25 gr
Cetearyl alcohol	2 gr	3.25 gr
Stearic acid	2 gr	6.5 gr
Shea butter	2 gr	6.5 gr
Beeswax	2 gr	6.5 gr
Tween 60	2 gr	3.25 gr

Table 2. Abbreviations of the prepared emulsions and their contents in Vitamins A, C and E.

	0.5% Vit A O/W	**1% Vit A O/W**	**2% Vit A O/W**	**0.5% Vit A W/O**	**1% Vit A W/O**	**2% Vit A W/O**	**0.5% Vit C O/W**	**1% Vit C O/W**	**2% Vit C O/W**	**0.5% Vit C W/O**	**1% Vit C W/O**	**2% Vit C W/O**	**0.5% Vit E O/W**	**1% Vit E O/W**	**2% Vit E O/W**	**0.5% Vit E W/O**	**1% Vit E W/O**	**2% Vit E W/O**
Vitamin A	0.5 gr			0.5 gr			0.5 gr			0.5 gr			0.5 gr			0.5 gr		
Vitamin C		1 gr			1 gr			1 gr			1 gr			1 gr			1 gr	
Vitamin E			2 gr			2 gr			2 gr			2 gr			2 gr			2 gr

2.3. *pH Measurement*

The stability of the prepared emulsions was studied via pH and viscosity measurements after 1, 7, 14, 30, 60 and 90 days of storage in RT after the preparation of the emulsions [29]. The pH measurements of the emulsions were read precisely from the instrument by dipping the pH sensor (Microprocessor, WTW, pH 535, Gemini BV, Apeldoorn, The Netherlands), and the mean value of three consecutive measurements was calculated. Variations in the results of pH measurements could not exceed 10% of the initial value for the emulsion to be considered stable.

2.4. *Viscosity Determination*

Viscosity measurements were performed using the R3 spindle of a Visco Star Plus viscometer (Controla S.A, Thessaloniki, Greece) under 50 and 100 rpm after 1, 7, 14, 30, 60 and 90 days of storage in RT after the fabrication of the emulsions. Apparent viscosity determinations were assessed by dipping the R3 spindle of the viscometer for 1 min in each of the studied emulsions [30], and the obtained values were expressed as Centipoise (cP). Variation in the results of viscosity measurements could not exceed 10% of the initial value for the emulsion to be considered stable.

2.5. *Sun Protection Factor (SPF)*

The diluted solution transmittance method was employed for the investigation of the SPF values of the emulsions. Briefly, 1% *w*/*v* from each sample was added in ethanol, completely homogenized using sonication and filtered through Whatman filters. 20% *v*/*v* of each solution was diluted with ethanol. The absorption values of the emulsions were achieved in the range of 290–320 nm (every 5 nm) using a UV-Vis spectrophotometer

(Shimadzu, Tokyo, Japan) [28]. Mansur equation was utilized to determine the SPF values of the emulsions:

$$\text{SPF in vitro} = \text{CF} * \sum_{290}^{320} \text{EE}(\lambda) * \text{I}(\lambda) * \text{abs}(\lambda) \quad (1)$$

where CF (correction factor) is 10), EE(λ) is the erythemogenic effect of radiation at wavelength λ, I(λ) denotes the intensity of solar light at wavelength λ and abs(λ) denotes the absorbance of the sample at wavelength λ. "EE" and × "I" are constants that were first determined in the work of Sayre, et al. [31].

2.6. Antioxidant Study

The antioxidant activity (AA) of the prepared emulsions was determined via the 2,2-Diphenyil-1-picrylhydrazyl (DPPH) method, according to the method developed by Blois in 1958 [32]. Specifically, 1 mL of each sample was dispersed in ethanol (1% v/v) and was added to 3 mL of a 5×10^{-3} mg/mL ethanol DPPH solution. The reference sample was composed of ethanol and the DPPH/EtOH solution in a 1/3 ratio. The samples were sonicated for 10 min, and their absorbance was recorded after 30 min at 517 nm using a UV-Vis spectrometer (UV Probe 1650, Shimadzu, Tokyo, Japan). The free radical scavenging activity was described according to the following equation [33]:

$$\text{Free radical scavenging activity } (\%) = \frac{\text{Absorbance of control} - \text{Absorbance of extracts}}{\text{Absorbance of control}} * 100 \quad (2)$$

All experiments for each formulation were performed in triplicate for the validation of the obtained results.

3. Results

3.1. pH Stability

Figure 1 presents the obtained pH values for each investigated emulsion during storage at room temperature after 1, 7, 14, 30, 60 and 90 days after their preparation. The pH stability, in general, was excellent, and this can be attributed to the citric acid used as a pH regulator. Obviously, the composition of the emulsion affects the pH. It is well known that olive oil possesses more acidic pH than deionized water; hence W/O type of emulsion is expected to exhibit higher values of pH. All other emulsifiers used in the current work have a neutral pH of around 7 and thus do not have a significant effect on the stability. Generally, the pH values of the studied samples varied in the range of 4–6. However, emulsions with 2% w/w Vitamin C exhibited pH values below 4. Consecutively, the specific emulsions are not appropriate for dermal application. For this reason, no further characterization tests (viscosity, SPF, antioxidant) were performed on these samples.

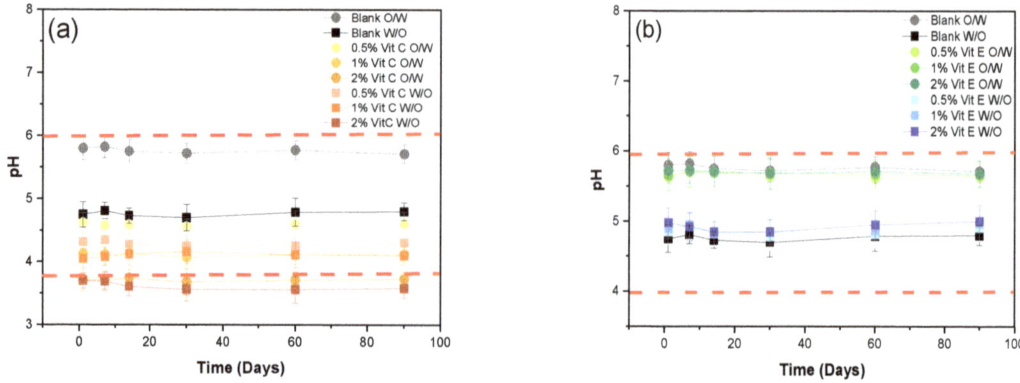

Figure 1. *Cont.*

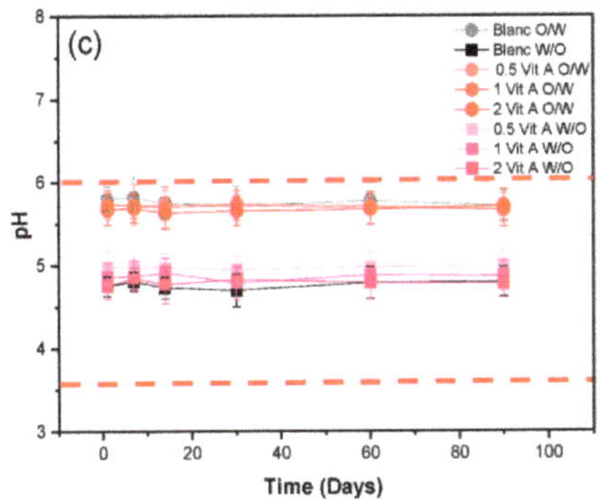

Figure 1. Change in pH values over time for O/W and W/O Blank emulsions, O/W and W/O emulsions with (**a**) Vitamin C; (**b**) Vitamin E; and (**c**) Vitamin A.

3.2. Rheological Study

Viscosity measurements were determined for all the prepared emulsions, and their stability profiles at 50 and 100 RPM during storage for up to 90 days are presented in Figures 2–4. It is clearly evident that no viscosity instabilities were observed during storage. This is due to the nature of the emulsifiers and the viscosity regulators used in the water and in the oil phase. The xanthate gum used in the water phase is an excellent polysaccharide-based viscosity stabilizer and thickening agent, producing very stable emulsions. Additionally, the Tween 60, possessing 60 moles of the long hydrophilic ethylene oxide chain, confers excellent emulsifying properties, thus stabilizing the emulsion. The increased amount of oily and fatty substances, such as oil and beeswax, resulted in increased viscosity values of the W/O blank. This is a result of the stickiness of fat particles, and the viscosity increase is attributed to the solidification of fat. Moreover, both at 50 and 100 RPM, similar values with the blank sample, varying in the range of 30–80 cP, were obtained for all O/W formulations containing the studied vitamins, with the ones at 50 RPM being higher. Regarding the W/O emulsions, significant changes were noticed. Specifically, at 50 RPM, Vit A samples showed decreased viscosity compared to blank W/O, with the one containing 0.5% Vit A having the lowest value around 200 cP. On the contrary, at 100 RPM, emulsions with 1% Vit A showed higher values compared to the blank sample. Vit C formulations, both at 50 and 100 RPM, exhibited similar and higher values than the blank W/O cream. Lastly, all samples containing Vit E exhibited higher values than the control sample. At 50 RPM, the viscosity profiles of all creams were around 350 cP. However, at 100 RPM, 0.5% Vit A sample was near 175 cP. 1 and 2% Vit A samples exhibited increased values near 300 cP.

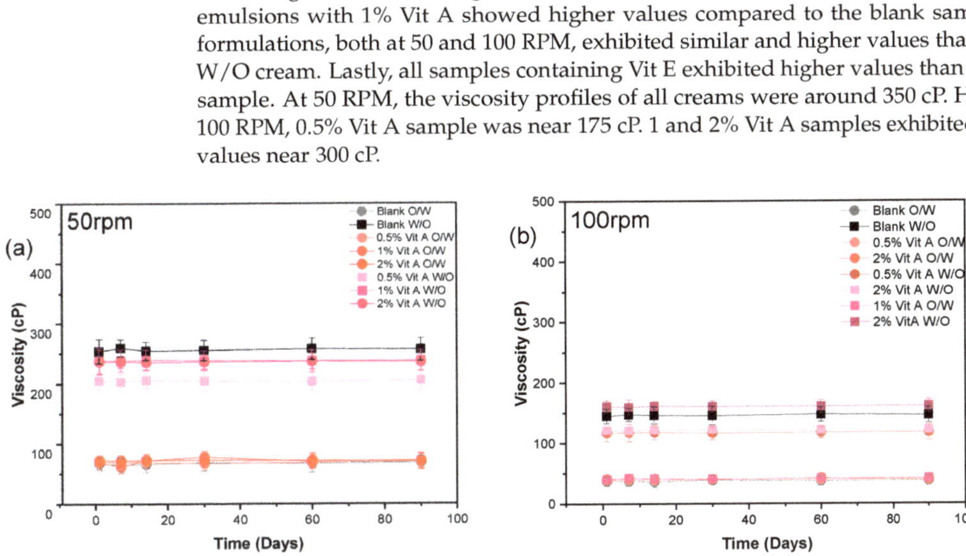

Figure 2. (**a**) Viscosity tests for Vitamin A emulsions at (**a**) 50 rpm and (**b**) 100 rpm.

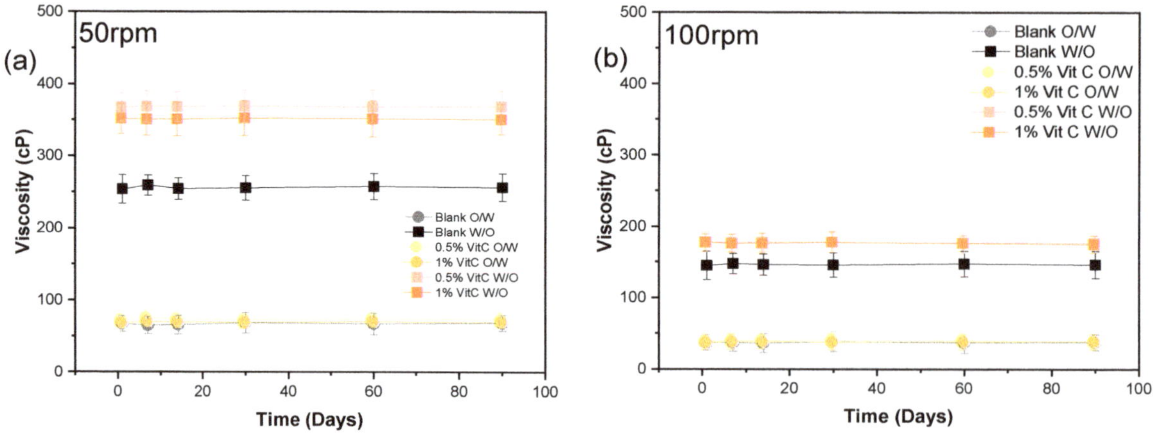

Figure 3. Viscosity tests for Vitamin C emulsions at (**a**) 50 rpm and (**b**) 100 rpm.

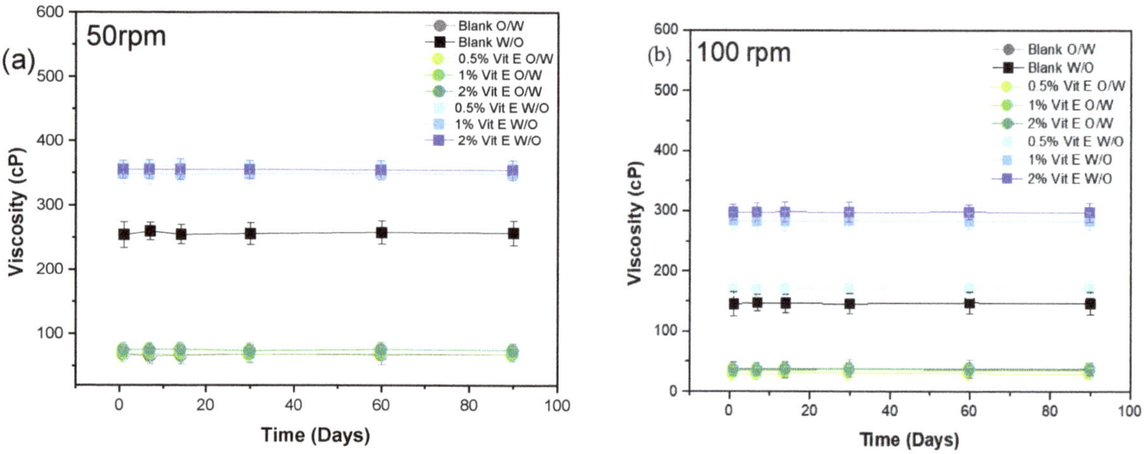

Figure 4. Viscosity tests for Vitamin E emulsions at (**a**) 50 rpm and (**b**) 100 rpm.

3.3. Assessment of Sunscreen Activity

Figure 5 presents the in vitro SPF tests conducted on the prepared emulsions. Blank O/W and W/O samples exhibited 3.72 and 3.96 SPF, respectively, with the W/O having slightly higher SPF, probably due to the increased amount of olive oil that is well-known for its UV-blocking properties [34]. O/W and W/O emulsions containing Vit A exhibited an increasing trend with increasing vitamin addition. The highest SPF value both amongst the Vit A and the overall formulations was shown by 2% Vit A W/O cream at 22.6. O/W 0.5 and 1% Vit C emulsions exhibited lower SPF values than the blank samples, indicating that the addition of Vit C has a negative effect on the anti-UV behavior. However, the addition of Vit C in W/O emulsions increased the SPF values up to 8.52 in the 1% Vit C W/O sample. A similar value (8.11) was obtained by the 1% Vit E W/O emulsion. The addition of 2% Vit E in both O/W and W/O samples seems to have an inhibitory effect when compared to 1% samples since an SPF decrease is observed.

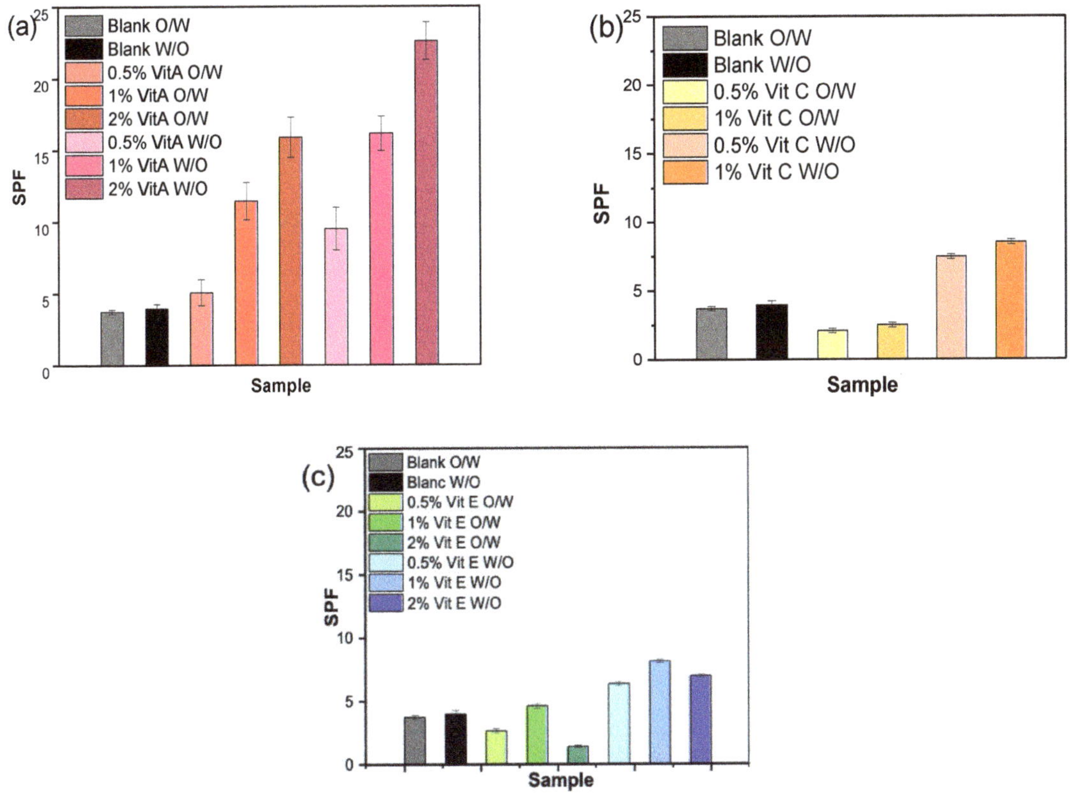

Figure 5. In vitro SPF measurements of O/W and W/O emulsions containing (**a**) Vitamin A, (**b**) Vitamin C, and (**c**) Vitamin E.

3.4. Antioxidant Properties

Figure 6 illustrates the in vitro antioxidant ability of the prepared emulsions. It was obtained that all W/O emulsions showed enhanced antioxidant capacity compared to O/W emulsions. This is due to the higher amount of olive oil [35] and shea butter [36] that present strong antioxidant properties. All samples containing vitamins exhibited improved antioxidant ability compared to the blank emulsions, which was enhanced with the increase of vitamin content from 0.5 to 2% *w*/*w*. However, the samples showed significant differences from each other in their antioxidant ability. Emulsion with 1% Vit C W/O exhibited the highest antioxidant capacity at 80.56% AA during the 8 days of this study. The emulsions containing Vitamin E exhibited the next-best results, as 2% Vit E W/O had 77.86% AA. The formulations with Vitamin A showed lower antioxidant capacity since the best result was shown by 2% Vit A W/O cream with 65.23% AA. All emulsions showed a progressive increase in the % antioxidant capacity up to 8 days after their preparation. This was also observed in previous works that studied the antioxidant properties of emulsions containing curcumin derivatives [27]. This may be attributed to the synergistic effect taking place during storage between vitamins and antioxidant ingredients (olive oil and shea butter) used in the oil phase of the prepared emulsions [37].

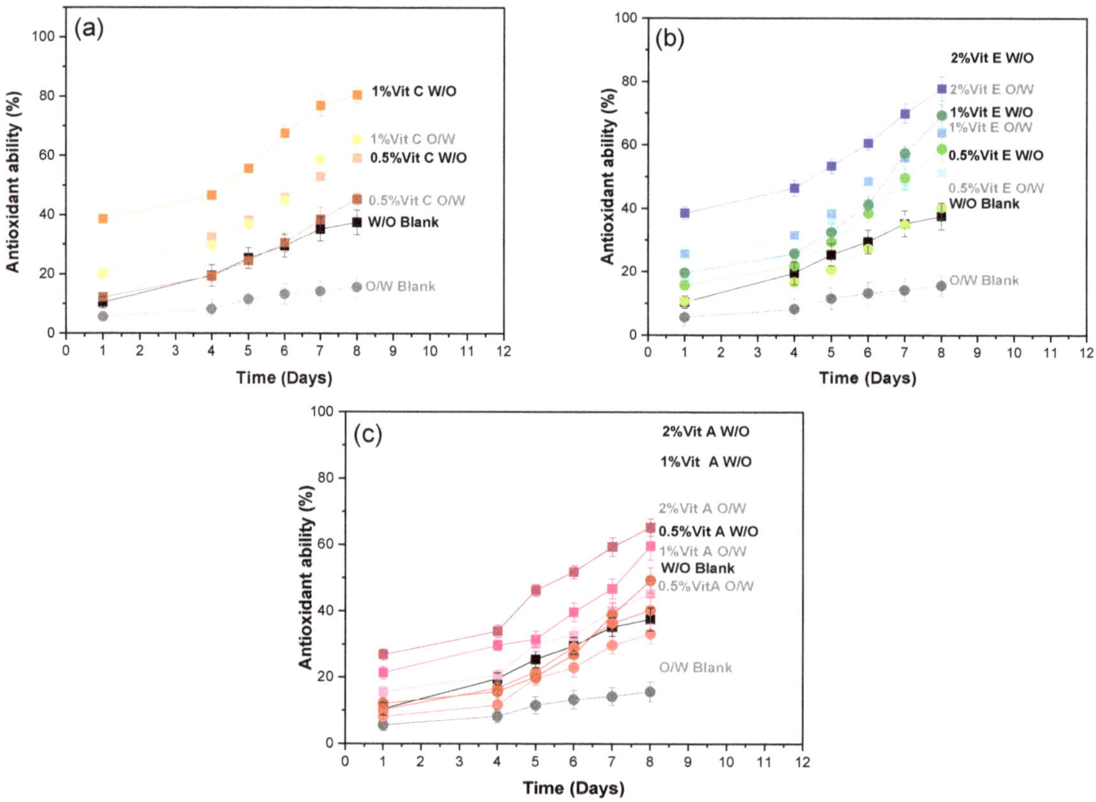

Figure 6. Study of the kinetic of the antioxidant ability of emulsions during the first 8 days. (**a**) represents the % antioxidant ability of the emulsions containing Vit C, (**b**) the emulsions containing Vit E and (**c**) the emulsions containing Vit A.

4. Discussion

The popularity of a cosmetic product varies not only on the efficiency of the active ingredient(s) but also on consumer approval, which is greatly affected by the sensory properties of the formulation [38]. The physical stability of the creams is an important parameter in the stabilization of the vitamins for topical dermal delivery and for ensuring the effectiveness of their usage. The stability of the formulation, temperature variations during transportation or storage and pH changes may lead to the degradation of the active compounds [21]. Consecutively, these factors in a formulation are critical and should be carefully investigated and monitored to determine the bio-functionality of a cosmetic product and to fulfill consumers' needs.

4.1. pH Stability

Since the beginning of its study, human skin has been believed to have an acidic surface [39]. In general, it was proposed by researchers that the skin pH varies between 5.4 and 5.9. Recent works have claimed that the pH of adult skin has an even more acidic nature, with the scientific community accepting that, depending on the part of the body, it ranges from 4.1 to 5.8 (mean 4.9) [40]. In pH values higher than 6, several biological functions of the SC, such as the formation of epidermal lipids and the keratinocyte differentiation process, are suspended, while the development of microbes is favored, resulting in skin disorders and disturbances [41]. Cosmetic formulations, due to their extensive daily application, should contribute to skin health conservation by maintaining the pH of the skin's surface [42]. Moreover, the pH of a topical product is vital for optimal bioavailability, as is the presence of a penetration enhancer. Taking into consideration the abovementioned, it is of great significance for a topical skin care product to retain pH values between 4 and 6.

In the current work, it was observed (Figure 1) that the pH values of almost all the prepared emulsions remained between 4 and 5.9, even after 90 days of storage at room temperature. These results demonstrate the stability of the creams in terms of pH values even after a long period of storage. Only the emulsions containing 2% w/v Vit C, both O/W and W/O, exhibited highly acetic values in the range of 3.1–3.8. As previously mentioned, the pH of a cosmetic product should vary in the range of 4–5.8; otherwise, it could cause skin irritations or suspension of biological functions. This could be attributed to the increased amount of ascorbic acid (Vit C) that was added to each emulsion. Similar results were reported by Cefali, et al. [43], who studied the stability of emulsions containing Vitamin C in acerola extracts. During the stability tests, the pH values of the emulsions dropped to 4.02 ± 0.3 due to the fact that Vit C is prone to thermal degradation at elevated temperatures and that light could decrease pH values during storage. Kim, et al. [44] also reported very low values of pH and a constant decrease (from 5 to 2) when studying the stability of L-ascorbic acid in O/W emulsions at 25 °C.

4.2. Rheological Study

The viscosity measurement of an emulsion may provide significant information on an emulsion's stability and inner microstructure and, thus, is an important factor to take into consideration. Any variations in viscosity through storage may result in several defects in the aesthetic appearance of the end product [27]. However, from the obtained values of viscosity (Figures 3–5), no apparent viscosity instabilities deriving from the addition of the vitamins are observed. This is a great indication of the excellent viscosity stability of the studied emulsions during storage in RT for up to 90 days. As expected, due to the increased number of oily substances and thick liquid nature, W/O emulsions exhibited very high values of viscosity. Moreover, in the measured samples, the viscosity dramatically decreased with increasing RPM, as shown in Figures 3–5, suggesting that the W/O emulsions exhibit a shear-thinning behavior as pseudoplastic fluids. This thixotropic behavior is highly necessary for topical skin formulations as they are considered to deform during application and become fluid, thus facilitating the spreading of the product on the surface of the skin [45]. Moreover, the recovery of the initial viscosity after application prevents the product from dripping [46].

4.3. Assessment of Sunscreen Activity

Exposure to UV irradiation is responsible for numerous destructive effects on human skin health, such as sunburn (erythema), blistering (edema), photoaging and carcinogenesis [47,48]. Thus, when considering the preparation of a cosmetic product, anti-UV properties should be employed. The activity of a sunscreen product is mostly rated and marketed by SPF, which measures the fraction of sunburn. This term may be explained as the fraction of the UV radiation required to induce nominal erythema on the skin after applying sunscreen to the amount of energy necessary to convey the same effect on the skin without any sunscreen. The higher the SPF, the higher the protection created against the sun's rays [49,50]. There are several factors affecting the determination of SPF values, such as the nature of the emulsion, the effects and interactions of vehicle components—such as emollients and emulsifiers inside the product as well as with the skin—the addition of any active ingredients, the pH system and the emulsion rheological properties, among other factors, which may decrease or enhance the UV absorption of a sunscreen formulation [51]. At present, a well-known method in the cosmetic industry is to incorporate antioxidants, such as vitamins [47], as additions to UV filters since almost all post-radiation reactions involve reactive oxygen species (ROS) [52]. These photo-oxidative reactions initiate the development of many disorders affecting the skin [53,54].

The in vitro assessment of SPF of the investigated samples in the current work showed that in all cases, W/O emulsions exhibit higher UV protection (higher SPF). This can be attributed to the presence of an increased amount of olive oil and oily substances. Several scientific works have highlighted the anti-UV efficacy of olive oil in the past [34,55,56]. Kaur, et al. [51] calculated an in vitro 7.549 SPF value using a UV-Vis spectrophotometer. This was the highest value amongst a variety of investigated natural oils such as coconut, almond, castor, peppermint, lavender and lemon grass oil. Regarding the investigated samples, W/O emulsions containing Vit with a 2% ratio exhibited a 22.6 SPF, the highest value amongst all creams. As previously reported by Antille, et al. [57], due to its physical properties, Vitamin A (retinyl esters and retinol) strongly absorbs ultraviolet radiation between 300 and 350 nm, with a maximum at 325 nm. This wavelength range is transmitted from the sun at the earth's level and is accountable for many of the deleterious biological effects of the sun. In a study conducted on mice, it was shown that the topical delivery of Vitamin A prevents UV-induced epidermal hypovitaminosis A, while topical Vitamin E inhibits oxidative stress and systemic immuno- suppression elicited by UV [58]. In the current work, an 8.11 SPF value was obtained by the 1% Vit E W/O emulsion. Additionally, an inhibitory influence was observed with the addition of 2% Vit E in both O/W and W/O samples when compared to 1% samples since SPF noticeably decreased. O/W 0.5 and 1% Vit C emulsions exhibited lower SPF values than the blank samples. This may be attributed to the composition of the emulsion vehicle in combination with the addition of Vit C [59]. Nevertheless, 1% Vit C W/O exhibited SPF values up to 8.52, indicating a potential UV-blocking ability.

4.4. Antioxidant Capability of Emulsions

A number of oxygen-derived free radicals (molecules with an unpaired electron, ROS) are induced through many endogenous sources, e.g., enzyme activity or activated neutrophils, and external sources, such as exposure to pollutants, drugs and solar ultraviolet radiation. Superoxide radicals [$O_2\cdot-$] and hydroxyl radicals [$OH\cdot$] are the most found free radicals in the body. These agents can cause structural alterations to DNA, lipid peroxidation, induce direct damage to lipids and proteins, modify the enzyme structures, secretion of inflammatory cytokines, and finally cause cell death. Free radicals are responsible for at least part of the degenerative changes leading to cutaneous aging and skin cancer [60]. The production of endogenous antioxidant compounds by the human body can protect against ROS. Antioxidants are molecules that stabilize free radicals by the donation of an electron without also becoming free radicals themselves (Figure 7). Their function is to prevent or delay the cellular damage induced by ROS and work by reducing local oxygen concentrations and impairing chain initiation reactions [61]. Nevertheless, the intake of external antioxidant agents that intercept and scavenge radicals to inhibit the initiation and break chain propagation can further enhance protection against free radicals [62,63].

Vitamin C, Vitamin E and Vitamin A are capable of attenuating the damaging effects of O_2 in lipid and nonlipid cellular sections, thereby acting as potential antioxidants [64]. Several reports have proposed novel therapeutic approaches with topical or systemic administration of these antioxidant agents [65,66]. According to these results, the emulsions with Vit C exhibited the highest antioxidant ability. Vitamin C can either react directly with chain-carrying peroxyl radicals (ROO′, where R = H, substituted alkyl, etc.) or indirectly, by the reduction of the alpha-tocopherol radical to renew Vitamin E [67,68]. Furthermore, it was found that the addition of Vit E (alpha-tocopherol) improved the antioxidant capacity of the prepared emulsions since it is also a potent biological chain-breaking antioxidant and effectively inhibits the peroxidation of lipids [69–71]. However, despite the fact that it has been reported that Vitamin A and its congeners act as chain-breaking antioxidants in biological membranes, in the present work, the addition of Vitamin A illustrated a lower antioxidant rate compared to formulations with vitamins C and E [64].

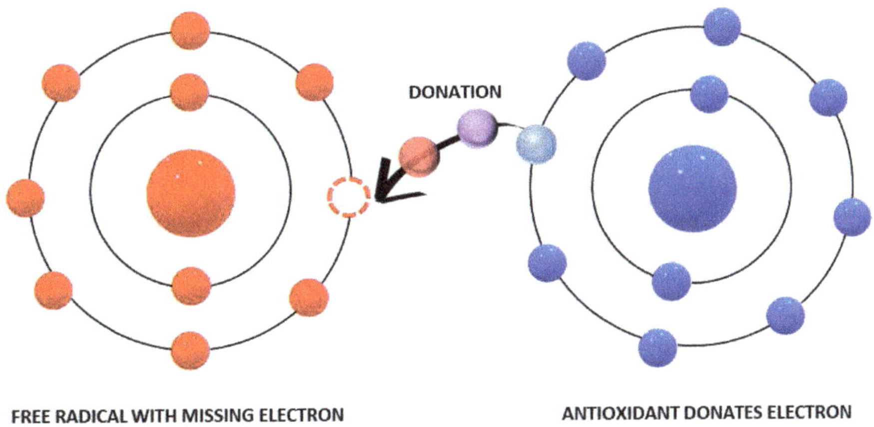

Figure 7. The mechanism of antioxidant agents' activity by the donation of an electron to a free radical.

5. Conclusions

In this study, O/W and W/O emulsions containing 0.5, 1 and 2% *w/w* Vitamins A, C and E for topical skin delivery applications are reported. Investigation of viscosity and pH stability at room temperature showed no deviations from the initial values and thus confirmed the stability of the samples over storage. 2% Vit C O/W and W/O samples exhibited pH values below 4, which is the acceptable value for human skin application, and hence their use is prohibited for dermal usage. In vitro SPF determination showed no insignificant values, suggesting that incorporated vitamins exhibit a mild sun protection activity. Great results were obtained by the performed antioxidants test, where a sample with 1% Vit C O/W presented the highest antioxidant ability, reaching up to 80.56% in the first 8 days. This was attributed to the excellent capability of vitamins to scavenge ROS such as [$O_2\cdot-$] and [OH]. All in all, the results obtained here are promising for the treatment of various skin disorders. Certainly, additional drug release and in vivo studies should be performed in future research since these emulsions are destined for human treatment. Lastly, it would be worth studying complex formulations that contain more than one vitamin for stability over time.

Author Contributions: Conceptualization, N.N.; Investigation, K.H. and S.L.; Methodology, S.L.; Project administration, N.N.; Resources, S.L.; Supervision, N.N.; Writing—original draft, N.D.B.; Writing—review and editing, I.K. and P.B. All authors have read and agreed to the published version of the manuscript.

Funding: This research received no external funding.

Institutional Review Board Statement: Not applicable.

Informed Consent Statement: Not applicable.

Data Availability Statement: Not applicable.

Conflicts of Interest: The authors declare no conflict of interest.

References

1. Abdo, J.M.; Sopko, N.A.; Milner, S.M. The Applied Anatomy of Human Skin: A Model for Regeneration. *Wound Med.* **2020**, *28*, 100179. [CrossRef]
2. Dąbrowska, A.K.; Spano, F.; Derler, S.; Adlhart, C.; Spencer, N.D.; Rossi, R.M. The Relationship between Skin Function, Barrier Properties, and Body-Dependent Factors. *Ski. Res. Technol.* **2018**, *24*, 165–174. [CrossRef] [PubMed]
3. Mccarley, K.D.; Bunge, A.L. Pharmacokinetic Models of Dermal Absorption. *J. Pharm. Sci.* **2001**, *90*, 1699–1719. [CrossRef] [PubMed]

4. Semenzato, A.; Costantini, A.; Meloni, M.; Maramaldi, G.; Meneghin, M.; Baratto, G. Formulating O/W Emulsions with Plant-Based Actives: A Stability Challenge for an Effective Product. *Cosmetics* **2018**, *5*, 59. [CrossRef]
5. Feng, J.; Chen, Q.; Wu, X.; Jafari, S.M.; McClements, D.J. Formulation of Oil-in-Water Emulsions for Pesticide Applications: Impact of Surfactant Type and Concentration on Physical Stability. *Environ. Sci. Pollut. Res.* **2018**, *25*, 21742–21751. [CrossRef]
6. Tang, C.; Li, Y.; Pun, J.; Osman, A.S.M.; Tam, K.C. Polydopamine Microcapsules from Cellulose Nanocrystal Stabilized Pickering Emulsions for Essential Oil and Pesticide Encapsulation. *Colloids Surf. A Physicochem. Eng. Asp.* **2019**, *570*, 403–413. [CrossRef]
7. Okuro, P.K.; Gomes, A.; Costa, A.L.R.; Adame, M.A.; Cunha, R.L. Formation and Stability of W/O-High Internal Phase Emulsions (HIPEs) and Derived O/W Emulsions Stabilized by PGPR and Lecithin. *Food Res. Int.* **2019**, *122*, 252–262. [CrossRef]
8. Torres, L.G.; Iturbe, R.; Snowden, M.J.; Chowdhry, B.Z.; Leharne, S.A. Preparation of o/w Emulsions Stabilized by Solid Particles and Their Characterization by Oscillatory Rheology. *Colloids Surf. A Physicochem. Eng. Asp.* **2007**, *302*, 439–448. [CrossRef]
9. Otto, A.; Du Plessis, J.; Wiechers, J.W. Formulation Effects of Topical Emulsions on Transdermal and Dermal Delivery. *Int. J. Cosmet. Sci.* **2009**, *31*, 1–19. [CrossRef]
10. Simovic, S.; Ghouchi-Eskandar, N.; Prestidge, C.A. Pickering Emulsions for Dermal Delivery. *J. Drug Deliv. Sci. Technol.* **2011**, *21*, 123–133. [CrossRef]
11. Shah, B.R.; Xu, W.; Mráz, J. Formulation and Characterization of Zein/Chitosan Complex Particles Stabilized Pickering Emulsion with the Encapsulation and Delivery of Vitamin D_3. *J. Sci. Food Agric.* **2021**, *101*, 5419–5428. [CrossRef]
12. Eskandar, N.G.; Simovic, S.; Prestidge, C.A. Nanoparticle Coated Submicron Emulsions: Sustained in-Vitro Release and Improved Dermal Delivery of All-Trans-Retinol. *Pharm. Res.* **2009**, *26*, 1764–1775. [CrossRef]
13. Lunter, D.J.; Daniels, R. New Film Forming Emulsions Containing Eudragit® NE and/or RS 30D for Sustained Dermal Delivery of Nonivamide. *Eur. J. Pharm. Biopharm.* **2012**, *82*, 291–298. [CrossRef]
14. Dickerson, E.B.; Dreaden, E.C.; Huang, X.; El-Sayed, I.H.; Chu, H.; Pushpanketh, S.; McDonald, J.F.; El-Sayed, M.A.; O'Neal, D.P.; Hirsch, L.R.; et al. Pharmaceutical Strategies for the Topical Dermal Delivery of Peptides/Proteins for Cosmetic and Therapeutic Applications. *Cancer Lett.* **2014**, *269*, 57–66. [CrossRef]
15. Gašperlin, M.; Gosenca, M. Main Approaches for Delivering Antioxidant Vitamins through the Skin to Prevent Skin Ageing. *Expert Opin. Drug Deliv.* **2011**, *8*, 905–919. [CrossRef]
16. Tozer, S.; O'Mahony, C.; Hannah, J.; O'Brien, J.; Kelly, S.; Kosemund-Meynen, K.; Alexander-White, C. Aggregate Exposure Modelling of Vitamin A from Cosmetic Products, Diet and Food Supplements. *Food Chem. Toxicol.* **2019**, *131*, 110549. [CrossRef]
17. Polcz, M.E.; Barbul, A. The Role of Vitamin A in Wound Healing. *Nutr. Clin. Pract.* **2019**, *34*, 695–700. [CrossRef]
18. Caritá, A.C.; Fonseca-Santos, B.; Shultz, J.D.; Michniak-Kohn, B.; Chorilli, M.; Leonardi, G.R. Vitamin C: One Compound, Several Uses. Advances for Delivery, Efficiency and Stability. *Nanomed. Nanotechnol. Biol. Med.* **2020**, *24*, 102117. [CrossRef]
19. Ngoc, L.T.N.; Van Tran, V.; Moon, J.Y.; Chae, M.; Park, D.; Lee, Y.C. Recent Trends of Sunscreen Cosmetic: An Update Review. *Cosmetics* **2019**, *6*, 64. [CrossRef]
20. Proksch, E. PH in Nature, Humans and Skin. *J. Dermatol.* **2018**, *45*, 1044–1052. [CrossRef]
21. Chu, C.C.; Nyam, K.L. Kenaf (*Hibiscus cannabinus* L.) Seed Oil: Application as Cosmetic Product Ingredients. *Ind. Crop. Prod.* **2020**, *156*, 112871. [CrossRef]
22. Thiele, J.J.; Hsieh, S.N.; Ekanayake-Mudiyanselage, S. Vitamin E: Critical Review of Its Current Use in Cosmetic and Clinical Dermatology. *Dermatol. Surg.* **2005**, *31*, 805–813. [CrossRef] [PubMed]
23. Yoshida, K.; Sekine, T.; Matsuzaki, F.; Yanaki, T.; Yamaguchi, M. Stability of Vitamin A in Oil-in-Water-in-Oil-Type Multiple Emulsions. *J. Am. Oil Chem. Soc.* **1999**, *76*, 1–6. [CrossRef]
24. Maia, A.M.; Baby, A.R.; Pinto, C.A.S.O.; Yasaka, W.J.; Suenaga, E.; Kaneko, T.M.; Velasco, M.V.R. Influence of Sodium Metabisulfite and Glutathione on the Stability of Vitamin C in O/W Emulsion and Extemporaneous Aqueous Gel. *Int. J. Pharm.* **2006**, *322*, 130–135. [CrossRef]
25. Fraj, J.; Petrović, L.; Đekić, L.; Budinčić, J.M.; Bučko, S.; Katona, J. Encapsulation and Release of Vitamin C in Double W/O/W Emulsions Followed by Complex Coacervation in Gelatin-Sodium Caseinate System. *J. Food Eng.* **2021**, *292*, 110353. [CrossRef]
26. Moyano, M.A.; Segall, A. Vitamin a Palmitate and α-Lipoic Acid Stability in O/W Emulsions for Cosmetic Application. *J. Cosmet. Sci.* **2011**, *62*, 405–415.
27. Dalla, E.; Koumentakou, I.; Bikiaris, N.; Balla, E.; Lykidou, S.; Nikolaidis, N. Formulation, Characterization and Evaluation of Innovative O/W Emulsions Containing Curcumin Derivatives with Enhanced Antioxidant Properties. *Antioxidants* **2022**, *11*, 2271. [CrossRef]
28. Ntohogian, S.; Gavriliadou, V.; Christodoulou, E.; Nanaki, S.; Lykidou, S.; Naidis, P.; Mischopoulou, L.; Barmpalexis, P.; Nikolaidis, N.; Bikiaris, D.N. Chitosan Nanoparticles with Encapsulated Natural and Uf-Purified Annatto and Saffron for the Preparation of UV Protective Cosmetic Emulsions. *Molecules* **2018**, *23*, 2107. [CrossRef]
29. Bikiaris, N.D.; Koumentakou, I.; Lykidou, S.; Nikolaidis, N. Innovative Skin Product O/W Emulsions Containing Lignin, Multiwall Carbon Nanotubes and Graphene Oxide Nanoadditives with Enhanced Sun Protection Factor and UV Stability Properties. *Appl. Nano* **2022**, *3*, 1–15. [CrossRef]
30. Bikiaris, N.D.; Michailidou, G.; Lazaridou, M.; Christodoulou, E.; Gounari, E.; Ofrydopoulou, A.; Lambropoulou, D.A.; Vergkizi-Nikolakaki, S.; Lykidou, S.; Nikolaidis, N. Innovative Skin Product Emulsions with Enhanced Antioxidant, Antimicrobial and UV Protection Properties Containing Nanoparticles of Pure and Modified Chitosan with Encapsulated Fresh Pomegranate Juice. *Polymers* **2020**, *12*, 1542. [CrossRef]

31. Sayre, R.M.; Agin, P.P.; LeVee, G.J.; Marlowe, E. A Comparison of in Vivo and in Vitro Testing of Sunscreening Formulas. *Photochem. Photobiol.* **1979**, *29*, 559–566. [CrossRef]
32. Blois, M.S. Antioxidant Determinations by the Use of a Stable Free Radical. *Nature* **1958**, *181*, 1199–1200. [CrossRef]
33. Brand-Williams, W.; Cuvelier, M.E.; Berset, C. Use of a Free Radical Method to Evaluate Antioxidant Activity. *LWT Food Sci. Technol.* **1995**, *28*, 25–30. [CrossRef]
34. Gause, S.; Chauhan, A. UV-Blocking Potential of Oils and Juices. *Int. J. Cosmet. Sci.* **2016**, *38*, 354–363. [CrossRef]
35. Kesen, S.; Kelebek, H.; Selli, S. Characterization of the Volatile, Phenolic and Antioxidant Properties of Monovarietal Olive Oil Obtained from Cv. Halhali. *J. Am. Oil Chem. Soc.* **2013**, *90*, 1685–1696. [CrossRef]
36. Adefegha, S.A.; Oboh, G.; Olasehinde, T.A. Alkaloid Extracts from Shea Butter and Breadfruit as Potential Inhibitors of Monoamine Oxidase, Cholinesterases, and Lipid Peroxidation in Rats' Brain Homogenates: A Comparative Study. *Comp. Clin. Pathol.* **2016**, *25*, 1213–1219. [CrossRef]
37. Scomoroscenco, C.; Teodorescu, M.; Burlacu, S.G.; Gîfu, I.C.; Mihaescu, C.I.; Petcu, C.; Raducan, A.; Oancea, P.; Cinteza, L.O. Synergistic Antioxidant Activity and Enhanced Stability of Curcumin Encapsulated in Vegetal Oil-Based Microemulsion and Gel Microemulsions. *Antioxidants* **2022**, *11*, 854. [CrossRef]
38. Montenegro, L.; Rapisarda, L.; Ministeri, C.; Puglisi, G. Effects of Lipids and Emulsifiers on the Physicochemical and Sensory Properties of Cosmetic Emulsions Containing Vitamin E. *Cosmetics* **2015**, *2*, 35–47. [CrossRef]
39. Rippke, F.; Schreiner, V.; Schwanitz, H.J. The Acidic Milieu of the Horny Layer: New Findings on the Physiology and Pathophysiology of Skin PH. *Am. J. Clin. Dermatol.* **2002**, *3*, 261–272. [CrossRef]
40. Lukić, M.; Pantelić, I.; Savić, S.D. Towards Optimal Ph of the Skin and Topical Formulations: From the Current State of the Art to Tailored Products. *Cosmetics* **2021**, *8*, 69. [CrossRef]
41. Menon, G.K.; Cleary, G.W.; Lane, M.E. The Structure and Function of the Stratum Corneum. *Int. J. Pharm.* **2012**, *435*, 3–9. [CrossRef] [PubMed]
42. Darmstadt, G.L.; Mao-Qiang, M.; Chi, E.; Saha, S.K.; Ziboh, V.A.; Black, R.E.; Santosham, M.; Elias, P.M. Impact of Topical Oils on the Skin Barrier: Possible Implications for Neonatal Health in Developing Countries. *Acta Paediatr. Int. J. Paediatr.* **2002**, *91*, 546–554. [CrossRef]
43. Cefali, L.C.; de Oliveira Maia, L.; Stahlschimidt, R.; Ataide, J.A.; Tambourgi, E.B.; Rosa, P.C.P.; Mazzola, P.G. Vitamin C in Acerola and Red Plum Extracts: Quantification via HPLC, in Vitro Antioxidant Activity, and Stability of Their Gel and Emulsion Formulations. *J. AOAC Int.* **2018**, *101*, 1461–1465. [CrossRef] [PubMed]
44. Kim, S.; Lee, T.G. Stabilization of L-Ascorbic Acid in Cosmetic Emulsions. *J. Ind. Eng. Chem.* **2018**, *57*, 193–198. [CrossRef]
45. Gaspar, L.R.; Campos, P.M. Rheological Behavior and the SPF of Sunscreens. *Int. J. Pharm.* **2003**, *250*, 35–44. [CrossRef]
46. Gonçalves, G.M.S.; Campos, P.M. Shelf Life and Rheology of Emulsions Containing Vitamin C and Its Derivatives. *Rev. Cienc. Farm. Basica Apl.* **2009**, *30*, 217–224.
47. Roberts, R.L.; Greene, J.A. High Efficiency Skin Protection Formulation With Sunscreen Agents and Antioxidants. *Geothermics* **1985**, *14*, 595–599.
48. He, H.; Li, A.; Li, S.; Tang, J.; Li, L.; Xiong, L. Natural Components in Sunscreens: Topical Formulations with Sun Protection Factor (SPF). *Biomed. Pharmacother.* **2021**, *134*, 111161. [CrossRef]
49. Dutra, E.A.; Oliveira, D.A.G.D.C.; Kedor-Hackmann, E.R.M.; Santoro, M.I.R.M. Determination of Sun Protection Factor (SPF) of Sunscreens by Ultraviolet Spectrophotometry. *Rev. Bras. Ciênc. Farm.* **2004**, *40*, 381–385. [CrossRef]
50. Donglikar, M.M.; Deore, S.L. Sunscreens: A Review. *Pharmacogn. J.* **2016**, *8*, 171–179. [CrossRef]
51. Kaur, C.D.; Saraf, S. In Vitro Sun Protection Factor Determination of Herbal Oils Used in Cosmetics. *Pharmacogn. Res.* **2010**, *2*, 22–25. [CrossRef]
52. Lin, Q.; Xu Xu, R.H.J.; Yang, N.; Karim, A.A.; Loh, X.J.; Zhang, K. UV Protection and Antioxidant Activity of Nanodiamonds and Fullerenes for Sunscreen Formulations. *ACS Appl. Nano Mater.* **2019**, *2*, 7604–7616. [CrossRef]
53. Fernández-García, E. Skin Protection against UV Light by Dietary Antioxidants. *Food Funct.* **2014**, *5*, 1994–2003. [CrossRef]
54. Mansuri, R.; Diwan, A.; Kumar, H.; Dangwal, K.; Yadav, D. Potential of Natural Compounds as Sunscreen Agents. *Pharmacogn. Rev.* **2021**, *15*, 47–56. [CrossRef]
55. Smith, G.J.; Miller, I.J.; Clare, J.F.; Diffey, B.L. The Effect of UV Absorbing Sunscreens on the Reflectance and the Consequent Protection of Skin. *Photochem. Photobiol.* **2002**, *75*, 122. [CrossRef]
56. Da Silva, A.C.P.; Paiva, J.P.; Diniz, R.R.; dos Anjos, V.M.; Silva, A.B.S.M.; Pinto, A.V.; dos Santos, E.P.; Leitão, A.C.; Cabral, L.M.; Rodrigues, C.R.; et al. Photoprotection Assessment of Olive (*Olea europaea* L.) Leaves Extract Standardized to Oleuropein: In Vitro and in Silico Approach for Improved Sunscreens. *J. Photochem. Photobiol. B Biol.* **2019**, *193*, 162–171. [CrossRef]
57. Antille, C.; Tran, C.; Sorg, O.; Carraux, P.; Didierjean, L.; Saurat, J.H. Vitamin A Exerts a Photoprotective Action in Skin by Absorbing Ultraviolet B Radiation. *J. Investig. Dermatol.* **2003**, *121*, 1163–1167. [CrossRef]
58. Sorg, O.; Tran, C.; Saurat, J.H. Cutaneous Vitamins A and E in the Context of Ultraviolet- or Chemically-Induced Oxidative Stress. *Ski. Pharmacol. Appl. Ski. Physiol.* **2001**, *14*, 363–372. [CrossRef] [PubMed]
59. Daher, C.C.; Fontes, I.S.; Rodrigues, R.D.O.; Damasceno, G.A.D.B.; Soares, D.D.S.; Aragão, C.F.S.; Gomes, A.P.B.; Ferrari, M. Development of O/W Emulsions Containing Euterpe Oleracea Extract and Evaluation of Photoprotective Efficacy. *Braz. J. Pharm. Sci.* **2014**, *50*, 639–652. [CrossRef]

60. Briganti, S.; Picardo, M. Antioxidant Activity, Lipid Peroxidation and Skin Diseases. What's New. *J. Eur. Acad. Dermatol. Venereol.* **2003**, *17*, 663–669. [CrossRef] [PubMed]
61. Hunina, L.M.; Oliĭnyk, S.A. Oxidative Stress and Its Role in Carcinogenesis. *Fiziolohichnyĭ Zhurnal* **2006**, *52*, 78–89.
62. Gülçin, I.; Huyut, Z.; Elmastaş, M.; Aboul-Enein, H.Y. Radical Scavenging and Antioxidant Activity of Tannic Acid. *Arab. J. Chem.* **2010**, *3*, 43–53. [CrossRef]
63. Gulcin, İ. Antioxidants and Antioxidant Methods: An Updated Overview. *Arch. Toxicol.* **2020**, *94*, 651–715. [CrossRef]
64. Livrea, M.A.; Tesoriere, L. Antioxidant Activity of Vitamin A within Lipid Environments. *Subcell. Biochem.* **1998**, *30*, 113–143. [CrossRef]
65. Fuchs, J. Potentials and Limitations of the Natural Antioxidants RRR-Alpha-Tocopherol, L-Ascorbic Acid and β-Carotene in Cutaneous Photoprotection. *Free Radic. Biol. Med.* **1998**, *25*, 848–873. [CrossRef]
66. Podda, M.; Grundmann-Kollmann, M. Low Molecular Weight Antioxidants and Their Role in Skin Ageing. *Clin. Exp. Dermatol.* **2001**, *26*, 578–582. [CrossRef]
67. Bendich, A.; Machlin, L.J.; Scandurra, O.; Burton, G.W.; Wayner, D.D.M. The Antioxidant Role of Vitamin C. *Adv. Free Radic. Biol. Med.* **1986**, *2*, 419–444. [CrossRef]
68. Uluata, S.; McClements, D.J.; Decker, E.A. How the Multiple Antioxidant Properties of Ascorbic Acid Affect Lipid Oxidation in Oil-in-Water Emulsions. *J. Agric. Food Chem.* **2015**, *63*, 1819–1824. [CrossRef]
69. Burton, G.W.; TRaber, M.G. Vitamin E: Antioxidant Activity, Biokinetics, and Bioavailability. *Annu. Rev. Nutr.* **1990**, *10*, 357–382. [CrossRef]
70. Yamauchi, R. Vitamin E: Mechanism of Its Antioxidant Activity. *Food Sci. Technol. Int. Tokyo* **1997**, *3*, 301–309. [CrossRef]
71. Wu, D.; Hayek, M.G.; Meydani, S.N. Symposium: Molecular Mechanisms of Protective Effects of Vitamin E in Atherosclerosis: Vitamin E and Macrophage Cyclooxygenase Regulation in the Aged. *J. Nutr.* **2001**, *131*, 382S–388S. [CrossRef] [PubMed]

Disclaimer/Publisher's Note: The statements, opinions and data contained in all publications are solely those of the individual author(s) and contributor(s) and not of MDPI and/or the editor(s). MDPI and/or the editor(s) disclaim responsibility for any injury to people or property resulting from any ideas, methods, instructions or products referred to in the content.

cosmetics

MDPI

Article

Protective Effects of Naringenin against UVB Irradiation and Air Pollution-Induced Skin Aging and Pigmentation

Christina Österlund *, Nina Hrapovic, Virginie Lafon-Kolb, Nahid Amini, Sandra Smiljanic and Lene Visdal-Johnsen

Swedish Research and Innovation Lab, Oriflame Cosmetics AB, Fleminggatan 14, 112 26 Stockholm, Sweden; lene.visdal-johnsen@oriflame.com (L.V.-J.)
* Correspondence: christina.osterlund@oriflame.com

Abstract: Both UVB irradiation and air pollution are major extrinsic factors causing premature aging of the skin, including sagging, wrinkles, and pigmentation spots. Naringenin, a naturally occurring flavanone, found in citrus fruits, and known for its good antioxidant and anti-inflammatory effects, was investigated for protective effects in human skin cells and reconstructed epidermis. The results showed that naringenin inhibits UVB-induced inflammation markers MMP1, MMP3, IL6, and GM-CSF, as well as pollution-induced MMP1 in human skin fibroblasts. Furthermore, naringenin inhibited the pollution-induced expression of the *CYP1A1* gene in human skin keratinocytes. In melanocytes and pigmented reconstructed epidermis, naringenin significantly downregulated several genes involved in melanogenesis, such as MITF, MLPH, and MYO5A. Additionally topical treatment with naringenin on pigmented reconstructed epidermis significantly decreased melanin production. In conclusion, this study demonstrates that naringenin could be a valuable ingredient in skincare products, protecting against the detrimental effects of both UVB and pollution on the skin.

Keywords: naringenin; natural ingredient; UV protection; pollution protection; pigmentation

Citation: Österlund, C.; Hrapovic, N.; Lafon-Kolb, V.; Amini, N.; Smiljanic, S.; Visdal-Johnsen, L. Protective Effects of Naringenin against UVB Irradiation and Air Pollution-Induced Skin Aging and Pigmentation. *Cosmetics* **2023**, *10*, 88. https://doi.org/10.3390/cosmetics10030088

Academic Editor: Agnieszka Feliczak Guzik

Received: 25 April 2023
Revised: 26 May 2023
Accepted: 30 May 2023
Published: 7 June 2023

Copyright: © 2023 by the authors. Licensee MDPI, Basel, Switzerland. This article is an open access article distributed under the terms and conditions of the Creative Commons Attribution (CC BY) license (https://creativecommons.org/licenses/by/4.0/).

1. Introduction

Skin, the largest organ of the human body, undergoes significant changes during the aging process, which is influenced by both genetic and environmental factors. Intrinsic aging, driven by genetic factors, results in the gradual breakdown of collagen and elastin fibers, leading to loss of skin elasticity and firmness. Extrinsic aging, on the other hand, is largely driven by environmental factors such as UV radiation and pollution [1–3].

The mechanisms underlying the effects of UV radiation and pollution on skin aging are complex and multifactorial. UV radiation generates reactive oxygen species (ROS) that contribute to the breakdown of collagen and elastin fibers in the skin, resulting in a loss of elasticity and firmness [4]. Exposure to pollution can also generate ROS and trigger inflammatory pathways, which can further exacerbate the effects of UV radiation and accelerate the aging process [5]. Notably, air pollution exposure has been shown to be correlated with the development of pigment spots [5,6].

Air pollution as well as UV radiation can activate the aryl hydrocarbon receptor (AhR) [7,8]. This, in turn, leads to the activation of downstream genes, such as *MMP1* and *CYP1A1*, resulting in increased production of ROS, inflammatory cytokines, and collagen breakdown [8–10].

Effective protection of the skin against the detrimental effects of both UV radiation and pollution is therefore essential to maintain healthy skin.

In recent years, there has been growing interest in the use of natural compounds in skin-care products. One class of compounds that has gained considerable attention is flavonoids. Flavonoids are a diverse group of phytochemicals, widely distributed in fruits, vegetables, herbs, and medicinal plants, and are known for their various biological activities. In the realm of skincare, flavonoids, such as quercitin, kaempferol, and rutin, are highly regarded for their potent antioxidant and anti-inflammatory properties [11–13]. Another promising area of application for flavonoids in skincare is their photoprotective potential. Flavonoids have demonstrated their capacity to absorb and scatter UV radiation, thus reducing its penetration into the skin and minimizing its harmful effects [14–16]. One particular flavonoid of interest is naringenin (5,7,4′-trihydroxyflavanone, depicted in Figure 1), which is abundantly found in citrus fruits. Several animal studies have indicated that naringenin holds immense potential in benefiting the skin through its photoprotective, anti-inflammatory, antioxidant, and skin-brightening effects [17–21]. These accumulated findings strongly suggest that naringenin may prove to be effective not only in shielding the skin against UV radiation-induced photoaging but also in protecting it from the harmful effects of pollution.

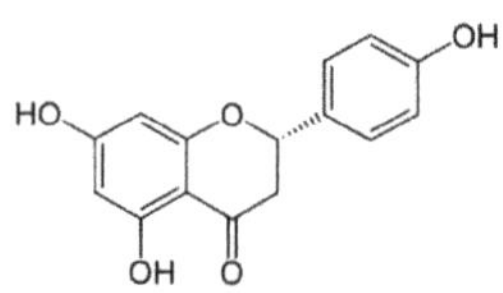

Figure 1. Chemical structure of naringenin.

The present study aims to (1) investigate the in vitro UV protective/anti-inflammatory effects of naringenin on human primary skin cells, (2) investigate naringenin's ability to protect against pollution-induced skin inflammation and oxidative stress in human skin cells, and (3) investigate naringenin's effect on melanogenesis genes and melanin production in human skin cells and reconstructed pigmented epidermis.

2. Materials and Methods

2.1. Viability

In parallel with all assays, viability testing was performed using CellTiterGlo (Promega, Madison, WI, USA) according to manufacturer's instructions.

2.2. Cell Culture

Primary human keratinocytes, fibroblasts, and melanocytes, all came from healthy female donors, 24–56 years of age (Cellsystems GmbH, Troisdorf, Germany). They were maintained in EpiLife supplemented with HKGS (Gibco, Grand Island, NY, USA), DMEM containing 10% fetal bovine serum (Gibco) or Media 254 supplemented with HMGS at 37 °C in a 5% CO_2 humidified incubator.

2.3. UVB Keratinocyte/Fibroblast Assay

The UVB keratinocyte/fibroblast assay is a paracrine assay where the media from UV stimulated primary keratinocytes is used to activate primary dermal fibroblasts (method adapted from Wang et al. [22]). Briefly, the keratinocytes were cultured until they reached approximately 80% confluence and then were irradiated with 50 mJ/cm^2 in the UVB range of 280–315 nm, using the UV-MAT Irradiation controller from Dr Gröbel GmbH, Ettlingen, Germany. The cells were then cultured at 37 °C in a 5% CO_2 humidified incubator, and the media was collected after 24 h. Fibroblasts were cultured to approximately 80% confluence and pretreated with naringenin, at 1.5 µM, 15 µM or 150 µM, (Sigma-Aldrich, St. Louis, MO, USA >95% purity) for 1 h before changing to the media from the UV-treated keratinocytes. After 24 h incubation the media was collected and analyzed using MMP1 ELISA (Sigma-Aldrich) or Luminex analysis (ProCartaPlex, Affymetrix, Santa Clara, CA, USA). Standard curve from the ELISA was plotted as log concentration vs. absorbance (-blank) and then analyzed with variable slope four parameter curve fit. Standard curve from the Luminex was plotted as log concentration vs. absorbance (-blank) and then analyzed with asymmetric five parameter curve fit according to manufacturer's instructions. Assay is considered valid when the biomarker in question is significantly higher after stimulation than the untreated vehicle control. Cells are stimulated in parallel with IL1-beta (Sigma-Aldrich) and inhibited with IL-1 receptor antagonist (Sigma-Aldrich) as a functional assay control.

2.4. Pollution-Induced Keratinocyte/Fibroblast Assay

The pollution-induced keratinocyte/fibroblast assay is a paracrine assay in which the media from keratinocytes stimulated with diesel particulate matter (DPM), is utilized to activate dermal fibroblasts.

Keratinocytes were cultured to approximately 80% confluence and then cultured in the presence of 10 µg/mL DPM (1650b, National Institute of Standards and Technology, Gaithersburg, MD, USA). The media was collected after 24 h. Fibroblasts were cultured to approximately 80% confluence and pretreated with 150 µM naringenin for 1 h before changing to the media from the DPM-treated keratinocytes in combination with 150 µM naringenin (Sigma-Aldrich). After 24 h incubation, the fibroblast media was collected and analyzed by MMP1 ELISA (Sigma-Aldrich) according to manufacturer's instructions. Standard curve from the ELISA was plotted as log concentration vs. absorbance (-blank) and then analyzed with variable slope four parameter curve fit. Assay is considered valid when the biomarker in question is significantly higher after stimulation than the untreated vehicle control. Cells are in parallel stimulated with IL1-beta, 100 ng/mL (Sigma-Aldrich), and inhibited with IL-1 receptor antagonist, 200 ng/mL (Sigma-Aldrich), as a functional assay control.

2.5. Pollution-Induced CYP1A1 Assay

The pollution induced CYP1A1 assay is a qPCR assay analyzing *CYP1A1* gene expression in keratinocytes after DPM stimulation. Keratinocytes were cultured to approximately 90% confluence followed by pretreatment with 150 µM naringenin (Sigma-Aldrich) for 1 h before the addition DPM, 10 µg/mL (1650b, National Institute of Standards and Technology). The keratinocytes were then cultured in the presence of DPM and 150 µM naringenin for an additional 5 h. Assay is considered valid when the biomarker in question is significantly higher after stimulation than the untreated vehicle control. Cells are in parallel treated with the inhibitor quercetin, 10 µM, as a functional assay control.

2.6. RNA Isolation and cDNA Synthesis

Total RNA was isolated using the RNeasy® Mini Kit (Qiagen, Hilden, Germany) and reverse transcription of RNA was performed with the iScript™ Advanced cDNA Synthesis Kit (BioRad, Hercules, CA, USA) all according to the manufacturer's instructions.

2.7. qPCR Analysis of CYP1A1 Gene Expression

The qPCR reaction was performed in a total volume of 20 µL containing 10 µL of SsoAdvanced™ Universal SYBR® Green Supermix (BioRad), cDNA and the PrimePCR™ SYBR® CYP1A1 (BioRad). The thermal cycling was carried out in a CFX96 ThermalCycler (BioRad) with a program of 95 °C for 3 min, followed by 40 cycles with denaturation at 95 °C for 10 s, and annealing and elongation at 55 °C for 30 s. The gene expression levels were normalized to the expression level of GAPDH housekeeping gene (PrimePCR™ SYBR® GAPDH, BioRad). Relative gene expression changes, calculated using the $2^{-\Delta\Delta CT}$ method, are reported as number-fold changes compared to those in the control samples.

2.8. Gene Expression Analysis of Pigmentation Genes in Melanocytes

The melanocytes were cultured to approximately 90% confluence and thereafter cultured in the presence of naringenin (Sigma-Aldrich), 150 µM, or vehicle control (water) for 24 h. Total RNA was isolated from primary melanocytes using the RNeasy® Mini Kit (Qiagen). Nanostring hybridization was set up according to manufacturer's instructions, 25 ng RNA/reaction (nCounter® XT assay, nanoString Technologies, Seattle, WA, USA). Nanostring analysis was performed at KIGene core facility, CMM, Karolinska Institutet, Sweden. Gene expression analysis of 27 selected melanogenesis genes (*MITF, ADRB2, CDH1, CTNNB1, DCT, DCTN1, EDN1, EDNRB, GPR143, GRIN1, GSTP1, HMOX1, KIT, MAPK1, MAPK3, MC1R, MLANA, MLPH, MYO5A, NFE2L2, PMEL, POMC, SFRP2, TYR, TYRP1, WNT3, WNT7A*) normalized to the two housekeeping genes, *GAPDH* and *PGK1*. Statistical analysis was performed by use of the software nSolver 3.0 (nanoString Technologies, USA)

2.9. Pigmentation Analysis, Reconstructed Human Epidermis

2.9.1. Cell Culture and Reconstruction of Epidermis

The study was conducted by Straticell, Gembloux, Belgium, using reconstructed human epidermis (RHE/MEL/001) that contained normal human keratinocytes and normal, darkly pigmented epidermal melanocytes. The RHE/MEL/001 was reconstructed at the air–liquid interface in EpiLife growth media supplemented with HKGS (Gibco) for 14 days in a humid atmosphere at 37 °C and 5% CO_2. Bovine pituitary extract (BPE, Gibco) was added to the media at a concentration of 50 µg/mL from day 0 to day 4, and from day 5 to day 14, the media was supplemented without BPE.

2.9.2. Pigmentation Assay, with or without UV Challenge

Naringenin (Sigma-Aldrich) was applied topically in butylene glycol to the reconstructed skin at 0.5% and 0.1%. Application was done at six occasions during 10 days on day 4, 7, 8, 9, 10 and 11. In the UV challenge model the RHE/MEL/001 was treated with α-MSH, 1µM; on day 7, 8 9,10 and 11 in combination with UVA, 1 J/ cm^2; and UVB, 50 mJ/cm^2, on day 8, 9, 10 and 11. Analysis took place on day 14. In parallel, the effect of kojic acid, 250 µM, (added systemically to the RHE/MEL/001) was measured as a reference compound to validate the analysis. At the end of treatments, melanin was extracted using Solvable (Perkin Elmer, Waltham, MA, USA), heated at 80 °C for 1 h and quantified spectrophotometrically at 490 nm. Synthetic melanin (Sigma-Aldrich) was used for standard curve. All data were compared to untreated control or vehicle control (butylene glycol) with or without UV challenge.

2.10. Statistical Analysis

Quantitative data are represented as mean ± SEM. To analyze the statistical difference between two groups, two-sided Student's *t*-tests were used for the comparisons. For two or more groups, one-way analysis of variance (ANOVA) with Tukey's multiple-comparison test was used to analyze the statistical difference. Two-way ANOVA was chosen for analyzing the statistical difference between data points in groups. GraphPad Prism (GraphPad Software Inc., San Diego, CA, USA) was used for data analysis and graphic presentations. $p^{*} \leq 0.05$, $p^{**} \leq 0.01$ and $p^{***} \leq 0.001$.

3. Results

3.1. Viability

No decrease in cell viability was observed in either keratinocytes or fibroblasts after 24 h of treatment with naringenin up to a concentration of 150 µM. Therefore, this concentration was selected for use throughout the entire study.

3.2. Naringenin Inhibits UVB Induced Inflammatory Response in Human Primary Skin Cells

Abnormal induction of MMP-1 plays an important role in UV-induced photoaging through its degradation of collagen in dermal skin [4,23]. UVB induced MMP1 production was significantly inhibited after 1 h pretreatment, followed by 24 h treatment with naringenin in human primary dermal fibroblasts, as shown in Figure 2a. This inhibition is dose-dependent, with an IC50 (50% inhibition concentration) of 1.4 µM of naringenin, as shown in Figure 2b.

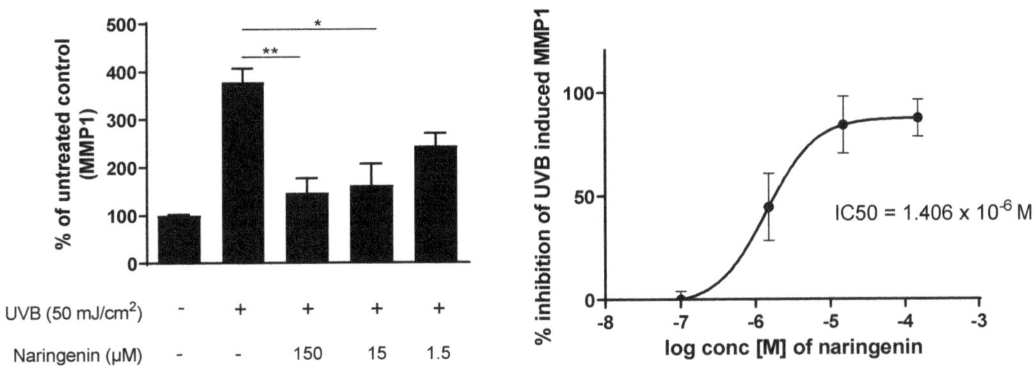

(a) (b)

Figure 2. (**a**) Human dermal fibroblasts pre-treated for 1 h with naringenin 150 µM, 15 µM or 1.5 µM and then treated for 24 h with naringenin and media from UVB irradiated keratinocytes (50 mJ/cm^2). MMP1 concentration presented as % of untreated control. Statistics were performed on raw data with unpaired student's *t*-test, where $p^* \leq 0.05$, and $p^{**} \leq 0.01$. $n = 3$, where n is the number of fibroblast donors. (**b**) Concentration of naringenin (log) was plotted against % inhibition of UVB-induced MMP1. The IC50 for naringenin's ability to inhibit UVB induced MMP1 in dermal human fibroblasts was calculated to be 1.4 µM.

In addition to inhibition of MMP1, naringenin also significantly inhibited UVB-induced MMP3, IL-6 and GM-CSF, as shown in Figure 3. They have all been shown to be upregulated in skin after UV irradiation and involved in the inflammation and photoaging process of the skin [24–26].

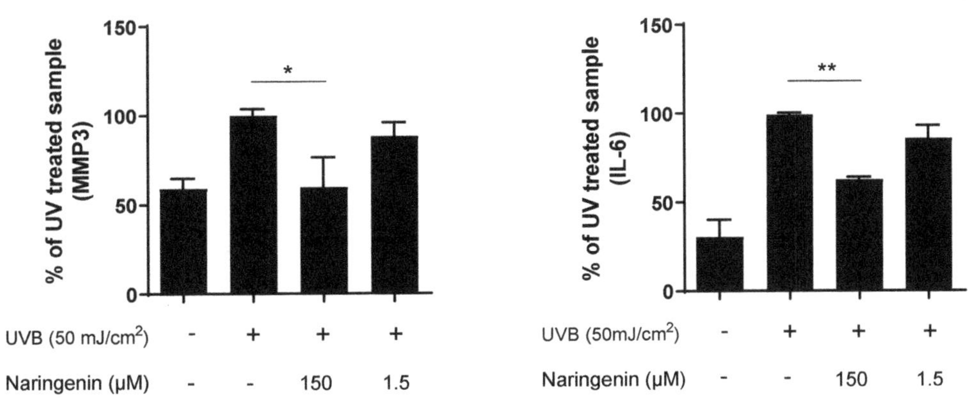

(a) (b)

Figure 3. *Cont.*

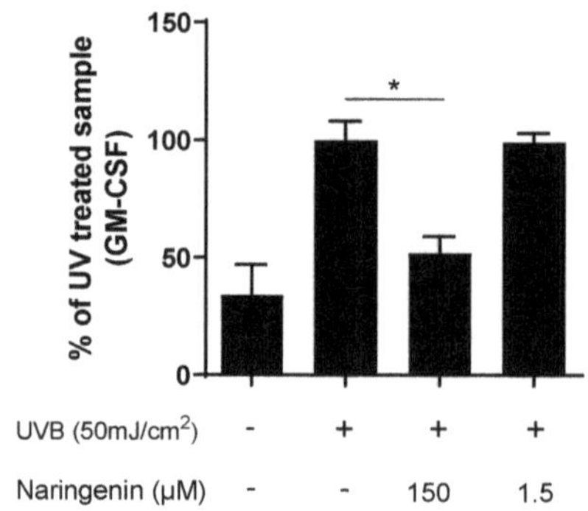

(c)

Figure 3. Human dermal fibroblasts pre-treated for 1 h with naringenin 150 µM or 1.5 µM and then treated for 24 h with naringenin in media from UVB irradiated keratinocytes (50 mJ/cm^2). (**a**) MMP3 concentration presented as % of UV-treated sample. (**b**) IL-6 concentration presented as % of UV-treated sample. (**c**) GM-CSF concentration presented as % of UV-treated sample. Statistics were performed on raw data with unpaired student's *t*-test, where p * $\leq$ 0.05 and p ** $\leq$ 0.01. n = 2 in duplicates where n is the number of fibroblast donors.

3.3. Naringenin Inhibits Pollution-Induced MMP1 in Human Dermal Fibroblasts

As shown in previous studies, pollution induced activation of the AhR pathway activate MMP-1 expression in human fibroblasts and keratinocytes [9], thus contributing to matrix degradation and extrinsic aging. Here, we show that naringenin significantly inhibits DPM-induced MMP1 expression in human primary fibroblasts, as shown in Figure 4.

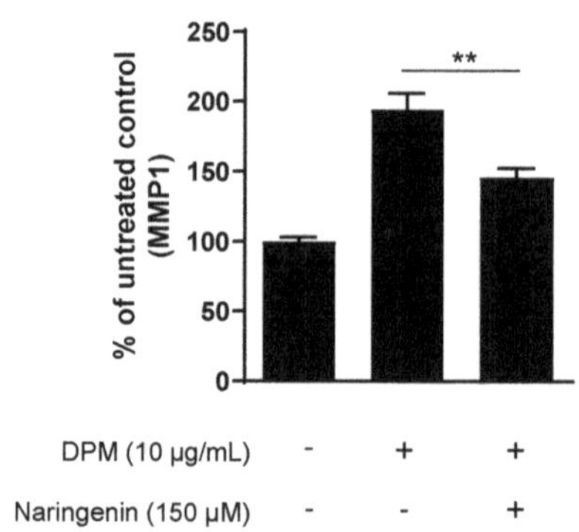

Figure 4. Human dermal fibroblasts pre-treated for 1 h with 150 µM naringenin, and then treated for 24 h with naringenin and media from DPM treated (10 µg/mL) keratinocytes. MMP1 concentration presented as % of untreated control. Statistics were performed on raw data with unpaired student´s *t*-test, where p ** $\leq$ 0.01. n = 3, where n is the number of fibroblast donors.

3.4. Naringenin Inhibits Pollution-Induced CYP1A1 in Human Keratinocytes

Air pollution has been shown to increase the expression of the *CYP1A1* gene, which belongs to the P450 family of enzymes. When activated, CYP1A1 metabolizes environmental pollutants into highly reactive intermediates that can cause DNA damage, oxidative stress, and inflammation [27,28]. In this study, we demonstrate that *CYP1A1* expression is upregulated after exposure to diesel particulate matter (DPM). Primary human keratinocytes were treated with DPM for 5 h, resulting in a significant 14-fold increase in *CYP1A1* expression compared to the control group; however, pre-treatment with 150 µM naringenin for 1 h followed by 5 h of combined naringenin and DPM exposure resulted in a significant downregulation of *CYP1A1* gene expression back to basal levels (Figure 5).

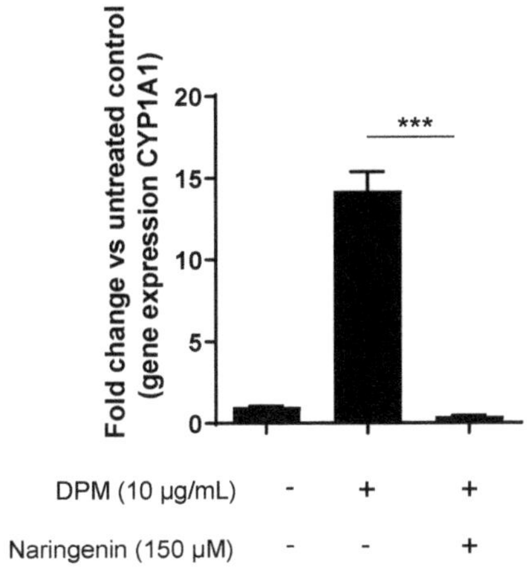

Figure 5. Human dermal keratinocytes pre-treated for 1 h with naringenin 150 µM, and then treated for 5 h with naringenin and DPM (10 µg/mL). Gene expression presented as fold change compared to untreated control. Statistics were performed on ΔCt. $n = 3$, where n is the number of keratinocyte donors. Unpaired student's t-test, where p *** ≤ 0.001.

3.5. Naringenin Downregulates Key Melanogenesis Genes

To investigate the effect of naringenin on genes involved in melanogenesis, we treated melanocytes with naringenin for 24 h and analyzed the gene expression of 27 melanogenesis-associated genes.

Three of these genes, *MITF*, *MLPH* and *MYO5A*, were significantly downregulated by naringenin compared to the vehicle (Table 1).

Table 1. Gene expression analysis in primary human melanocytes treated with 150 µM naringenin for 24 h. Statistical analysis was performed by use of the software nSolver 3.0 (NanoString Technologies, USA), $n = 1$ in duplicates, where n is the number of melanocyte donors.

Gene ID		Naringenin	Vehicle	Ratio	*p* Value
MLPH	XM_006712737.1	192.88	226.91	0.85	0.049
		159.27	226.91	0.7	0.028
MYO5A	NM_000259.3	544.97	808.06	0.67	0.03
		549.92	808.06	0.68	0.024
MITF	NM_000248.3	298.35	401.2	0.74	0.026
		302.99	401.2	0.76	0.049

3.6. Naringenin Inhibits Basal and UV-Induced Melanogenesis in Human Pigmented Reconstructed Epidermis

3.6.1. Visual Analysis

The macroscopic pictures show that systemic treatment with the control compound kojic acid results in whitening of the tissue compared to untreated, both with and without UV. It can also be concluded that topical treatment with naringenin results in a visual whitening effect compared to vehicle both with and without UV, Figure 6.

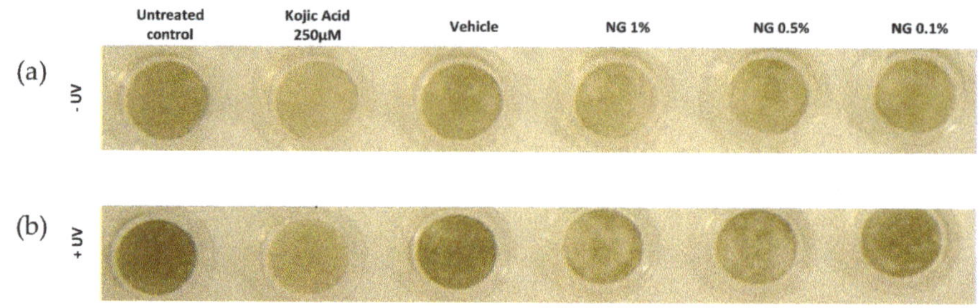

Figure 6. Macroscopic picture of RHE/MEL tissues treated systemically with kojic acid or topically with naringenin (NG) or vehicle (butylene glycol) for 10 days (from day 4 to 14), without (**a**) or with UVA (1 J/cm^2), UVB (50 mJ/cm^2) and α-MSH (1 µM) (**b**).

3.6.2. Melanin Content Analysis after Chemical Extraction

Following the visual analysis, melanin was extracted from the RHE/MEL tissues at the end of treatments on day 14 and quantified spectrophotometrically. The level of melanin in the untreated samples without UV was as expected lower than in the UVA/UVB + α-MSH-treated tissues, 33 µg/mL compared to 53 µg/mL (Figure 7a and 7b, respectively). Treatment with the control compound kojic acid significantly decreased the melanin content in epidermis both with and without UV challenge, validating the experimental setup.

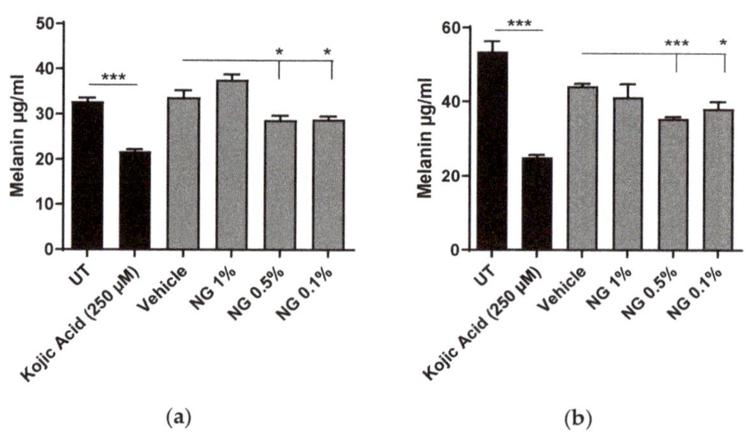

Figure 7. Quantification of extracted melanin from reconstructed human epidermis containing melanocytes. Topically treated for 10 days with naringenin (NG) or systemically with the control compound kojic acid (KA), without (**a**) or with UVA/B (**b**). The histograms represent values obtained from three tissues (n = 3) for each treatment. For statistical analysis, a student's t-test was performed, where p * $\leq$ 0.05, and p *** $\leq$ 0.001.

Topical treatment with naringenin, 0.5% and 0.1%, also results in a significant decrease in melanin both with and without UV. At 1%, visual effects are observed from the macroscopic pictures but could not be confirmed after melanin extraction (Figure 6a,b).

4. Discussion

Naringenin, a flavonoid that is commonly found in citrus fruits, has been demonstrated to possess various beneficial properties for the skin. These include antioxidant, anti-inflammatory, and anticancer effects, as reported in previous studies [29–31]. Furthermore, topical application of naringenin has been shown to reduce skin inflammation and oxidative stress caused by UVB radiation in mice [19].

The primary aim of this study was to thoroughly investigate the protective and restorative effects of naringenin on human primary skin cells and reconstructed skin when exposed to external stressors, thus shedding light on its potential in promoting overall skin health. Throughout this study, we specifically utilized primary skin cells due to their inherent advantages, such as better functionality and the ability to exhibit more accurate physiological responses, as compared to cell lines commonly used in research. Additionally, the utilization of reconstructed skin models in this study further enhanced the relevance and translatability of our findings.

UV radiation is a known cause of skin damage, resulting in premature aging, wrinkles, and an increased risk of skin cancer. In the context of skin aging and photoaging, an increase in matrix metalloproteinases plays a critical role in the unbalanced turnover or rapid breakdown of collagen and elastin fibers in the skin, leading to the development of fine lines, wrinkles, and loss of elasticity [4].

In this study, we investigated the effect of naringenin on two of these matrix metalloproteinases (MMPs), MMP1 and MMP3, which are known to be upregulated in response to UV radiation in human skin cells. MMP1 is considered the major collagenase involved in photoaging caused by UV irradiation [23,24]. In line with previous results showing MMP1 inhibition in rat, mice, and human cell lines, [32–35], we showed a significant inhibition of UVB induced MMP1 production in human primary fibroblasts, as shown in Figure 2a. Moreover, this downregulation was found to be dose-dependent, and an IC50 value of 1.4 μM was established for MMP1, as shown in Figure 2b.

In addition to MMP1 inhibition, we also demonstrate that naringenin significantly reduces the UVB-induced increase in MMP3, Figure 3a, in human primary fibroblasts. MMP3, also plays a crucial role in tissue remodeling and degradation of the extracellular matrix [36]. MMP3 has also been linked to inflammatory processes in the skin and excessive MMP3 activity can promote the release of pro-inflammatory cytokines and chemokines, triggering an inflammatory response [37].

Interleukin-6 (IL-6) and granulocyte-macrophage colony-stimulating factor (GM-CSF) are both cytokines involved in immune responses and inflammation. While these cytokines play important roles in the immune system, excessive or dysregulated levels of IL-6 and GM-CSF can have negative effects on the skin. They are both pro-inflammatory cytokines that can contribute to chronic inflammation when their levels are elevated and prolonged inflammation in the skin can lead to tissue damage, impaired skin barrier function, and the development of various skin disorders. Elevated levels of IL-6 and GM-CSF have also been associated with skin aging, by promoting the production of MMPs and elastase, leading to reduced skin elasticity, increased wrinkles, and loss of skin firmness [38,39].

Both IL-6 and GM-CSF are known to be upregulated by UVB radiation in the skin [25,26]. In this study we demonstrate that following treatment with naringenin, the UVB-induced levels of IL-6 and GM-CSF were normalized (Figure 3b and 3c, respectively). Collectively, these findings in human primary skin fibroblasts support previous studies conducted in animal models and in a human cell line that have shown the photoprotective properties of naringenin [19,30,32,40].

Like UV radiation, air pollution also leads to an increase of intracellular reactive oxygen species (ROS), inflammatory mediators, pro-inflammatory cytokines, and matrix metalloproteinases in the skin. Moreover, pollution can interact with UV radiation and have synergistic effects on the skin, leading to more pronounced damage [41]. The effect of pollution to the skin is primarily due to the direct activation of the aryl hydrocarbon receptor (AhR) [7–10]. One of the downstream genes that is directly activated by pollution via the AhR is *CYP1A1*, which belongs to the cytochrome P450 family of enzymes that play a role in the metabolism of various xenobiotics, including polycyclic aromatic hydrocarbons (PAHs) that are present in cigarette smoke and air pollution. The activation of CYP1A1 increases the metabolism of these compounds, resulting in their detoxification and elimination from the body. However, CYP1A1 also catalyzes the conversion of the common exhaust pollutant benzo[a]pyrene (B[a]P) into the highly toxic epoxide (BP-7,8-

dihydrodiol-9,10-epoxide). Thus, inhibiting *CYP1A1* may be of interest to protect the skin from pollution-induced damage [8,42]. In this study, we demonstrated that naringenin inhibits both pollution-induced upregulation of MMP1 in human fibroblasts (Figure 4) and AhR-mediated upregulation of *CYP1A1* expression (Figure 5).

In addition to causing oxidative stress, inflammation, and premature skin aging, UV and polluted air can lead to uneven skin tone and hyperpigmentation [5,43,44]. Previous studies reported that naringenin can both stimulate and inhibit melanogenesis [45–48]; however, they were conducted in murine melanoma cells. In our study, we demonstrate that in human primary melanocytes, naringenin acts as an inhibitor of several key melanogenesis genes (Table 1). Microphthalmia-associated transcription factor (MITF) is a master regulator of melanocyte development, differentiation, and function. It regulates the expression of several genes involved in melanogenesis, including TYR, TYRP1, DCT, and PMEL, which are necessary for melanin production [49]. Overexpression or hyperactivation of MITF can lead to hyperpigmentation. Melanophilin (MLPH) and myosin 5 (MYO5A) both play essential roles in melanogenesis by regulating the maturation and transport of melanosomes and maintaining the proper distribution of melanin in the skin [50,51]. Furthermore, we show that when topically applied to reconstructed pigmented epidermis with human primary cells, naringenin significantly reduces pigmentation (Figures 6 and 7).

5. Conclusions

In conclusion, this study confirms naringenins photoprotective properties in human primary skin cells and, to our knowledge, for the first time, demonstrates its ability to protect against pollution-induced skin damage by inhibiting MMP1 as well as CYP1A1. In combination with naringenin's ability to reduce pigmentation, it counteracts some of the major hallmarks of extrinsic aging caused by UVB and pollution. Overall, given the protective effects of naringenin against both UVB radiation and pollution-induced skin damage, it holds great promise as a candidate for the development of skincare products.

Author Contributions: Conceptualization and supervision, L.V.-J.; methodology, N.H., V.L.-K., N.A., S.S. and C.Ö.; formal analysis, C.Ö.; investigation, C.Ö., N.H., V.L.-K. and N.A.; data curation, C.Ö.; writing—original draft preparation, C.Ö.; writing—review and editing, C.Ö., N.H., N.A., S.S. and L.V.-J. All authors have read and agreed to the published version of the manuscript.

Funding: This research received no external funding.

Institutional Review Board Statement: Not applicable.

Informed Consent Statement: Not applicable.

Data Availability Statement: The data presented in this study are available upon request from the corresponding author.

Conflicts of Interest: The authors declare no conflict of interest.

References

1. Krutmann, J.; Liu, W.; Li, L.; Pan, X.; Crawford, M.; Sore, G.; Seite, S. Pollution and Skin: From Epidemiological and Mechanistic Studies to Clinical Implications. *J. Dermatol. Sci.* **2014**, *76*, 163–168. [CrossRef] [PubMed]
2. Velasco, M.V.R.; Sauce, R.; de Oliveira, C.A.; de Oliveira Pinto, C.A.S.; Martinez, R.M.; Baah, S.; Almeida, T.S.; Rosado, C.; Baby, A.R. Active Ingredients, Mechanisms of Action and Efficacy Tests of Antipollution Cosmetic and Personal Care Products. *Brazilian J. Pharm. Sci.* **2018**, *54*, e01003. [CrossRef]
3. Khmaladze, I.; Leonardi, M.; Fabre, S.; Messaraa, C.; Mavon, A. The Skin Interactome: A Holistic "Genome-Microbiome-Exposome" Approach to Understand and Modulate Skin Health and Aging. *Clin. Cosmet. Investig. Dermatol.* **2020**, *13*, 1021–1040. [CrossRef] [PubMed]
4. Fisher, G.J.; Kang, S.; Varani, J.; Bata-Csorgo, Z.; Wan, Y.; Datta, S.; Voorhees, J.J. Mechanisms of Photoaging and Chronological Skin Aging. *Arch. Dermatol.* **2002**, *138*, 1462–1470. [CrossRef] [PubMed]
5. Vierkötter, A.; Schikowski, T.; Ranft, U.; Sugiri, D.; Matsui, M.; Krämer, U.; Krutmann, J. Airborne Particle Exposure and Extrinsic Skin Aging. *J. Investig. Dermatol.* **2010**, *130*, 2719–2726. [CrossRef]

6. Ahn, Y.; Lee, E.J.; Luo, E.; Choi, J.; Kim, J.Y.; Kim, S.; Kim, S.H.; Bae, Y.J.; Park, S.; Lee, J.; et al. Particulate Matter Promotes Melanin Production through Endoplasmic Reticulum Stress–Mediated IRE1α Signaling. *J. Investig. Dermatol.* **2022**, *142*, 1425–1434.e6. [CrossRef]
7. Nguyen, L.P.; Bradfield, C.A. The Search for Endogenous Activators of the Aryl Hydrocarbon Receptor. *Chem. Res. Toxicol.* **2008**, *21*, 102–116. [CrossRef]
8. Costa, C.; Catania, S.; De Pasquale, R.; Stancanelli, R.; Scribano, G.M.; Melchini, A. Exposure of Human Skin to Benzo[a]Pyrene: Role of CYP1A1 and Aryl Hydrocarbon Receptor in Oxidative Stress Generation. *Toxicology* **2010**, *271*, 83–86. [CrossRef]
9. Ono, Y.; Torii, K.; Fritsche, E.; Shintani, Y.; Nishida, E.; Nakamura, M.; Shirakata, Y.; Haarmann-Stemmann, T.; Abel, J.; Krutmann, J.; et al. Role of the Aryl Hydrocarbon Receptor in Tobacco Smoke Extract–Induced Matrix Metalloproteinase-1 Expression. *Exp. Dermatol.* **2013**, *22*, 349–353. [CrossRef]
10. Tigges, J.; Haarmann-Stemmann, T.; Vogel, C.F.A.; Grindel, A.; Hübenthal, U.; Brenden, H.; Grether-Beck, S.; Vielhaber, G.; Johncock, W.; Krutmann, J.; et al. The New Aryl Hydrocarbon Receptor Antagonist E/Z-2-Benzylindene-5,6- Dimethoxy-3,3-Dimethylindan-1-One Protects against UVB-Induced Signal Transduction. *J. Investig. Dermatol.* **2014**, *134*, 556–559. [CrossRef]
11. Barreca, D.; Mandalari, G.; Calderaro, A.; Smeriglio, A.; Trombetta, D.; Felice, M.R.; Gattuso, G. Citrus Flavones: An Update on Sources, Biological Functions, and Health Promoting Properties. *Plants* **2020**, *9*, 288. [CrossRef]
12. de Lima Cherubim, D.J.; Buzanello Martins, C.V.; Oliveira Fariña, L.; da Silva de Lucca, R.A. Polyphenols as Natural Antioxidants in Cosmetics Applications. *J. Cosmet. Dermatol.* **2020**, *19*, 33–37. [CrossRef]
13. Panche, A.N.; Diwan, A.D.; Chandra, S.R. Flavonoids: An Overview. *J. Nutr. Sci.* **2016**, *5*, e47. [CrossRef]
14. Nunes, A.R.; Vieira, Í.G.P.; Queiroz, D.B.; Leal, A.L.A.B.; Maia Morais, S.; Muniz, D.F.; Calixto-Junior, J.T.; Coutinho, H.D.M. Use of Flavonoids and Cinnamates, the Main Photoprotectors with Natural Origin. *Adv. Pharmacol. Sci.* **2018**, *2018*, 5341487. [CrossRef]
15. Takayama, K.S.; Monteiro, M.C.; Saito, P.; Pinto, I.C.; Nakano, C.T.; Martinez, R.M.; Thomaz, D.V.; Verri, W.A.; Baracat, M.M.; Arakawa, N.S.; et al. Rosmarinus Officinalis Extract-Loaded Emulgel Prevents UVB Irradiation Damage to the Skin. *An. Acad. Bras. Cienc.* **2022**, *94*, e20201058. [CrossRef]
16. Velasco, M.V.R.; Sarruf, F.D.; Salgado-Santos, I.M.N.; Haroutiounian-Filho, C.A.; Kaneko, T.M.; Baby, A.R. Broad Spectrum Bioactive Sunscreens. *Int. J. Pharm.* **2008**, *363*, 50–57. [CrossRef]
17. Lee, C.-H.; Jeong, T.-S.; Choi, Y.-K.; Hyun, B.-H.; Oh, G.-T.; Kim, E.-H.; Kim, J.-R.; Han, J.-I.; Bok, S.-H. Anti-Atherogenic Effect of Citrus Flavonoids, Naringin and Naringenin, Associated with Hepatic ACAT and Aortic VCAM-1 and MCP-1 in High Cholesterol-Fed Rabbits. *Biochem. Biophys. Res. Commun.* **2001**, *284*, 681–688. [CrossRef]
18. Al-Roujayee, A.S. Naringenin Improves the Healing Process of Thermally-Induced Skin Damage in Rats. *J. Int. Med. Res.* **2017**, *45*, 570–582. [CrossRef]
19. Martinez, R.M.; Pinho-Ribeiro, F.A.; Steffen, V.S.; Silva, T.C.C.; Caviglione, C.V.; Bottura, C.; Fonseca, M.J.V.; Vicentini, F.T.M.C.; Vignoli, J.A.; Baracat, M.M.; et al. Topical Formulation Containing Naringenin: Efficacy against Ultraviolet B Irradiation-Induced Skin Inflammation and Oxidative Stress in Mice. *PLoS ONE* **2016**, *11*, e0146296. [CrossRef]
20. Pinho-Ribeiro, F.A.; Zarpelon, A.C.; Fattori, V.; Manchope, M.F.; Mizokami, S.S.; Casagrande, R.; Verri, W.A. Naringenin Reduces Inflammatory Pain in Mice. *Neuropharmacology* **2016**, *105*, 508–519. [CrossRef]
21. Ali, R.; Shahid, A.; Ali, N.; Hasan, S.K.; Majed, F.; Sultana, S. Amelioration of Benzo[a]Pyrene-Induced Oxidative Stress and Pulmonary Toxicity by Naringenin in Wistar Rats: A Plausible Role of COX-2 and NF-KB. Hum. *Exp. Toxicol.* **2016**, *36*, 349–364. [CrossRef] [PubMed]
22. Wang, X.; Bi, Z.; Chu, W.; Wan, Y. IL-1 Receptor Antagonist Attenuates MAP Kinase/AP-1 Activation and MMP1 Expression in UVA-Irradiated Human Fibroblasts Induced by Culture Medium from UVB-Irradiated Human Skin Keratinocytes. *Int. J. Mol. Med.* **2005**, *16*, 1117–1124. [CrossRef] [PubMed]
23. Brennan, M.; Bhatti, H.; Nerusu, K.C.; Bhagavathula, N.; Kang, S.; Fisher, G.J.; Varani, J.; Voorhees, J.J. Matrix Metalloproteinase-1 Is the Major Collagenolytic Enzyme Responsible for Collagen Damage in UV-Irradiated Human Skin. *Photochem. Photobiol.* **2007**, *78*, 43–48. [CrossRef]
24. Lu, J.; Guo, J.H.; Tu, X.L.; Zhang, C.; Zhao, M.; Zhang, Q.W.; Gao, F.H. Tiron Inhibits UVB-Induced AP-1 Binding Sites Transcriptional Activation on MMP-1 and MMP-3 Promoters by MAPK Signaling Pathway in Human Dermal Fibroblasts. *PLoS ONE* **2016**, *11*, e0159998. [CrossRef]
25. Choi, Y.J.; Uehara, Y.; Park, J.Y.; Kim, S.J.; Kim, S.R.; Lee, H.W.; Moon, H.R.; Chung, H.Y. MHY884, a Newly Synthesized Tyrosinase Inhibitor, Suppresses UVB-Induced Activation of NF-KB Signaling Pathway through the Downregulation of Oxidative Stress. *Bioorg. Med. Chem. Lett.* **2014**, *24*, 1344–1348. [CrossRef]
26. Sharma, S.D.; Katiyar, S.K. Dietary Grape Seed Proanthocyanidins Inhibit UVB-Induced Cyclooxygenase-2 Expression and Other Inflammatory Mediators in UVB-Exposed Skin and Skin Tumors of SKH-1 Hairless Mice. *Pharm. Res.* **2010**, *27*, 1092–1102. [CrossRef]
27. Nebert, D.W.; Dalton, T.P.; Okey, A.B.; Gonzalez, F.J. Role of Aryl Hydrocarbon Receptor-Mediated Induction of the CYP1 Enzymes in Environmental Toxicity and Cancer. *J. Biol. Chem.* **2004**, *279*, 23847–23850. [CrossRef]
28. Furue, M.; Uchi, H.; Mitoma, C.; Hashimoto-Hachiya, A.; Tanaka, Y.; Ito, T.; Tsuji, G. Implications of Tryptophan Photoproduct FICZ in Oxidative Stress and Terminal Differentiation of Keratinocytes. *G. Ital. Dermatol. Venereol.* **2019**, *154*, 37–41. [CrossRef]

29. Manchope, M.F.; Casagrande, R.; Verri, W.A. Naringenin: An Analgesic and Anti-Inflammatory Citrus Flavanone. *Oncotarget* **2017**, *8*, 3766–3767. [CrossRef]
30. Kumar, R.; Bhan Tiku, A. Naringenin Suppresses Chemically Induced Skin Cancer in Two-Stage Skin Carcinogenesis Mouse Model. *Nutr. Cancer* **2020**, *72*, 976–983. [CrossRef]
31. Sun, R.; Liu, C.; Liu, J.; Yin, S.; Song, R.; Ma, J.; Cao, G.; Lu, Y.; Zhang, G.; Wu, Z.; et al. Integrated Network Pharmacology and Experimental Validation to Explore the Mechanisms Underlying Naringenin Treatment of Chronic Wounds. *Sci. Rep.* **2023**, *13*, 132. [CrossRef]
32. Jung, S.K.; Ha, S.J.; Jung, C.H.; Kim, Y.T.; Lee, H.K.; Kim, M.O.; Lee, M.H.; Mottamal, M.; Bode, A.M.; Lee, K.W.; et al. Naringenin Targets ERK2 and Suppresses UVB-Induced Photoaging. *J. Cell. Mol. Med.* **2016**, *20*, 909–919. [CrossRef]
33. Martinez, R.M.; Pinho-Ribeiro, F.A.; Steffen, V.S.; Caviglione, C.V.; Vignoli, J.A.; Barbosa, D.S.; Baracat, M.M.; Georgetti, S.R.; Verri, W.A.; Casagrande, R. Naringenin Inhibits UVB Irradiation-Induced Inflammation and Oxidative Stress in the Skin of Hairless Mice. *J. Nat. Prod.* **2015**, *78*, 1647–1655. [CrossRef]
34. Chang, H.L.; Chang, Y.M.; Lai, S.C.; Chen, K.M.; Wang, K.C.; Chiu, T.T.; Chang, F.H.; Hsu, L.S. Naringenin Inhibits Migration of Lung Cancer Cells via the Inhibition of Matrix Metalloproteinases-2 and-9. *Exp. Ther. Med.* **2017**, *13*, 739–744. [CrossRef]
35. Wang, C.C.; Guo, L.; Tian, F.D.; An, N.; Luo, L.; Hao, R.H.; Wang, B.; Zhou, Z.H. Naringenin Regulates Production of Matrix Metalloproteinases in the Knee-Joint and Primary Cultured Articular Chondrocytes and Alleviates Pain in Rat Osteoarthritis Model. Brazilian. *J. Med. Biol. Res.* **2017**, *50*, e5714. [CrossRef]
36. Quan, T.; Little, E.; Quan, H.; Qin, Z.; Voorhees, J.J.; Fisher, G.J. Elevated Matrix Metalloproteinases and Collagen Fragmentation in Photodamaged Human Skin: Impact of Altered Extracellular Matrix Microenvironment on Dermal Fibroblast Function. *J. Investig. Dermatol.* **2013**, *133*, 1362. [CrossRef]
37. Fingleton, B. Matrix Metalloproteinases as Regulators of Inflammatory Processes. *Biochim. Biophys. Acta Mol. Cell. Res.* **2017**, *1864*, 2036–2042. [CrossRef]
38. Onodera, S.; Kaneda, K.; Mizue, Y.; Koyama, Y.; Fujinaga, M.; Nishihira, J. Macrophage Migration Inhibitory Factor Up-Regulates Expression of Matrix Metalloproteinases in Synovial Fibroblasts of Rheumatoid Arthritis. *J. Biol. Chem.* **2000**, *275*, 444–450. [CrossRef]
39. Imokawa, G.; Nakajima, H.; Ishida, K. Biological Mechanisms Underlying the Ultraviolet Radiation-Induced Formation of Skin Wrinkling and Sagging II: Over-Expression of Neprilysin Plays an Essential Role. *Int. J. Mol. Sci.* **2015**, *16*, 7776–7795. [CrossRef]
40. El-Mahdy, M.A.; Zhu, Q.; Wang, Q.E.; Wani, G.; Patnaik, S.; Zhao, Q.; Arafa, E.S.; Barakat, B.; Mir, S.N.; Wani, A.A. Naringenin Protects HaCaT Human Keratinocytes against UVB-Induced Apoptosis and Enhances the Removal of Cyclobutane Pyrimidine Dimers from the Genome. *Photochem. Photobiol.* **2008**, *84*, 307–316. [CrossRef]
41. Kim, K.E.; Cho, D.; Park, H.J. Air Pollution and Skin Diseases: Adverse Effects of Airborne Particulate Matter on Various Skin Diseases. *Life Sci.* **2016**, *152*, 126–134. [CrossRef] [PubMed]
42. Shimizu, Y.; Nakatsuru, Y.; Ichinose, M.; Takahashi, Y.; Kume, H.; Mimura, J.; Fujii-Kuriyama, Y.; Ishikawa, T. Benzo[a]Pyrene Carcinogenicity Is Lost in Mice Lacking the Aryl Hydrocarbon Receptor. *Proc. Natl. Acad. Sci. USA* **2000**, *97*, 779. [CrossRef] [PubMed]
43. Grether-Beck, S.; Felsner, I.; Brenden, H.; Marini, A.; Jaenicke, T.; Aue, N.; Welss, T.; Uthe, I.; Krutmann, J. Air Pollution-Induced Tanning of Human Skin. *Br. J. Dermatol.* **2021**, *185*, 1026–1034. [CrossRef] [PubMed]
44. Shi, Y.; Zeng, Z.; Liu, J.; Pi, Z.; Zou, P.; Deng, Q.; Ma, X.; Qiao, F.; Xiong, W.; Zhou, C.; et al. Particulate Matter Promotes Hyperpigmentation via AhR/MAPK Signaling Activation and by Increasing α-MSH Paracrine Levels in Keratinocytes. *Environ. Pollut.* **2021**, *278*, 116850. [CrossRef] [PubMed]
45. Kim, S.S.; Kim, M.J.; Choi, Y.H.; Kim, B.K.; Kim, K.S.; Park, K.J.; Park, S.M.; Lee, N.H.; Hyun, C.G. Down-Regulation of Tyrosinase, TRP-1, TRP-2 and MITF Expressions by Citrus Press-Cakes in Murine B16 F10 Melanoma. *Asian Pac. J. Trop. Biomed.* **2013**, *3*, 617–622. [CrossRef]
46. Murata, K.; Takahashi, K.; Nakamura, H.; Itoh, K.; Matsuda, H. Search for Skin-Whitening Agent from Prunus Plants and the Molecular Targets in Melanogenesis Pathway of Active Compounds. *Nat. Prod. Commun.* **2014**, *9*, 185–188. [CrossRef]
47. Huang, Y.C.; Yang, C.H.; Chiou, Y.L. Citrus Flavanone Naringenin Enhances Melanogenesis through the Activation of Wnt/β-Catenin Signalling in Mouse Melanoma Cells. *Phytomedicine* **2011**, *18*, 1244–1249. [CrossRef]
48. Nasr Bouzaiene, N.; Chaabane, F.; Sassi, A.; Chekir-Ghedira, L.; Ghedira, K. Effect of Apigenin-7-Glucoside, Genkwanin and Naringenin on Tyrosinase Activity and Melanin Synthesis in B16F10 Melanoma Cells. *Life Sci.* **2016**, *144*, 80–85. [CrossRef]
49. Vance, K.W.; Goding, C.R. The Transcription Network Regulating Melanocyte Development and Melanoma. *Pigment. Cell. Res.* **2004**, *17*, 318–325. [CrossRef]
50. Wu, X.; Bowers, B.; Rao, K.; Wei, Q.; Hammer, J.A. Visualization of Melanosome Dynamics within Wild-Type and Dilute Melanocytes Suggests a Paradigm for Myosin V Function In Vivo. *J. Cell Biol.* **1998**, *143*, 1899–1918. [CrossRef]
51. Hume, A.N.; Ushakov, D.S.; Tarafder, A.K.; Ferenczi, M.A.; Seabra, M.C. Rab27a and MyoVa Are the Primary Mlph Interactors Regulating Melanosome Transport in Melanocytes. *J. Cell Sci.* **2007**, *120*, 3111–3122. [CrossRef]

Disclaimer/Publisher's Note: The statements, opinions and data contained in all publications are solely those of the individual author(s) and contributor(s) and not of MDPI and/or the editor(s). MDPI and/or the editor(s) disclaim responsibility for any injury to people or property resulting from any ideas, methods, instructions or products referred to in the content.

cosmetics

MDPI

Review

The Biological Role of Dead Sea Water in Skin Health: A Review

Daoxin Dai *, Xiaoyu Ma, Xiaojuan Yan and Xijun Bao

Fosun Cosmetics (Shanghai) Bio-Technology Co., Ltd., 333 GuiPing Road, Shanghai 200233, China
* Correspondence: daidaoxin2022@163.com

Abstract: Applying natural mineral water to skin care is a popular tendency and many cosmetics products based on thermal spring water have been developed. The special location and environmental conditions provide Dead Sea water (DSW) with unique ion composition and concentrations, which bring comprehensive positive effects on skin health. This article reviews two potential action modes of DSW, and the biological function of DSW and its related complex in dermatology and skin care. Previous studies have proved the functions of skin moisturization, anti-inflammation, skin barrier repair, and anti-pollution. Especially, the anti-aging effect of DSW and related complexes can act in three different ways: keratinocyte rejuvenation, photo-protection, and cellular energy elevation. Additionally, the issues that need further investigation are also discussed. We hope that this review will help to improve the understanding of DSW and its related complex, and further contribute to product development in the skincare industry.

Keywords: Dead Sea; natural mineral water; skin health; cosmetics; molecular mechanism

Citation: Dai, D.; Ma, X.; Yan, X.; Bao, X. The Biological Role of Dead Sea Water in Skin Health: A Review. *Cosmetics* **2023**, *10*, 21. https://doi.org/10.3390/cosmetics10010021

Academic Editor: Agnieszka Feliczak-Guzik

Received: 21 November 2022
Revised: 23 December 2022
Accepted: 27 December 2022
Published: 19 January 2023

Copyright: © 2023 by the authors. Licensee MDPI, Basel, Switzerland. This article is an open access article distributed under the terms and conditions of the Creative Commons Attribution (CC BY) license (https://creativecommons.org/licenses/by/4.0/).

1. Introduction

The Dead Sea, located on the border of Israel, Palestine, and Jordan, is the lowest point on the continent and one of the three most saline lakes in the world, with a salinity of about 300‰. The extreme environmental conditions have shaped the Dead Sea into a forbidden area for higher plants and animals. Only a few salt-tolerant plants and microorganisms can survive on the shore or around the lake. Nevertheless, the skincare benefits of Dead Sea water (DSW) have been known since biblical times [1], mainly due to its unique ionic concentration and composition.

Compared to other natural waters, such as ordinary seawater and hot springs, DSW has a very high ratio of divalent to monovalent cation concentrations. The main divalent cations are magnesium, calcium, and strontium, and the main monovalent cations are sodium and potassium. Additionally, the highest concentration of anions is not chloride ion, but bromine ion, and DSW also contain some trace metal elements, such as zinc and manganese. The elemental composition of DSW is shown in Table 1.

Numerous experimental and cohort studies have proved the therapeutic properties of DSW in dermatological conditions. Therapeutic bathing in the Dead Sea can significantly improve skin dryness, peeling, itching, and pain, and alleviate the related inflammation [2]. All these are the common symptoms caused by chronic skin diseases, such as psoriasis, atopic dermatitis, seborrheic dermatitis, and vitiligo [3]. Therefore, Dead Sea climate therapy has become a recognized adjunctive treatment for skin diseases recommended by dermatologists [4]. Thanks to these properties, scientists endeavored to investigate the skincare benefits and potential cosmetics application of DSW through both in vitro and in vivo evaluations. The results showed that DSW and related complexes can protect the skin comprehensively via moisturization, barrier repair, anti-inflammation, and anti-aging.

This review is aimed to summarize the reports on the mechanism of action, the skincare properties of the DSW and its related complex, as well as research directions in the future.

Table 1. Major elements composition in Dead Sea water.

Main Elements	Name	Dead Sea Water (mg/L)
Na^+	Sodium	2295
K^+	Potassium	1440
Ca^{2+}	Calcium	27,620
Mg^{2+}	Magnesium	67,120
Sr^{2+}	Strontium	516
Cl^-	Chlorine	2300
Br^-	Bromine	38,000
SiO_2	Silicate	<20
Li^+	Lithium	30
Mn^{2+}	Manganese	6
Zn^{2+}	Zinc	≤ 2

2. Mechanism of Action of DSW

2.1. Direct Action

There is a long history of trying to incorporate or directly use mineral-rich water for skin care. Massive scientific studies have proved the positive skin conditioning effect as a result of the different types, concentrations, and ratios of mineral elements [5]. For example, using hairless mice as experimental subjects, the scientist found that applying magnesium chloride solution alone can accelerate skin barrier restoration [6]. The exocytosis function of lipid-containing platelet vesicles within upper epidermal keratinocytes was elevated after using chloride ion carriers [7]. Another experiment found that the application of K^+ channel blockers inhibited skin recovery, whereas treatment with molecules that open the same channels (or are classified as K^+ carriers) can accelerate skin recovery by regulating platelet vesicle secretion. This suggested that the alterations of K^+ channel activity can significantly affect skin barrier homeostasis [8]. Moreover, Ca^{2+} gradient and signaling are also key for a healthy skin barrier and barrier homeostasis. During aging, as well as in diabetic skin, dysregulated calcium signaling occurs and the Ca^{2+} gradient is flattened [9]. Besides driving keratinocyte differentiation, the Ca^{2+} gradient also plays a role in cell migration and wound healing [10]. In summary, it can be assumed that the abundance of Mg^{2+}, Ca^{2+}, Cl^-, and K^+ in DSW can prominently improve the barrier function of the skin. At the same time, the experiment has revealed that the application of 5% $MgCl_2$ can specifically inhibit the TNF-α production by epidermal cells, and also the antigen-presenting capacity of Langerhans cells [11]. Due to the high concentration of Mg^{2+} in DSW, we can conclude that DSW has good anti-inflammation potential.

In addition to different types of elements, the concentrations and ratios of elements can also impose various effects on skin health. For example, Avène Thermal Spring Water, which is low in minerals but rich in bicarbonate and silicates, could serve as a regulator for cell membrane fluidity, antioxidants, and an anti-inflammatory agent [12,13]. Oppositely, the mineral-rich Vichy Thermal Spring Water exhibited more diverse skincare effects, including the increase of stratum corneum peroxidase activity and the promotion of skin homeostasis-related gene expression [14,15]. Thus the unique ionic composition and ratio in DSW may also confer powerful potential for skin care applications.

2.2. Indirect Action

In previous research, Ca^{2+}, due to its high inspecting sensitivity, was selected as the biomarker to represent the ion transport in DSW. No change was observed in the calcium concentration of the culture medium, which suggests that the ions in DSW might not function via transdermal transport [16]. Moderate ionic osmotic stress (MIOS), induced by applying hypersaline materials like Dead Sea water and mud, has been proven to have beneficial contributions to skin health. The positive effects include an impact on the modulation of cell-cycle dynamics, which further leads to a stronger epidermal barrier function, skin hydration elevation, and inflammatory response reduction [17]. The fundamental principle is that the dissolved mineral salts act through the induction of the cell osmosis, and participate in osmotic stress mechano-transduction via piezo-electric ion channels [18].

Although these conclusions have been proven by many experiments at the molecular level, the fundamental mechanism remains unknown. To verify the conjecture, Cohen et al. performed experiments on internal ROS-elevated skin cells and organ models. They found that enhanced Nrf2 (nuclear factor erythroid-2- related factor 2) translocation into the nucleus, upregulation of phase-II antioxidant enzymes, and downregulation of NF-κB-related inflammatory proteins (cytokines IL-1β, IL-8, and caspase-3) are witnessed after MIOS exposure. Taken together, MIOS can result in modulating intracellular ROS generation, which activates the physiological redox homeostasis of the skin and evokes the induction of various biochemical pathways, such as the Nrf2 pathway [19].

3. Biological Function of DSW and Related Complexes (Figure 1)

3.1. Dermatological Treatment

Atopic dermatitis, psoriasis, vitiligo, ichthyosis, and granuloma annulare, are typical chronic skin diseases with a high rate of relapse [20]. The treatment of these diseases often involves the topical use of drugs, such as corticosteroids, antihistamines, immunomodulators, and antibacterial agents, which can lead to great side effects and drug dependence [21]. Dead Sea climatotherapy, as an effective, cost-effective, and safer method, has been consolidated in many studies for its therapeutic efficacy [22,23].

In 2000, Elkayam et al. conducted Dead Sea climatotherapy (DSC) in psoriatic arthritis patients [24]. Within four weeks, several clinical indicators were measured by dermatologists at regular intervals to evaluate the efficacy. PASI (Psoriasis Area and Severity Index), patient self-assessment, and the Schober test showed statistically significant improvements after treatment. These variables are valid and reliable in psoriatic arthritis severity definition [25,26]. The cohort study of 1718 patients with atopic dermatitis showed that over 95% clearance could be achieved in 4 weeks and more after DSC [27]. Also, DSW-containing cream has been demonstrated to improve skin parameters associated with atopic dermatitis in children, particularly in transepidermal water loss (TEWL) and objective severity assessment of atopic dermatitis (OSAAD) values [28].

TEWL measurement is widely used to assess skin barrier function [29] and has a significant correlation with the clinical severity of chronic dermatosis [30]. Plaque psoriasis is the most common form of psoriasis and its treatment effect can be partly reflected by the PASI value. Harari et al. found a positive effect on PASI after DSC, particularly in the early stages of the disease [31]. Previous studies have also revealed increased levels of encephalin, an opioid peptide known to modulate inflammatory responses and keratinocyte proliferation, in psoriatic skin tissues [32]. After four-week DSC treatments, the clinical symptoms of patients disappeared and enkephalin levels in keratinocytes decreased by 21% [33]. Meanwhile, the mean SCORAD (atopic dermatitis score) value was found to decrease from 50.5 to 11 after around 30-day DSC treatments [34]. The potential therapeutic effect of DSC on vitiligo was also confirmed by analyzing the clinical statistical parameters of 436 patients. At the end of treatment, more than 80% of the patients showed improved repigmentation, which was better than the typical narrow-band ultraviolet B treatment [35]. DSC treatment in psoriasis disease can not only bring immediate alleviation but also exert a long-lasting effect [36]. Also, human trials suggested that the mean PASI decreased from 31.7 to 1.42 after the four-week DSC treatment with an improvement of 95.5%. All patients achieved PASI 50, which was thought to be a clinically meaningful improvement and primary endpoint in psoriasis [37], and the therapeutic effect can last up to 33.6 weeks [38]. These results together revealed the good applicate potential of DSW in chronic skin disease treatment.

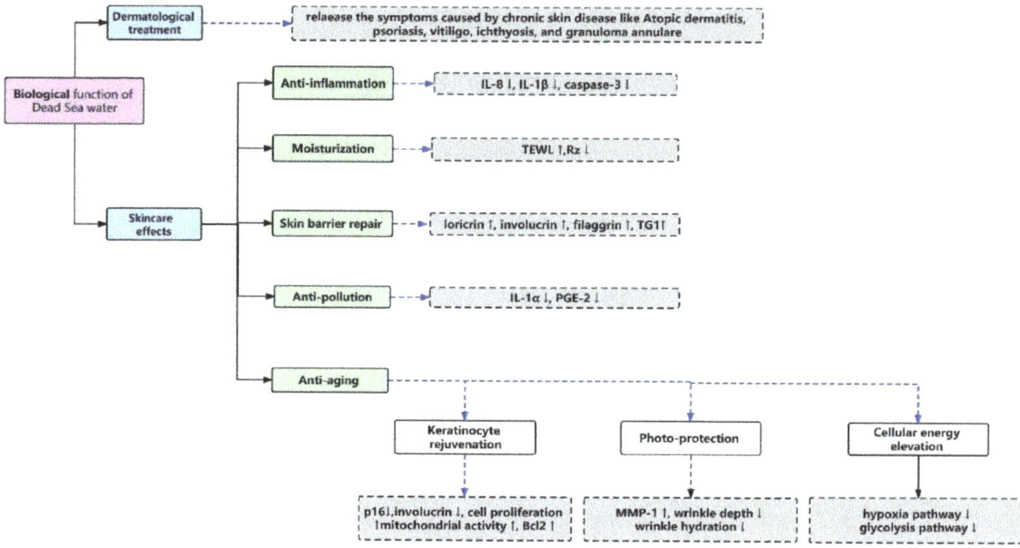

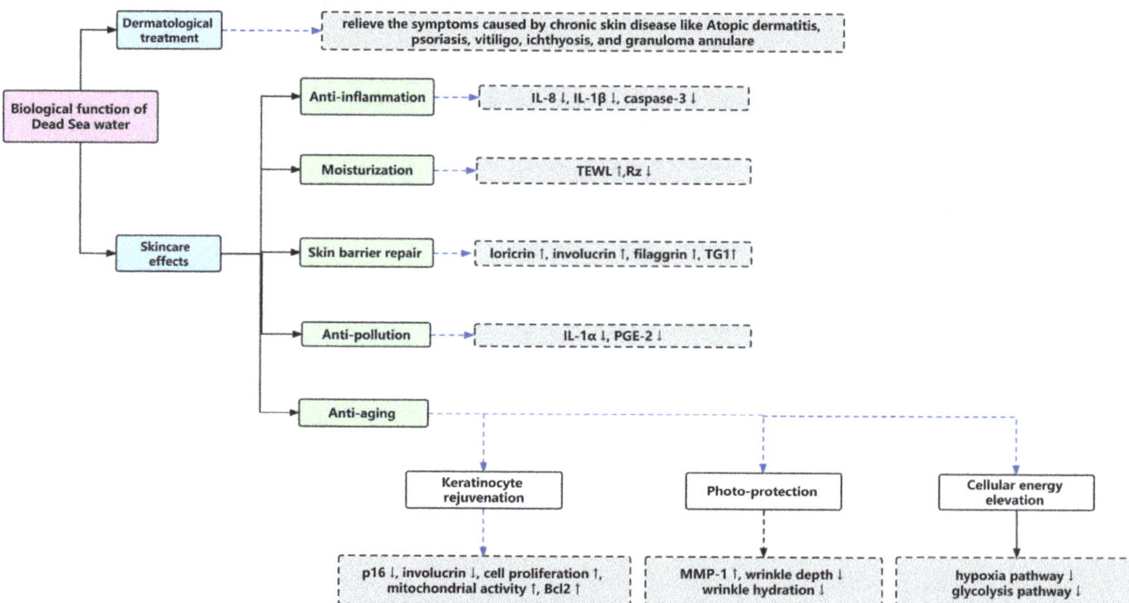

Figure 1. Overview of the biological function of Dead Sea water in dermatology and skin care.

3.2. *Skincare Effects*

3.2.1. Moisturization

To clarify the effect of bathing in DSW, especially the biophysical characteristics in atopic dry skin, transepidermal water loss (TEWL), skin hydration, skin roughness, and skin redness were measured. Eitan et al. compared the indexes before the study and at weeks 1–6, then they observed the elevation of basal TEWL and enhanced skin hydration. Meanwhile, the common skin inflammation markers, such as roughness and redness, were also significantly reduced after treatment [39]. By evaluating skin roughness using computer-aided laser profilometry, Ma'Or et al. investigated the cutaneous smoothing effects of three different liquid gels, one of which contained Dead Sea minerals. After four weeks, the gel containing 1% Dead Sea mineral solution reduced skin roughness by 40.7%, which boosted a better moisturizing effect than the other two gels [40].

In addition, the complexes that combine DSW with other active substances also exhibits good moisturizing properties. Triple D Complex^{TM}, the mixture of DSW, *Dunaliella salina* algae extract, and desert plants, was developed and added into a cosmetic cream. *Dunaliella salina*, a unicellular halophilic microalga with a red colony, is the commercialized producer of many compounds within the carotenoid family. Its extract showed good antiglycation, anti-aging, and anti-inflammatory capabilities in the ex vivo human skin explants model [41,42]. After four-week treatments, they found an average reduction of the skin roughness parameter Rz by 43%, which is measured by the silicone impression. The skin surface hydration state was also assessed by a corneometer (a skin capacitance-based instrument) and the appearance improvement was also observed after application [43].

3.2.2. Anti-Inflammation

Cohen et al. used UVB irradiation-induced skin explant as the model and co-cultured it with the DSW. A downregulated level of interleukins secretion levels, such as IL-8 and IL-1β, as well as a lower level of caspase-3 proapoptotic enzyme, were observed after DSW treatment [19]. Furthermore, Portugal-Cohen et al. also constructed another inflammation-induced model by lipopolysaccharides (LPS) on Human skin organ culture. They found that treatment with DSW at concentrations of 0.1% and 0.5% can significantly attenuate IL-1β induction by 46% and 54%, respectively [16]. From the in vitro tests, it is obvious that DSW exerts inflammation inhibition ability.

Clinically, the DSW also demonstrated an anti-inflammation effect on chronic inflammatory skin diseases, such as psoriasis or atopic dermatitis. As reported by Proksch [39], a magnesium-rich Dead Sea salt solution was applied to patients with atopic dry skin. A 15 min treatment in bath solution containing 5% Dead Sea salt for 6 weeks could greatly improve skin hydration, roughness, and reduce skin redness. It was proposed that the anti-inflammatory property of the DSW might originate from the modulation of interleukins, as well as the antigen-presenting capacity of Langerhans' cells [44].

Magnesium in DSW might contribute the most to inflammation inhibition. DSW is known for the abundance of magnesium ions. As summarized by Tarnowska [45], magnesium could reduce TNF-α production in epidermal cells, thus bringing skin soothing effects. Additionally, magnesium-rich Dead Sea therapy has been proven to downregulate nuclear factor κB (NFκB) to avoid further inducement of proinflammatory factors and integrin. In Kim's study [46], a Korean Sea derivative with similar mineral composition showed similar anti-inflammatory properties, consolidating the hypothesis that magnesium-enriched seawater has a protective effect against skin inflammation from different sources.

3.2.3. Skin Barrier Repair

Skin, as the largest organ in the human body, forms the first barrier against different exogenous stress, performing a barrier function maintenance in skin homeostasis. To clarify the internal mechanism, four barrier function-related proteins: loricrin, involucrin, filaggrin, and transglutaminase 1 (TG1), were measured on skin equivalents after applying DSW at concentrations of 0.8% and 2% [16]. The results suggested that the latter three structural-related biomarkers were upregulated after topical DSW application. Involucrin is thought to be one of the precursors cross-linked during the resistant cornified envelope assembly process, which is partly driven by TG1 [47]. Filaggrin is another important protein during the final stages of keratinocyte differentiation, which can promote keratin filaments aggregation and further form into tight bundles [48]. Further tests on SDS-induced (sodium dodecyl sulfate) human skin irritation organ culture (HSOC) showed that topical DSW application alleviated the reduced epidermal viability and decreased IL-1α and PGE2 levels, which are in line with previous reports of skin response to minerals in animal models. DSW can also attenuate LPS-induced IL-1β secretion, which is probably related to its skin barrier-restoring effect. The complete skin barrier can prevent skin from the harmful stimulation of the external environment, and from the release of inflammatory markers, such as cytokine and PGE [49]. Former research has found that osmotic pressure can stimulate TRPV4, an important signal molecule in the regulation of calcium gradient and barrier homeostasis in the epidermis [50]. Considering the central role of Ca^{2+} in epidermic protein synthesis and its high concentration in DSW, it can be conjectured that DSW might function in two parallel ways, namely the mediation of TRPV4 and the calcium-pump activation by osmotic pressure, thus regulating the expression and activity of the epidermal-related protein.

3.2.4. Anti-Pollution

With the increasing deterioration of the natural environment, the negative skin impact of air pollution attracts the attention of dermatologists [51]. In a recent study, the common pollution models, ozone and a mixture of pollutants (MOP) composed of heavy metal and atmospheric particulate matter [52], were selected to induce the oxidatively-stressed state in 3D skin cell culture. Using epidermal viability and inflammatory biomarkers as the indicator of effect, it was found that DSW can inhibit the IL-1α overproduction following MOP exposure. When mixed with another active ingredient, for example, anionic polysaccharide PolluStop® (bio-saccharide gum-4 or 1,2-hexanediol), the release of IL-1α and PGE-2 induced by ozone exposure can be further reduced [53].

3.2.5. Anti-Aging

Keratinocyte Rejuvenation

Skin aging often leads to increased wrinkles, decreased elasticity, and reduced skin thickness [54]. These phenotypes can not only bring a negative appearance but also deteriorate the individual's confidence. Accordingly, the anti-aging function of DSW was studied and verified by a series of lab experiments. First, the researchers developed the biological model of aged epidermal keratinocytes by characterizing the cellular and molecular properties, including the morphological, fluorometric, and biochemical parameters on both skin cell and organ cultures [55]. Then, the altered expression of 16 biochemical molecules in both aged cultured cells and tissues was observed and selected as the aging biomarkers, including caspases-1 and 3, beta-galactosidases, p16, Ki67, 20S proteasome, and effectors of the Fas-dependent apoptotic pathway [56]. After applying DSW to both models, they found that mitochondrial activity and cell proliferation are increased at subtoxic doses. The DSW treatment group also showed significantly reduced p16 and involucrin signals and increased Bcl2 levels, as observed in non-senescent cells. A possible explanation is that DSW can eliminate poorly proliferative and aged cells, to increase the activity of the whole keratinocyte population. In summary, DSW can stimulate proliferation and mitochondrial activity, decrease the expression of aging biomarkers, and limit apoptotic damage after UVB irradiation.

Apart from the use of pure DSW, a combination of DSW and traditional anti-aging actives may present a synergistic effect. Retinol is a commonly-employed ingredient to improve age-related skin issues, but its safety profile is controversial. The scientists designed a new complex named "pRetinolTM" (PRE), which contains β-carotene, niacinamide, the extract of *Dunaliella salina*, and DSW. The former two can serve as the precursors to synthesizing retinol [57]. The extract of *Dunaliella salina* alga has been proven to contain multiple anti-antioxidant substances, which include fatty acids and pigments, such as β-carotene and chlorophyll. The biological function of this complex was evaluated and compared with retinol only by using three different models: in vitro human dermal fibroblasts cell, reconstructed 3D skin equivalent, and ex vivo human skin organ culture. The measurement of hyaluronic acid, TNF-α, and IL-1α expression levels revealed it is of retinol-like skin activity, yet it led to less skin irritation. The whole-genome microarray was also performed to compare the different expression pathways between PRE complex and retinol treatment. The enrichment analysis showed that PRE can reduce many pathways involving inflammation, such as NOD-like receptor signaling [58], TNF signaling, nuclear factor-kappa B (NF-κB) signaling [59], and apoptosis. The complex can additionally up-regulate the BASE excision repair-related gene expression. As an important part of the oxidative DNA damage defense mechanism [60], this complex might provide a protective effect on retinol, which has side effects and toxicity [61].

Photo-Protection

Among the multiple exogenous sources that can result in skin photo-damage and aging, UV exposure is the most important and well-known factor [62]. A mixture of DSW and three plants (Tibetan goji berry, Himalayan raspberry root extract, and Iceland moss (lichen)), namely Extreme Complex^{TM}, was developed in the previous report [63]. The UVB-induced ex vivo human skin model was reconstructed and used to assess its protection function against light radiation. A reduced caspase-3 activity and pro-inflammatory cytokine TNF-α secretion were shown, which suggested its anti-apoptotic and anti-inflammatory functions [64]. Additionally, the complex can also decrease the activity of collagen balance-related biomarkers, degrading enzymes, and collagen maturation byproducts. Within the matrix metalloproteinase (MMP) family, MMP-1 is the most injured enzyme in collagen damage when facing photoaging [65]. MMP-1 activation can further induce the increased expression level of MMP-3 and MMP-9, accelerating collagen degradation [66]. In clinical tests on human subjects, skin wrinkle depth and hydration were common metrics that can reflect the degree of skin photoaging, which can be measured using the PRIMOS optical 3D measuring device and corneometer, respectively [67]. A significant improvement in skin moisture and wrinkle depth is observed after application. Altogether, the antioxidant, anti-apoptotic, and anti-inflammatory characteristics of this complex might bring comprehensive alleviation effects of skin photo-damage and appearance improvement.

Under certain conditions, adding metal ions into the fermentation medium has been proven to improve the structure and function of products [68]. Since Dead Sea water is rich in minerals and trace elements, scientists developed a fermentation supplement by mixing the water and mud of the Dead Sea, which acts as a positive stress supplement during yeast Pichia pastoris (aka *Komagataella phaffii*) fermentation. This kind of methylotrophic microorganism is one of the most commonly used cell factories for heterologous protein production and has been used in many industries [69]. To evaluate the biological functions of this complex, Portugal-Cohen et al. first measured a series of skin elasticity biomarkers on 3D human skin equivalents and performed the whole-genome DNA microarray test at the same time, to investigate its effect on both gene and protein levels, respectively [70]. After treatment, the experimental group had significant alleviation of abnormal UVB-induced alterations; for example, both elastin and fibulin are the main components of the extracellular matrix, which is the damage target of UV solar [71]. Tight junction protein (TJ) can connect multiple parallel intramembrane strands and neighboring cells into a regulatory and structural network [72], which can contribute to the skin's UV resistance capacity [73]. The elevation of these proteins indicates the skincare potential of this complex.

The UV-protection effect of DSW is also consolidated in DSW-containing cosmetic creams [74]. A cream composed of DSW, zinc oxide, aloe vera extract, pro-vitamin B5, and vitamin E, was formulated and topically applied to human skin organ cultures exposed to UVB. A severe mitochondrial activity loss was observed after UV radiation, as activated by caspase-3 and cytokine secretion, which was well accorded with previous studies [75]. On the contrary, the DSW-incorporated formulation significantly promoted the anti-oxidative capacity and reduced cell damage and apoptosis. It was inferred that the DSW could minimize oxidative stress and inflammatory signs after UV exposure.

Cellular Energy Elevation

Organisms that live in extreme environments usually evolved many special mechanisms for survival [76]. *Calotropis procera*, the traditional American medicinal plant, naturally grows in the flora of the Dead Sea region [77]. The previous study has found many pharmacological actions of its extracts, such as anti-inflammatory, antibacterial, and antioxidant [78]. After mixing with DSW, various biological activities are further amplified and enhanced, which was verified by the RNA microarray experiment on the reconstructed full-thickness skin tissues. The GSEA analysis results showed that the biological processes of hypoxia, glycolysis, and epithelial-mesenchymal transition pathway were significantly down-regulated after treatment. These results indicated that this complex had an unexpected biological potential for energy production, resistance to hypoxia, and ECM balance [79].

4. Current Challenges for the Use of DSW

Though DSW shows a unique biological role in maintaining skin health, there are still tremendous issues to be addressed.

The core problem is how to strike the balance between Dead Sea resources exploitation and eco-protection. Environmental factors, such as climate alteration, ongoing sinkholes, and geochemistry variation, could lead to the scarcity of water resources [80]. In addition to the natural influences, human intervention also threatens the preservation of Dead Sea resources. Water pollution, exhaustive exploitation, and changes in biodiversity could all aggravate the exhaustion of Dead Sea resources. As reported, the Dead Sea's sea level has dropped at a speed of 1 m/year in the last 5 decades, bringing the challenge of sustainable supply of DSW.

The exterior variation of the Dead Sea also contributes to another issue: quality control. Apart from seasonal and locational differentiation, the ion concentration, the ratio between different minerals, and the metabolites from microbials could be affected by environmental changes, which might eventually lead to unpredictable biological effects.

The application of DSW in cosmetic products also evokes great challenges, not only because a higher concentration of DSW in products might cause potential skin irritation and discomfort, but also due to its relatively high ion strength. This would hamper the stability of the cosmetic formula by sabotaging emulsion thermo-dynamic homogeneity and altering the rheological properties of the whole system. Additionally, DSW might also impose an antagonistic effect with other actives in formula, resulting in unwanted precipitation or efficacy invalidation. However, there is a cosmetic brand that has produced a series of related products with broad effects and diverse forms, like hydrating sprays, lotions, and creams. The DSW here functions more as the actives instead of the main ingredient.

Considering this, many efforts have been made to improve their utilization in formulations. For example, scientists dispersed the nanosized Dead Sea mineral in mixed oil (Crystal Osmoter™) and achieved six times higher concentrations DSW on skin, for which the clinical tests showed a better performance on wrinkle reduction, firming, and radiance than normal DSW with no irritation. At the same time, new delivery systems were also developed, like liposomes (LipOsmoter™) or strontium hexaferrite ($SrFe_{12}O_{19}$) nanomagnets, which enhance the safety of DSW and provide a longer-term skincare benefit.

Although there are some inventions about the DSW application in cosmetics products, still many issues need to be explored, like how to complement the synergy of DSW and how to stabilize the system even if added in a high DSW concentration.

Since the DSW is a complex mixture of different minerals and other trace elements, it is of great urgency to determine the exact composition of DSW. Especially for the trace elements, more advanced analytic techniques are needed to quantitatively evaluate their contents. Facing resource shortages and environment pollution, the artificial DSW (ADSW) consisting of the main ion compositions are formulated in labs. Unfortunately, it has shown weaker effects on both cell models and human skin organ cultures than natural DSW (unpublished data). It is probable that the trace element and microbe-secreted bioactive substances can make up part of the skincare effects of natural DSW. Thus a more delicate choice of DSW content should be studied first.

In addition, the interaction between different ions and dose-function relation of DSW should be further investigated. Each type of ion might play complex roles in modulating biological processes and molecular functions on skin, thus DSW might bring even more complicated cellular interplays.

To sum up, the current challenges for the use of DSW lie in the short supply, unpredictable quality variation, compatibility concerns in the formulas, as well as lack of understanding of DSW components and interrelated mechanism.

5. Future Direction

The skincare benefits of Dead Sea water and its related complex have been demonstrated in many studies and cover a wide range of functions, including moisturization, barrier repair, anti-inflammation, anti-aging, photo-protection, cellular energy elevation, and relief of skin disease symptoms. However, several issues are still worthy of further investigation, which are listed below.

Firstly, although the extreme hypersaline environmental conditions of Dead Sea water hamper the growth of higher plants and animals, microorganisms like fungi and bacteria can still survive [81–83]. Extremophiles are often thought to be a huge reservoir of active substances [84]. For example, an extremely salt-tolerant *Bacillus* strain isolated from Dead Sea had significant antibacterial and fungal activity in its aqueous extract. One of them can even resist all the tested bacteria strains [85,86]. *Haloarcula vallismortis* is a kind of halophilic archaeon with reddish colonies [87]. Its extract has a significant anti-inflammatory activity that can resist the DNA damage induced by UV exposure, which suggests its potential for use as a biological or natural sunscreen [88]. Although some Dead Sea source strains have been isolated and purified, studies for their bioactive product utilization are still scarce and need to be further investigated experimentally.

Secondly, the number of skin microbiome-related studies is increasing rapidly in recent years and the relationship between the skin microbiome and skin health is gradually being revealed [89–91]. The effects of DSW and related complexes on skin microbiomes are less evaluated, and most of them are involved in skin diseases. Based on a healthy population, a previous experiment compared the change in the skin microbiome after DSW application [83]. They found that bacterial community diversity is almost constant, while fungal diversity was significantly lower than before, the variation of which is driven primarily by *Malassezia* spp. It's known that the abnormal abundance of *Malassezia* can lead to free radicals release and inflammatory skin issues [92]. The variation pattern of AD (Atopic Dermatitis) patients has also been explored after using DSW. The results show that the unbalance of skin microbial ecology, which occurs at both lesion and non-lesion sites, was significantly attenuated after DSW application [93]. The most significant changes were seen in severe AD, mainly reflected in the relative abundance of *Staphylococcus epidermidis*, *Streptococcus mitis*, and *Micrococcus luteus*. Current studies usually used the high-throughput 16S rRNA or ITS (internally transcribed spacer) amplicon sequencing method, which is of low resolution and limited information availability. A more comprehensive metagenome approach should be established subsequently to obtain more accurate and profound findings [94].

Thirdly, in contrast to the extensive exploration of DSW, the research involved in the skincare effect of Dead Sea mud is rare [2,95–98]. Previous experimental results have proven that both short-term and long-term application of Dead Sea-derived mud has a high safety profile. It causes no damage to skin barrier integrity but has a firming effect instead [99,100]. Furthermore, Dead Sea black mud-derived masks have been shown to accelerate the wound healing process in mice skin by promoting granulation, angiogenesis, and collagen deposition [101]. The mud is also found to inactivate common microorganisms and produced an obvious growth inhibition area, suggesting its significant antibacterial effect [102]. As the seed of Dead Sea minerals, the studies of Dead Sea mud should not be limited to the basic skincare functions, but also more to its applications in cosmetics.

6. Conclusions

Dead Sea water has a unique ion composition and its benefits on skin health have been well-known since ancient times. From previous research, we summarize two potential action modes of DSW. The first one is the direct penetration of mineral ions, and the second one is the moderate ionic osmotic stress mechanism, which can activate the cellular osmotic stress-related pathway via ion channels. The chronic skin disease improvements and comprehensive skincare efficacy of DSW and its related complex are also illustrated. Specifically, they can resist skin senescence from three different perspectives (keratinocyte rejuvenation promotion, photo-protection, and cellar energy elevation), which indicates their strong application potential in anti-aging cosmetics product development. However, many other aspects of the Dead Sea resource are still unknown and need to be studied, such as the Dead Sea mud, secondary metabolites of Dead Sea bacteria and fungi, and also their effects on the skin microbiome.

Author Contributions: Conceptualization, D.D. and X.M.; writing—original draft preparation, D.D; writing—review and editing, D.D., X.M., X.Y. and X.B. All authors have read and agreed to the published version of the manuscript.

Funding: This research received no external funding.

Institutional Review Board Statement: Not applicable.

Informed Consent Statement: Not applicable.

Data Availability Statement: Not applicable.

Conflicts of Interest: The authors declare no conflict of interest.

References

1. Even-Paz, Z.; Shani, J. The Dead Sea and psoriasis: Historical and geographic background. *Int. J. Dermatol.* **1989**, *28*, 1–9. [CrossRef] [PubMed]
2. Buskila, D.; Abu-Shakra, M.; Neumann, L.; Odes, L.; Shneider, E.; Flusser, D.; Sukenik, S. Balneotherapy for fibromyalgia at the Dead Sea. *Rheumatol. Int.* **2001**, *20*, 105–108. [CrossRef] [PubMed]
3. Yao, Y.; Ravn Jørgensen, A.-H.; Thomsen, S.F. Biologics for chronic inflammatory skin diseases: An update for the clinician. *J. Dermatol. Treat.* **2020**, *31*, 108–130. [CrossRef] [PubMed]
4. Huang, A.; Seité, S.; Adar, T. The use of balneotherapy in dermatology. *Clin. Dermatol.* **2018**, *36*, 363–368. [CrossRef]
5. Cacciapuoti, S.; Luciano, M.A.; Megna, M.; Annunziata, M.C.; Napolitano, M.; Patruno, C.; Scala, E.; Colicchio, R.; Pagliuca, C.; Salvatore, P. The role of thermal water in chronic skin diseases management: A review of the literature. *J. Clin. Med.* **2020**, *9*, 3047. [CrossRef]
6. Maarouf, M.; Hendricks, A.J.; Shi, V.Y. Bathing additives for atopic dermatitis—A systematic review. *Dermatitis* **2019**, *30*, 191–197. [CrossRef]
7. Denda, M.; Fuziwara, S.; Inoue, K. Influx of calcium and chloride ions into epidermal keratinocytes regulates exocytosis of epidermal lamellar bodies and skin permeability barrier homeostasis. *J. Investig. Dermatol.* **2003**, *121*, 362–367. [CrossRef]
8. Bäsler, K.; Brandner, J.M. Tight junctions in skin inflammation. *Pflügers Arch.-Eur. J. Physiol.* **2017**, *469*, 3–14. [CrossRef]
9. Hudson, L.E.M. *Integration of Wound-Induced Calcium Signals to Transcriptional Activation and Regulation of Cutaneous Wound Healing Responses*; Newcastle University: Tyne, UK, 2015.
10. Lee, S.E.; Lee, S.H. Skin barrier and calcium. *Ann. Dermatol.* **2018**, *30*, 265–275. [CrossRef]
11. Schempp, C.M.; Dittmar, H.C.; Hummler, D.; Simon-Haarhaus, B.; Schöpf, E.; Simon, J.C.; Schulte-Mönting, J. Magnesium ions inhibit the antigen-presenting function of human epidermal Langerhans cells in vivo and in vitro. Involvement of ATPase, HLA-DR, B7 molecules, and cytokines. *J. Investig. Dermatol.* **2000**, *115*, 680–686. [CrossRef]
12. Eliasse, Y.; Redoules, D.; Espinosa, E. Impact of Avène Thermal Spring Water on immune cells. *J. Eur. Acad. Dermatol. Venereol.* **2020**, *34*, 21–26. [CrossRef] [PubMed]
13. Nocera, T.; Jean-Decoster, C.; Georgescu, V.; Guerrero, D. Benefits of Avène thermal hydrotherapy in chronic skin diseases and dermatological conditions: An overview. *J. Eur. Acad. Dermatol. Venereol.* **2020**, *34*, 49–52. [CrossRef] [PubMed]
14. Tacheau, C.; Weisgerber, F.; Fagot, D.; Bastien, P.; Verdier, M.; Liboutet, M.; Sore, G.; Bernard, B. Vichy Thermal Spring Water (VTSW), a cosmetic ingredient of potential interest in the frame of skin ageing exposome: An in vitro study. *Int. J. Cosmet. Sci.* **2018**, *40*, 377–387. [CrossRef] [PubMed]
15. Rasmont, V.; Valois, A.; Gueniche, A.; Sore, G.; Kerob, D.; Nielsen, M.; Berardesca, E. Vichy volcanic mineralizing water has unique properties to strengthen the skin barrier and skin defenses against exposome aggressions. *J. Eur. Acad. Dermatol. Venereol.* **2022**, *36*, 5–15. [CrossRef]

16. Portugal-Cohen, M.; Cohen, D.; Ish-Shalom, E.; Laor-Costa, Y.; Ma'or, Z.e. Dead Sea minerals: New findings on skin and the biology beyond. *Exp. Dermatol.* **2019**, *28*, 585–592. [CrossRef]
17. Levi-Schaffer, F.; Shani, J.; Politi, Y.; Rubinchik, E.; Brenner, S. Inhibition of proliferation of psoriatic and healthy fibroblasts in cell culture by selected Dead-sea salts. *Pharmacology* **1996**, *52*, 321–328. [CrossRef]
18. Carbajo, J.M.; Maraver, F. Salt water and skin interactions: New lines of evidence. *Int. J. Biometeorol.* **2018**, *62*, 1345–1360. [CrossRef]
19. Cohen, D.; Ma'or, Z.e.; Cohen, M.P.; Oron, M.; Kohen, R. Nrf2 Pathway Involvement in the Beneficial Skin Effects of Moderate Ionic Osmotic Stress–the Case of the Dead Sea Water. *J. Cosmet. Dermatol. Sci. Appl.* **2022**, *12*, 109–130.
20. Sawada, Y.; Saito-Sasaki, N.; Mashima, E.; Nakamura, M. Daily Lifestyle and Inflammatory Skin Diseases. *Int. J. Mol. Sci.* **2021**, *22*, 5204. [CrossRef]
21. Duong, T.A.; Valeyrie-Allanore, L.; Wolkenstein, P.; Chosidow, O. Severe cutaneous adverse reactions to drugs. *Lancet* **2017**, *390*, 1996–2011. [CrossRef]
22. Marsakova, A.; Kudish, A.; Gkalpakiotis, S.; Jahn, I.; Arenberger, P.; Harari, M. Dead Sea climatotherapy versus topical steroid treatment for atopic dermatitis children: Long-term follow-up study. *J. Dermatol. Treat.* **2019**, *31*, 711–715. [CrossRef] [PubMed]
23. Emmanuel, T.; Petersen, A.; Houborg, H.I.; Rønsholdt, A.B.; Lybæk, D.; Steiniche, T.; Bregnhøj, A.; Iversen, L.; Johansen, C. Climatotherapy at the Dead Sea for psoriasis is a highly effective anti-inflammatory treatment in the short term: An immunohistochemical study. *Exp. Dermatol.* **2022**, *31*, 1136–1144. [CrossRef] [PubMed]
24. Elkayam, O.; Ophir, J.; Brener, S.; Paran, D.; Wigler, I.; Efron, D.; Even-Paz, Z.; Politi, Y.; Yaron, M. Immediate and delayed effects of treatment at the Dead Sea in patients with psoriatic arthritis. *Rheumatol. Int.* **2000**, *19*, 77–82. [CrossRef]
25. Bożek, A.; Reich, A. The reliability of three psoriasis assessment tools: Psoriasis area and severity index, body surface area and physician global assessment. *Adv. Clin. Exp. Med.* **2017**, *26*, 851–856. [CrossRef]
26. Elewski, B.E.; Puig, L.; Mordin, M.; Gilloteau, I.; Sherif, B.; Fox, T.; Gnanasakthy, A.; Papavassilis, C.; Strober, B.E. Psoriasis patients with psoriasis Area and Severity Index (PASI) 90 response achieve greater health-related quality-of-life improvements than those with PASI 75–89 response: Results from two phase 3 studies of secukinumab. *J. Dermatol. Treat.* **2017**, *28*, 492–499. [CrossRef]
27. Harari, M.; Shani, J.; Seidl, V.; Hristakieva, E. Climatotherapy of atopic dermatitis at the Dead Sea: Demographic evaluation and cost-effectiveness. *Int. J. Dermatol.* **2000**, *39*, 59–69. [CrossRef] [PubMed]
28. Portugal-Cohen, M.; Oron, M.; Merrik, E.; Ben-Amitai, D.; Yogev, H.; Zvulunov, A. A dead sea water-enriched body cream improves skin severity scores in children with atopic dermatitis. *J. Cosmet. Dermatol. Sci. Appl.* **2011**, *1*, 71. [CrossRef]
29. Alexander, H.; Brown, S.; Danby, S.; Flohr, C. Research techniques made simple: Transepidermal water loss measurement as a research tool. *J. Investig. Dermatol.* **2018**, *138*, 2295–2300.e1. [CrossRef]
30. Montero-Vilchez, T.; Segura-Fernández-Nogueras, M.-V.; Pérez-Rodríguez, I.; Soler-Gongora, M.; Martinez-Lopez, A.; Fernández-González, A.; Molina-Leyva, A.; Arias-Santiago, S. Skin barrier function in psoriasis and atopic dermatitis: Transepidermal water loss and temperature as useful tools to assess disease severity. *J. Clin. Med.* **2021**, *10*, 359. [CrossRef]
31. Harari, M.; Czarnowicki, T.; Fluss, R.; Ruzicka, T.; Ingber, A. Patients with early-onset psoriasis achieve better results following Dead Sea climatotherapy. *J. Eur. Acad. Dermatol. Venereol.* **2012**, *26*, 554–559. [CrossRef]
32. Bigliardi, P.L.; Bigliardi-Qi, M.; Buechner, S.; Rufli, T. Expression of μ-opiate receptor in human epidermis and keratinocytes. *J. Investig. Dermatol.* **1998**, *111*, 297–301. [CrossRef] [PubMed]
33. Nissen, J.; Avrach, W.; Hansen, E.; Stengaard-Pedersen, K.; Kragballe, K. Increased levels of enkephalin following natural sunlight (combined with salt water bathing at the Dead Sea) and ultraviolet A irradiation. *Br. J. Dermatol.* **1998**, *139*, 1012–1019. [CrossRef] [PubMed]
34. Harari, M.; Dreiher, J.; Czarnowicki, T.; Ruzicka, T.; Ingber, A. SCORAD 75: A new metric for assessing treatment outcomes in atopic dermatitis. *J. Eur. Acad. Dermatol. Venereol.* **2012**, *26*, 1510–1515. [CrossRef] [PubMed]
35. Czarnowicki, T.; Harari, M.; Ruzicka, T.; Ingber, A. Dead Sea climatotherapy for vitiligo: A retrospective study of 436 patients. *J. Eur. Acad. Dermatol. Venereol.* **2011**, *25*, 959–963. [CrossRef] [PubMed]
36. Emmanuel, T.; Lybæk, D.; Johansen, C.; Iversen, L. Effect of Dead Sea climatotherapy on psoriasis; a prospective cohort study. *Front. Med.* **2020**, *7*, 83. [CrossRef] [PubMed]
37. Carlin, C.S.; Feldman, S.R.; Krueger, J.G.; Menter, A.; Krueger, G.G. A 50% reduction in the Psoriasis Area and Severity Index (PASI 50) is a clinically significant endpoint in the assessment of psoriasis. *J. Am. Acad. Dermatol.* **2004**, *50*, 859–866. [CrossRef] [PubMed]
38. Harari, M.; Novack, L.; Barth, J.; David, M.; Friger, M.; Moses, S.W. The percentage of patients achieving PASI 75 after 1 month and remission time after climatotherapy at the Dead Sea. *Int. J. Dermatol.* **2007**, *46*, 1087–1091. [CrossRef]
39. Proksch, E.; Nissen, H.P.; Bremgartner, M.; Urquhart, C. Bathing in a magnesium-rich Dead Sea salt solution improves skin barrier function, enhances skin hydration, and reduces inflammation in atopic dry skin. *Int. J. Dermatol.* **2005**, *44*, 151–157. [CrossRef]
40. Ma'Or, Z.; Yehuda, S.; Voss, W. Skin smoothing effects of Dead Sea minerals: Comparative profilometric evaluation of skin surface. *Int. J. Cosmet. Sci.* **1997**, *19*, 105–110. [CrossRef]
41. Havas, F.; Krispin, S.; Cohen, M.; Loing, E.; Farge, M.; Suere, T.; Attia-Vigneau, J. A Dunaliella salina Extract Counteracts Skin Aging under Intense Solar Irradiation Thanks to Its Antiglycation and Anti-Inflammatory Properties. *Mar. Drugs* **2022**, *20*, 104. [CrossRef]

42. Xu, Y.; Harvey, P.J. Carotenoid production by Dunaliella salina under red light. *Antioxidants* **2019**, *8*, 123. [CrossRef] [PubMed]
43. Ma'Or, Z.; Meshulam-Simon, G.; Yehuda, S.; Gavrieli, J.; Sea, D. Antiwrinkle and skin-moisturizing effects of a mineral-algal-botanical complex. *J. Cosmet. Sci.* **2000**, *51*, 27–36.
44. Wang, B.; Amerio, P.; Sauder, D.N. Role of cytokines in epidermal Langerhans cell migration. *J. Leukoc. Biol.* **1999**, *66*, 33–39. [CrossRef] [PubMed]
45. Tarnowska, M.; Briançon, S.; Resende de Azevedo, J.; Chevalier, Y.; Bolzinger, M.A. Inorganic ions in the skin: Allies or enemies? *Int. J. Pharm.* **2020**, *591*, 119991. [CrossRef] [PubMed]
46. Kim, J.H.; Lee, J.; Lee, H.B.; Shin, J.H.; Kim, E.K. Water-retentive and anti-inflammatory properties of organic and inorganic substances from Korean sea mud. *Nat. Prod. Commun.* **2010**, *5*, 395–398. [CrossRef] [PubMed]
47. Sevilla, L.M.; Nachat, R.; Groot, K.R.; Klement, J.F.; Uitto, J.; Djian, P.; Määttä, A.; Watt, F.M. Mice deficient in involucrin, envoplakin, and periplakin have a defective epidermal barrier. *J. Cell Biol.* **2007**, *179*, 1599–1612. [CrossRef]
48. Proksch, E.; Brandner, J.M.; Jensen, J.-M. The skin: An indispensable barrier. *Exp. Dermatol.* **2008**, *17*, 1063–1072. [CrossRef]
49. Song, C.; Liu, L.; Chen, J.; Hu, Y.; Li, J.; Wang, B.; Bellusci, S.; Chen, C.; Dong, N. Evidence for the critical role of the PI3K signaling pathway in particulate matter-induced dysregulation of the inflammatory mediators COX-2/PGE2 and the associated epithelial barrier protein Filaggrin in the bronchial epithelium. *Cell Biol. Toxicol.* **2020**, *36*, 301–313. [CrossRef]
50. Rosenbaum, T.; Benítez-Angeles, M.; Sánchez-Hernández, R.; Morales-Lázaro, S.L.; Hiriart, M.; Morales-Buenrostro, L.E.; Torres-Quiroz, F. TRPV4: A physio and pathophysiologically significant ion channel. *Int. J. Mol. Sci.* **2020**, *21*, 3837. [CrossRef]
51. Richard, F.; Creusot, T.; Catoire, S.; Egles, C.; Ficheux, H. Mechanisms of pollutant-induced toxicity in skin and detoxification: Anti-pollution strategies and perspectives for cosmetic products. In *Annales Pharmaceutiques Françaises*; Elsevier: Amsterdam, The Netherlands, 2019.
52. McDaniel, D.; Farris, P.; Valacchi, G. Atmospheric skin aging—Contributors and inhibitors. *J. Cosmet. Dermatol.* **2018**, *17*, 124–137. [CrossRef]
53. Portugal-Cohen, M.; Oron, M.; Cohen, D.; Ma'or, Z. Antipollution skin protection–a new paradigm and its demonstration on two active compounds. *Clin. Cosmet. Investig. Dermatol.* **2017**, *10*, 185. [CrossRef] [PubMed]
54. Mohiuddin, A.K. Skin aging & modern age anti-aging strategies. *PharmaTutor* **2019**, *7*, 22–70.
55. Soroka, Y.; Ma'or, Z.; Leshem, Y.; Verochovsky, L.; Neuman, R.; Brégégère, F.M.; Milner, Y. Aged keratinocyte phenotyping: Morphology, biochemical markers and effects of Dead Sea minerals. *Exp. Gerontol.* **2008**, *43*, 947–957. [CrossRef] [PubMed]
56. Chervonsky, A.V. Apoptotic and effector pathways in autoimmunity. *Curr. Opin. Immunol.* **1999**, *11*, 684–688. [CrossRef]
57. Karaźniewicz-Łada, M.; Główka, A. A review of chromatographic methods for the determination of water-and fat-soluble vitamins in biological fluids. *J. Sep. Sci.* **2016**, *39*, 132–148. [CrossRef]
58. Danis, J.; Mellett, M. Nod-like receptors in host defence and disease at the epidermal barrier. *Int. J. Mol. Sci.* **2021**, *22*, 4677. [CrossRef]
59. Wang, Y.; Wang, L.; Wen, X.; Hao, D.; Zhang, N.; He, G.; Jiang, X. NF-κB signaling in skin aging. *Mech. Ageing Dev.* **2019**, *184*, 111160. [CrossRef]
60. Lee, T.-H.; Kang, T.-H. DNA oxidation and excision repair pathways. *Int. J. Mol. Sci.* **2019**, *20*, 6092. [CrossRef]
61. Cohen, D.; Portugal-Cohen, M. Safe Retinol-Like Skin Biological Effect by a New Complex, Enriched with Retinol Precursors. *J. Cosmet. Dermatol. Sci. Appl.* **2020**, *10*, 59.
62. Wang, M.; Charareh, P.; Lei, X.; Zhong, J.L. Autophagy: Multiple Mechanisms to Protect Skin from Ultraviolet Radiation-Driven Photoaging. *Oxidative Med. Cell. Longev.* **2019**, *2019*, 8135985. [CrossRef]
63. Wineman, E.; Portugal-Cohen, M.; Soroka, Y.; Cohen, D.; Schlippe, G.; Voss, W.; Brenner, S.; Milner, Y.; Hai, N.; Ma'or, Z. Photo-damage protective effect of two facial products, containing a unique complex of Dead Sea minerals and Himalayan actives. *J. Cosmet. Dermatol.* **2012**, *11*, 183–192. [CrossRef] [PubMed]
64. Lee, C.-H.; Wu, S.-B.; Hong, C.-H.; Yu, H.-S.; Wei, Y.-H. Molecular mechanisms of UV-induced apoptosis and its effects on skin residential cells: The implication in UV-based phototherapy. *Int. J. Mol. Sci.* **2013**, *14*, 6414–6435. [CrossRef] [PubMed]
65. Cole, M.A.; Quan, T.; Voorhees, J.J.; Fisher, G.J. Extracellular matrix regulation of fibroblast function: Redefining our perspective on skin aging. *J. Cell Commun. Signal.* **2018**, *12*, 35–43. [CrossRef] [PubMed]
66. Gutop, E.; Diatlova, A.; Linkova, N.; Orlova, O.; Trofimova, S.; Khavinson, V. Aging of skin fibroblasts: Genetic and epigenetic factors. *Adv. Gerontol. Uspekhi Gerontol.* **2019**, *32*, 908–914.
67. Cook, M.K.; Kaszycki, M.A.; Richardson, I.; Taylor, S.L.; Feldman, S.R. Comparison of two devices for facial skin analysis. *J. Cosmet. Dermatol.* **2022**, *21*, 7001–7006. [CrossRef]
68. Overy, D.; Correa, H.; Roullier, C.; Chi, W.-C.; Pang, K.-L.; Rateb, M.; Ebel, R.; Shang, Z.; Capon, R.; Bills, G. Does osmotic stress affect natural product expression in fungi? *Mar. Drugs* **2017**, *15*, 254. [CrossRef]
69. Juturu, V.; Wu, J.C. Heterologous protein expression in Pichia pastoris: Latest research progress and applications. *ChemBioChem* **2018**, *19*, 7–21. [CrossRef]
70. Portugal-Cohen, M.; Dominguez, M.F.; Oron, M.; Holtz, R. Dead Sea minerals-induced positive stress as an innovative resource for skincare actives. *J. Cosmet. Dermatol. Sci. Appl.* **2015**, *5*, 22. [CrossRef]
71. Eckersley, A.; Ozols, M.; O'Connor, C.; Bell, M.; Sherratt, M.J. Predicting and characterising protein damage in the extracellular matrix. *J. Photochem. Photobiol.* **2021**, *7*, 100055. [CrossRef]
72. Heinemann, U.; Schuetz, A. Structural features of tight-junction proteins. *Int. J. Mol. Sci.* **2019**, *20*, 6020. [CrossRef]

73. Seo, S.H.; Kim, S.-E.; Lee, S.E. ER stress induced by ER calcium depletion and UVB irradiation regulates tight junction barrier integrity in human keratinocytes. *J. Dermatol. Sci.* **2020**, *98*, 41–49. [CrossRef] [PubMed]
74. Portugal-Cohen, M.; Soroka, Y.; Ma'or, Z.; Oron, M.; Zioni, T.; Brégégère, F.M.; Neuman, R.; Kohen, R.; Milner, Y. Protective effects of a cream containing Dead Sea minerals against UVB-induced stress in human skin. *Exp. Dermatol.* **2009**, *18*, 781–788. [CrossRef] [PubMed]
75. Birch-Machin, M.A.; Russell, E.V.; Latimer, J.A. Mitochondrial DNA damage as a biomarker for ultraviolet radiation exposure and oxidative stress. *Br. J. Dermatol.* **2013**, *169*, 9–14. [CrossRef]
76. Singh, P.; Jain, K.; Desai, C.; Tiwari, O.; Madamwar, D. Microbial community dynamics of extremophiles/extreme environment. In *Microbial Diversity in the Genomic Era*; Elsevier: Amsterdam, The Netherlands, 2019; pp. 323–332.
77. Yaniv, Z.; Koltai, H. Calotropis procera, Apple of Sodom: Ethnobotanical review and medicinal activities. *Isr. J. Plant Sci.* **2018**, *65*, 55–61. [CrossRef]
78. Imosemi, I.O. Evaluation of the toxicity, medicinal use and pharmacological actions of Calotropis procera. *Ejpmr* **2016**, *3*, 28–36.
79. Portugal-Cohen, M.; Ish-Shalom, E.; Mallon, R.; Corral, P.; Michoux, F. Apple of Sodom (*Calatropis procera*) callus extract, a novel skincare active and its biological activity in skin models when combined with Dead Sea water. *J. Cosmet. Dermatol. Sci. Appl.* **2018**, *8*, 73–91. [CrossRef]
80. Kottmeier, C.; Agnon, A.; Al-Halbouni, D.; Alpert, P.; Corsmeier, U.; Dahm, T.; Eshel, A.; Geyer, S.; Haas, M.; Holohan, E.; et al. New perspectives on interdisciplinary earth science at the Dead Sea: The DESERVE project. *Sci. Total Environ.* **2016**, *544*, 1045–1058. [CrossRef] [PubMed]
81. Anton, B.P.; DasSarma, P.; Martinez, F.L.; DasSarma, S.L.; Al Madadha, M.; Roberts, R.J.; DasSarma, S. Genome Sequence of Salarchaeum sp. strain JOR-1, an extremely halophilic archaeon from the Dead Sea. *Microbiol. Resour. Announc.* **2020**, *9*, e01505-19. [CrossRef] [PubMed]
82. Yin, W.; Wang, Y.; Liu, L.; He, J. Biofilms: The microbial "protective clothing" in extreme environments. *Int. J. Mol. Sci.* **2019**, *20*, 3423. [CrossRef]
83. Lebre, P.H.; De Maayer, P.; Cowan, D.A. Xerotolerant bacteria: Surviving through a dry spell. *Nat. Rev. Microbiol.* **2017**, *15*, 285–296. [CrossRef]
84. Patel, A.; Matsakas, L.; Rova, U.; Christakopoulos, P. A perspective on biotechnological applications of thermophilic microalgae and cyanobacteria. *Bioresour. Technol.* **2019**, *278*, 424–434. [CrossRef] [PubMed]
85. Obeidat, M. Isolation and characterization of extremely halotolerant Bacillus species from Dead Sea black mud and determination of their antimicrobial and hydrolytic activities. *Afr. J. Microbiol. Res.* **2017**, *11*, 1303–1314.
86. Al-Karablieh, N. Antimicrobial Activity of Bacillus Persicus 24-DSM Isolated from Dead Sea Mud. *Open Microbiol. J.* **2017**, *11*, 372–383. [CrossRef]
87. Oren, A.; Ginzburg, M.; Ginzburg, B.; Hochstein, L.; Volcani, B. *Haloarcula marismortui* (Volcani) sp. nov., nom. rev., an extremely halophilic bacterium from the Dead Sea. *Int. J. Syst. Evol. Microbiol.* **1990**, *40*, 209–210. [CrossRef] [PubMed]
88. Kim, J.H.; Shin, J.Y.; Hwang, S.J.; Kim, Y.S.; Kim, Y.M.; Gil, S.Y.; Jin, M.H.; Lee, S.H. Effect of Halophilic bacterium, Haloarcula vallismortis, extract on UV-induced skin change. *J. Soc. Cosmet. Sci. Korea* **2015**, *41*, 341–350.
89. Callewaert, C.; Ravard Helffer, K.; Lebaron, P. Skin Microbiome and its Interplay with the Environment. *Am. J. Clin. Dermatol.* **2020**, *21*, 4–11. [CrossRef]
90. Dimitriu, P.A.; Iker, B.; Malik, K.; Leung, H.; Mohn, W.; Hillebrand, G.G. New insights into the intrinsic and extrinsic factors that shape the human skin microbiome. *MBio* **2019**, *10*, e00839-19. [CrossRef]
91. Prescott, S.L.; Larcombe, D.-L.; Logan, A.C.; West, C.; Burks, W.; Caraballo, L.; Levin, M.; Etten, E.V.; Horwitz, P.; Kozyrskyj, A. The skin microbiome: Impact of modern environments on skin ecology, barrier integrity, and systemic immune programming. *World Allergy Organ. J.* **2017**, *10*, 29. [CrossRef]
92. Dawson, T.L., Jr. Malassezia: The forbidden kingdom opens. *Cell Host Microbe* **2019**, *25*, 345–347. [CrossRef]
93. Brandwein, M.; Fuks, G.; Israel, A.; Sabbah, F.; Hodak, E.; Szitenberg, A.; Harari, M.; Steinberg, D.; Bentwich, Z.; Shental, N. Skin microbiome compositional changes in atopic dermatitis accompany Dead Sea climatotherapy. *Photochem. Photobiol.* **2019**, *95*, 1446–1453. [CrossRef]
94. Chng, K.R.; Tay, A.S.L.; Li, C.; Ng, A.H.Q.; Wang, J.; Suri, B.K.; Matta, S.A.; McGovern, N.; Janela, B.; Wong, X.F.C.C. Whole metagenome profiling reveals skin microbiome-dependent susceptibility to atopic dermatitis flare. *Nat. Microbiol.* **2016**, *1*, 1–10. [CrossRef] [PubMed]
95. Sukenik, S.; Giryes, H.; Halevy, S.; Neumann, L.; Flusser, D.; Buskila, D. Treatment of psoriatic arthritis at the Dead Sea. *J. Rheumatol.* **1994**, *21*, 1305–1309. [PubMed]
96. Chadzopulu, A.; Adraniotis, J.; Theodosopoulou, E. The therapeutic effects of mud. *Prog. Health Sci.* **2011**, *1*, 132–136.
97. Sukenik, S.; Buskila, D.; Neumann, L.; Kleiner-Baumgarten, A. Mud pack therapy in rheumatoid arthritis. *Clin. Rheumatol.* **1992**, *11*, 243–247. [CrossRef] [PubMed]
98. Codish, S.; Abu-Shakra, M.; Flusser, D.; Friger, M.; Sukenik, S. Mud compress therapy for the hands of patients with rheumatoid arthritis. *Rheumatol. Int.* **2005**, *25*, 49–54. [CrossRef]
99. Hamed, S.; Almalty, A.-M. Skin Tolerance of Three Types of Dead Sea Mud on Healthy Skin: A Short-Term Study. *J. Cosmet. Sci.* **2018**, *69*, 269–278.

100. Hamed, S.; Almalty, A.M.; Alkhatib, H.S. The cutaneous effects of long-term use of Dead Sea mud on healthy skin: A 4-week study. *Int. J. Dermatol.* **2021**, *60*, 332–339. [CrossRef]
101. Abu-Al-Basal, M.A. Histological evaluation of the healing properties of Dead Sea black mud on full-thickness excision cutaneous wounds in BALB/c mice. *Pak. J. Biol. Sci. PJBS* **2012**, *15*, 306–315. [CrossRef]
102. Ma'or, Z.; Henis, Y.; Alon, Y.; Orlov, E.; Sørensen, K.B.; Oren, A. Antimicrobial properties of Dead Sea black mineral mud. *Int. J. Dermatol.* **2006**, *45*, 504–511. [CrossRef]

Disclaimer/Publisher's Note: The statements, opinions and data contained in all publications are solely those of the individual author(s) and contributor(s) and not of MDPI and/or the editor(s). MDPI and/or the editor(s) disclaim responsibility for any injury to people or property resulting from any ideas, methods, instructions or products referred to in the content.

cosmetics

MDPI

Article

Potential of Icariin–Glucosamine Combination in the Treatment of Osteoarthritis by Topical Application: Development of Topical Formulation and In Vitro Permeation Study

Katarzyna Pikosz [1,2], Izabela Nowak [1] and Agnieszka Feliczak-Guzik [1,*]

1 Faculty of Chemistry, Adam Mickiewicz University in Poznań, Uniwersytetu Poznańskiego 8, 61-614 Poznań, Poland
2 Invanto Sp. z o.o., ul. Hodowlana 5, 61-680 Poznań, Poland
* Correspondence: agaguzik@amu.edu.pl

Abstract: The aim of this study was to develop a topically applied formulation with the potential to alleviate arthritis ailments. A combination of two active ingredients, icariin from *Epimedium* L. (Species: *Epimedium Koreanum*) extract as a potential promoter of chondrogenesis and glucosamine sulfate as a precursor of cartilage tissues, was tested. In permeation studies, the potential for skin permeation of both substances was confirmed; however, the in vitro release test did not accurately reflect the degree of skin permeation. The in vitro release of icariin was at a level of 15.0–19.0% for the plant-extract-derived icariin and 29.0–35.0% for the pure substance. The level of glucosamine sulfate release was 38.4% (on average). For icariin of both origins, the release results were higher than those obtained via oral administration (about 12.0%), which shows the potential superiority of topical application. In addition, the physicochemical parameters that affect the in vitro release and performance of topical formulations were addressed. This preliminary research and permeation analysis of the formulation produced a promising picture of its prospects regarding arthritis treatment, although further investigation is needed.

Keywords: icariin; *Epimedium* L. extract; glucosamine sulfate; osteoarthritis; topical application; in vitro release

Citation: Pikosz, K.; Nowak, I.; Feliczak-Guzik, A. Potential of Icariin–Glucosamine Combination in the Treatment of Osteoarthritis by Topical Application: Development of Topical Formulation and In Vitro Permeation Study. *Cosmetics* **2023**, *10*, 36. https://doi.org/10.3390/cosmetics10010036

Academic Editor: Adeyemi Oladapo Aremu

Received: 12 January 2023
Revised: 6 February 2023
Accepted: 12 February 2023
Published: 16 February 2023

Copyright: © 2023 by the authors. Licensee MDPI, Basel, Switzerland. This article is an open access article distributed under the terms and conditions of the Creative Commons Attribution (CC BY) license (https://creativecommons.org/licenses/by/4.0/).

1. Introduction

Musculoskeletal disorders are the second most common cause of disability in the elderly worldwide [1]. Among them, the most common is osteoarthritis (OA), which affects approximately 527.81 million people worldwide [2]. One of the effects of this disease is the progressive, permanent loss of cartilage within the joint, whose function is the protection of joint surfaces against frictional, compressive, shear, and tensile loads during movement. The cartilage tissue is composed of chondrocytes suspended in the intercellular matrix consisting of water, hyaluronic acid, minerals, proteoglycan, and collagen fibers, mainly of type II [3]. Due to the lack of blood vessels and nerves, damaged cartilage has only limited self-healing ability [4,5]. The excessive mechanical overloading of cartilage tissue stimulates chondrocytes to produce cytokines and tumor necrosis factor α (TNF-α), which activate the synthesis of metalloproteinases, tissue plasminogen activator, plasmins, and other proteases involved in tissue destruction [6]. As a consequence, a shift in the balance between the production of enzymes and their inhibitors, which regulate the process of cartilage formation, in favor of the destructive effect is observed, which precludes effective regeneration. Attempts to regenerate articular cartilage without removing mechanical loads lead to the formation of tissue with a changed biochemical composition. The decrease in the content of proteoglycans and change in their composition, leading to a shortening of the glycosaminoglycan chains and a weaker ability to bind hyaluronic acid, together with a lack of the ability to aggregate, have the greatest impact on the deterioration of the biomechanical properties of the tissue. The collagen content usually remains the same, but the fibers are much thinner and, therefore, form looser nets. With time, the volume of the cartilage decreases, and significant defects appear within it; additionally, the bones under the cartilage can also be damaged, which may lead to their remodeling, sclerotization, and the formation of osteophytes [7]. The whole process is accompanied by inflammation of the synovial membrane, which is associated with lympho- and monocytic infiltrates and the phagocytosis of mechanically damaged fragments of articular cartilage. Then, there is general inflammation of all surrounding structures, i.e., the joint capsule, ligaments, tendons, and muscles belonging to the affected joint. The clinical symptoms include, first and foremost, pain associated especially with the commencement of movement, but also limited mobility of the joint, stiffness after immobility, swelling and thickening, and visible changes in the joint's outline [8]. Pain may be caused by damage to various tissues and may be related to the irritation of the nerve endings of the periosteum, damage to the subchondral layer of the bone, prolonged muscle tension, or improper tensioning of other structures to stabilize the joint and may also result from the presence of inflammation [9]. Many years of epidemiological studies conducted on various populations from around the world have allowed for the determination of differences in the incidence and factors increasing the risk of the disease, such as age, gender, or the environment. It has been repeatedly shown that the incidence of osteoarthritis increases with age, and more often affects women than men, while other factors increasing the risk of developing this disease include obesity, smoking, and injuries [2].

This disease leads to the progressive, permanent loss of cartilage within the joint. Due to the lack of blood vessels and nerves, damaged cartilage has only limited self-healing ability [5]. The excessive mechanical overloading of cartilage tissue stimulates chondrocytes to produce cytokines and tumor necrosis factor α (TNF-α), which activate the synthesis of substances involved in tissue destruction [6]. The result is a shift in the balance between the production of enzymes and their inhibitors, which regulate the process of cartilage formation, in favor of a destructive effect; thus, effective regeneration is precluded. The clinical symptoms include, first and foremost, pain associated especially with the commencement of movement, but also limited mobility of the joint, stiffness after immobility, swelling and thickening, and visible changes in the joint's outline. Pain may be caused by damage to various tissues and may be related to the irritation of the nerve endings of the periosteum, damage to the subchondral layer of the bone, prolonged muscle tension, or improper tensioning of other structures to stabilize the joint and may also result from the presence of inflammation. To this day, no effective treatment for causal OA has been developed. The most commonly used therapy includes pain reduction, the inhibition of the disease's progression, and joint function improvement, which are achieved through a combination of pharmacological and non-pharmacological methods and surgical treatment. For this purpose, generally available painkillers are used, and in the case of more severe pain and inflammation, drugs from the group of non-steroidal anti-inflammatory drugs (NSAIDs) are recommended [10]. However, their long-term use is not recommended due to the possibility of a number of side effects [11]. In pharmacological treatment, especially because of low toxicity, an alternative group of substances is used, namely, that of the symptomatic, slow-acting drugs for osteoarthritis (SYSADOA) [12–14]. This group includes both natural and synthetic substances that can be administered orally, intraarticularly, or topically. It has been shown that *Epimedium* L. (Species: *Epimedium Koreanum*) extract has an excellent beneficial impact on bone regeneration and can be involved in the regulation of multiple signaling pathways and biochemical processes [15]. *Epimedium* L. is one of the most commonly used herbs to treat bone fractures and osteoporosis in traditional Chinese medicine, wherein its assumed that *Epimedium* L. has the ability to "strengthen and regenerate bones" [16]. The *Epimedium* L. herb contains over 60 types of flavonoids, mainly glycosidic flavonoids, phenylpropanoids, alkaloids, polysaccharides, lignins, and sesquiterpenes [17–20], of which epimedin A, B, and C; icarizide_II; and icariin are considered the most important biologically active ingredients and constitute about 52%

of the total content of flavonoids in the plant [21,22]. Pharmacological studies have shown that *Epimedium* L. has a number of anti-inflammatory, antioxidant, and tumor-inhibitory properties [23,24], and clinical studies have suggested a beneficial effect on the treatment of diabetes, depression, cardiovascular diseases, rheumatoid arthritis, osteoporosis, and osteoathritis [25–28]. Regarding the treatment of diseases of the musculoskeletal system, icariin has been shown to have the greatest impact on bone regeneration from among other constituents of *Epimedium* L. and can be involved in the regulation of multiple signaling pathways and biochemical processes [29]. In a mouse model of osteoarthritis, icariin reduced the destruction of cartilage, promoted chondrocyte differentiation, upregulated the expression of parathyroid hormone-related protein, and down-regulated the expression of Indian hedgehog [30]. Icariin belongs to the glycosyl flavonoids with phytoestrogen activity, and it is the 8-prenyl derivative of kaempferol 3,7-O-diglucoside with the molecular formula of $C_{33}H_{40}O_{15}$. It is also safe, nontoxic, and its price is moderate, which makes it more attractive [31].

Another potentially very helpful substance in OA treatment is glucosamine, which is a naturally occurring aminomonosacharide and acts as a precursor in the biochemical synthesis of glycosylated proteins or lipids. Of particular importance to this study, it is a constituent of glycosaminoglycans in cartilage and synovial fluids [32]. Glucosamine is an endogenous substance, i.e., a typical constituent of the polysaccharide chains of the cartilage matrix and of glycosaminoglycans in the synovial fluid. In vitro and in vivo studies have shown that glucosamine stimulates the synthesis of physiological glycosaminoglycans and proteoglycans via chondrocytes and hyaluronic acid via synoviocytes [33]. Glucosamine's mechanism of action in humans is still not fully known, which makes it impossible to estimate the time required for the body to respond to the administered preparation. Glucosamine is a relatively small molecule (with a molecular weight of 179.00, while that of glucosamine sulfate potassium chloride is 605.50); it is easily soluble in water and in hydrophilic organic solvents. Only limited information is available on the pharmacokinetics of glucosamine. The commonly sold form of glucosamine that offers significant therapeutic effects is glucosamine sulfate [34–37]; in particular, crystalline glucosamine sulfate is considered highly bioavailable with a proven pharmacological effect [38–40]. The purpose of this study was to design and develop a topical formulation for the treatment of osteoarthritis. This formulation contains a combination of two active ingredients: icariin from *Epimedium* L. (Species: *Epimedium Koreanum*) extract, which may be a potential promoter of chondrogenesis, and glucosamine sulfate as a precursor of cartilage tissue; together, these two ingredients may exhibit multidirectional effects and offer excellent potential with respect to the treatment of osteoarthritis. To test the validity of further studies, including application and clinical trials, permeation tests were conducted as the first step. An in vitro release test was conducted to mirror the permeation study of the active ingredients to confirm their potential for topical application while avoiding the side effects accompanying long-term oral intake [41].

2. Materials and Methods

2.1. Reagents

D-glucosamine sulfate*2KCl (100%) was provided by Biomus Company (Białystok, Poland), Icariin (98.0%) was purchased from Chengdu Biopurity Phytochemicals Ltd. (Sichuan, China), and *Epimedium* L. extract (based on the dried aerial parts of *Epimedium Koreanum* plant) was obtained from The Garden of Naturalsolution Co., Ltd. (Osan-si, Republic of Korea). Basal substances used were as follows: sodium polyacrylate (BASF Care Creations, Germany), *vitis vinifera* L. seed oil, shea butter, xanthan gum, and PEG-40 hydrogenated castor oil (Brenntag Poland Sp. z.o.o., Poland); glyceryl stearate, stearyl alcohol, and Ceteareth-25 were purchased from Evonik—Personal Care (Poland). The following permeation promoters were used: propylene glycol (Chemitec, Sosnowiec, Poland); essential oils of rosemary, marjoram, clove, and cinnamon from Avicenna Oil (Wrocław, Poland); and d-menthol and d-camphor, which were provided by PPH Standard sp. z.o.o.

(Lublin, Poland). Phosphate-buffered saline was purchased from Thermo Fisher Scientific (Waltham, MA, USA), methanol clean for analysis was obtained from StanLab (Lublin, Poland), and glacial acetic acid (99.5% clean) for analysis was provided by Chempur (Piekary Śląskie, Poland).

2.2. Preparation of a Formulation Containing Icariin and Glucosamine

The compositions of the formulations are shown in Table 1.

Table 1. The compositions of the formulations proposed.

Prepared Formulations		
INGREDIENT	**CREAM (wt.%)**	**GEL (wt.%)**
Water	up to 100.00	up to 100.00
Epimedium L. extract (Species: *Epimedium Koreanum*)	70.00	70.00
Propylene glycol	3.00	3.00
D-Glucosamine sulfate potassium chloride salt	10.00	10.00
Xanthan gum	0.20	-
Sodium polyacrylate	-	3.00
Vitis vinifera L. seed oil	3.00	-
Glyceryl stearate	3.00	-
Stearyl alcohol	2.50	-
Ceteareth 25	2.00	-
Butyrospermum parkii (shea) butter	1.00	-
PEG-40 hydrogenated castor oil	-	1.00
D-Camphor	0.50	0.50
D-Menthol	0.20	0.20
Alcohol denat.	1.00	1.00
Rosemary essential oil	0.50	0.50
Marjoram essential oil	0.20	0.20
Cinnamon essential oil	0.05	0.05
Clove essential oil	0.20	0.20
Dehydroacetic acid (and) Benzyl alcohol	1.00	1.00

2.2.1. Preparation of Cream

Solid ingredients of the oil phase were melted at a temperature of 75 °C, and the liquid portion of the oil phase was added after melting of the solid portion. Glucosamine sulfate was dissolved in water phase with stirring (IKA RW 20 digital). The aqueous phase containing bioactive components was gradually added to the oil phase, with homogenization (IKA ULTRA TURRAX T25) at 9500–10,000 rpm for 2 min and continuous stirring at 320 rpm for 40 min until the temperature decreased to 40 °C.

2.2.2. Preparation of Gel

Sodium polyacrylate gel was prepared by a cold process. D-glucosamine sulfate potassium chloride salt was dissolved in water phase with stirring (IKA RW 20 digital) at 320 rpm for 10 min. Sodium polyacrylate was added to the water phase and mixed with continuous stirring at 450 rpm for 30 min. Sodium was added until homogenous. The oil phase was gradually added to the water phase upon continuous agitation.

2.2.3. Preparation of Ointment

The constituents of the ointment base were placed together and allowed to melt at 70 °C in water bath. After melting, the ingredients were stirred while gently maintaining temperature of 70 °C for 10 min, and then cooled with continuous stirring at 350 rpm. The active ingredients were added after lowering the temperature of the base to 40 °C and continuing to mix at 500 rpm for 20 min until a homogeneous mass was obtained.

2.2.4. Acceptance Criteria and Specifications

According to specific cosmetic guidelines, each topical preparation must satisfy certain important criteria, such as obtaining/maintaining a specific color, certain pH range, density, viscosity, and homogeneity, and producing no skin irritation effects.

2.3. *Physicochemical Tests*

2.3.1. Appearance and Homogeneity

The cream and gel bases prepared were inspected visually for clarity, homogeneity, color, and presence of any particles/aggregates.

2.3.2. Determination of pH of Formulations

Metrohm 827pH Lab pH meter (Metrohm Polska, Opacz-Kolonia, Poland) was used for determination of the pH of each formulation in triplicate. The pH meter was calibrated with standard buffer solutions before each use.

2.3.3. Density

Steel pycnometer (Quantachrome, Boynton Beach, FL, USA) with a capacity of 100 mL was used to determine density of the formulations. The pycnometer was filled with the product and measurements were repeated three times. Densities of the formulations were calculated using the following equation (Equation (1)):

$$P_s = 0.99985 \times m_s/v_p + 0.0012 \quad (1)$$

where P_s—formulation density (kg/m^3); m_s—substance mass (kg); v_p—pycnometer volume (m^3) 0.99985—atmospheric pressure factor; and 0.0012—coefficient accounting for air density.

Equation (1) Equation for calculating formulation density.

2.3.4. Dynamic Viscosity

Brookfield LVT rotational viscometer was used to determine dynamic viscosity of the prepared formulations. Viscosity was measured at room temperature (approximately 23 °C) with a spindle no. 3 for creams and no. 4 for gels, and spindle rotation speed of 12 rpm for gels and 1.5 rpm for emulsions.

2.3.5. Assay of Icariin Content in *Epimedium* L. (Species: Epimedium Koreanum) Extract—LC-MS

The icariin concentration in the commercially available extract of *Epimedium* L. (Species: *Epimedium Koreanum*) was determined using a Kinetex column 2.6 µm C18 (2.1-id x100 mm). Analysis conditions were as follows: column temperature of 35 °C, injection volume of 3 µL, and mobile phase flow of 0.3 mL/min. The mobile phases were A: water + 0.1 wt.% formic acid, and B: acetonitrile + 0.1 wt.% formic acid. Elution was performed using a stepwise gradient elution shown in Table 2.

Table 2. Gradient mobile phase elution for icariin determination in *Epimedium* L. (Species: *Epimedium Koreanum*) extract via LC-MS.

Time (min)	A (% *v/v*)	B (% *v/v*)
0	75	25
1	75	25
15	55	45
16	55	45
17	10	90
22	10	90
23	75	25
28	75	25

2.3.6. Release of Icariin

In vitro release test for icariin from *Epimedium* L. extract (Species: *Epimedium Koreanum*) was conducted with 708-DS Dissolution Apparatus Agilent Technologies Inc. (Santa Clara, CA, USA), synthetic membrane, Acceptor Solution phosphate buffer pH 5.8, and UV–VIS spectrophotometer $\lambda = 270$ nm.

2.3.7. Release of Glucosamine

In vitro release test for glucosamine was conducted with Franz Vertical Diffusion Cell Test System (Copley Scientific TM, Notinghma, United Kingdom). Synthetic membrane was soaked for 30 min prior to the in vitro assay in phosphate-buffered saline (PBS) of pH 7.4. The amount of formulation in each Franz cell and the time it was assembled were recorded. Conditions used were as follows: 32 °C, agitation of 50 rpm, acceptor solution—phosphate buffer of pH 7.4, and the amount of formulation—0.1 g. Samples were taken from the sampling port every hour for five hours. After each sample was collected, the phosphate buffer volume was restored to maintain conditions.

Concentrations of d-glucosamine sulfate in the samples were determined using HPLC-UV (Agilent Technologies) and C18 Column (3.5 μm, 4.6 mm id × 150 mm Waters) at temp. of 30 °C and wavelength of $\lambda = 254$ nm. The mobile phase consisted of methanol: water: glacial acetic acid (10:89, 96:0.04, pH 3.5); flow—1 mL/min; and injection volume of 20 μL. Accumulated amount of glucosamine sulfate was measured after derivatization of glucosamine to phenylthiocarbamyl glucosamine with phenyl isothiocyanate.

3. Results and Discussion

As a chronic degenerative and progressive disease, osteoarthritis affects the lives of people around the world, and to this day, no effective treatment has been developed. The main purpose of this study was to create dermocosmetic products with the potential to alleviate arthritis ailments. Glucosamine is an important precursor in the biochemical synthesis of glycosaminoglycans, which are major components of cartilage [32]. Moreover, icariin, due to its chondrogenic and osteoinductive potential, may be an effective promoter of bone regeneration [42–44]. In vitro studies have shown that icariin has the potential to promote bone engineering by initiating direct and stable chondrocyte differentiation, reducing its apoptosis, and creating an extracellular matrix by chondrocytes—without invoking a hypertrophy effect—due to the fact that it might be an effective accelerant of growth factors for cartilage [42,43]. A potent chondrogenic effect might be caused by the upregulation of the expression levels of aggrecan, collagen II, and SOX9 genes and the downregulation of the collagen I gene. Sun et al. have noted that the protection against cartilage and bone degradation is connected to the ability of icariin to inhibit the protease activity of cathepsin K [44]. Wiu et al. have established that icariin regulates articular bone loss, in part by regulating the receptor activator of the nuclear factor-B ligand RANKL (reduction) and osteoprotegrin OPG (enhancement) expression [45,46]. The data from clinical trials concerning the use of icariin in humans are limited. Only one blinded placebo control study has been reported. A randomized, double-blind study of *Epimedium* L. flavonoids that involved the administration of icariin (60 mg/day) to 100 postmenopausal women showed a significantly higher bone mineral density in the treatment group compared to the placebo group in 24 months [47]. It is important to remember that icariin has phytoestrogen activity, so it may disturb the endocrine system and connected pathways through oral administration, but this fact also indicates the superiority of topical application [48,49]. Unfortunately, there are just a few clinical studies that suggest that icariin might be helpful in OA treatment, and especially via topical administration, so this area needs to be further investigated. Both substances (icariin and glucosamine sulfate) are known to exhibit excellent therapeutic effects when used separately; due to their completely different mechanisms of action [33,50], they may also show a synergistic effect, which would be worth examining, especially due to the fact that glucosamine sulfate presents synergistic effects with a few NSAIDs and SYSADOA [40,51]. The determination of the icariin content in *Epimedium* L. (Species: *Epimedium Koreanum*) extract was essential in order to set the extract concentration in the target product formulation for further tests such that it is within the detection limits of the analytical devices used. The predicted concentration of icariin in the extract was very low based on the information from the manufacturer's specification; the mixture that was provided contained only 5 wt.% of *Epimedium* L. (Species: *Epimedium Koreanum*) extract and a much smaller amount of icariin. The LC-MS method was selected to conduct content analysis due to its high specificity, as only selected ions produced from the analytes of interest are monitored. This analysis allowed for the determination of the quantitative composition of the formulation such that it fell within the limits of detection of the devices used. The calibration function for a standard was made by injecting five different concentrations of reference solution in triplicate. The regression coefficient was 0.999 for the calibration curve, which demonstrates good linearity over the linear range (Figure 1). The LC-MS analysis showed an average of 168.76 ng of icariin per 1 mL of *Epimedium* L. (Species: *Epimedium Koreanum*) extract, which would be under the detection limit while using the standard of 1–5 wt.% of the extract. This result enabled the content of the extract in the product mass, which was set at the level of 70%, to be within the detection limits of icariin for further analysis, and faclilitated the determination of the amount of icariin corresponding to the extract in the in vitro permeation study.

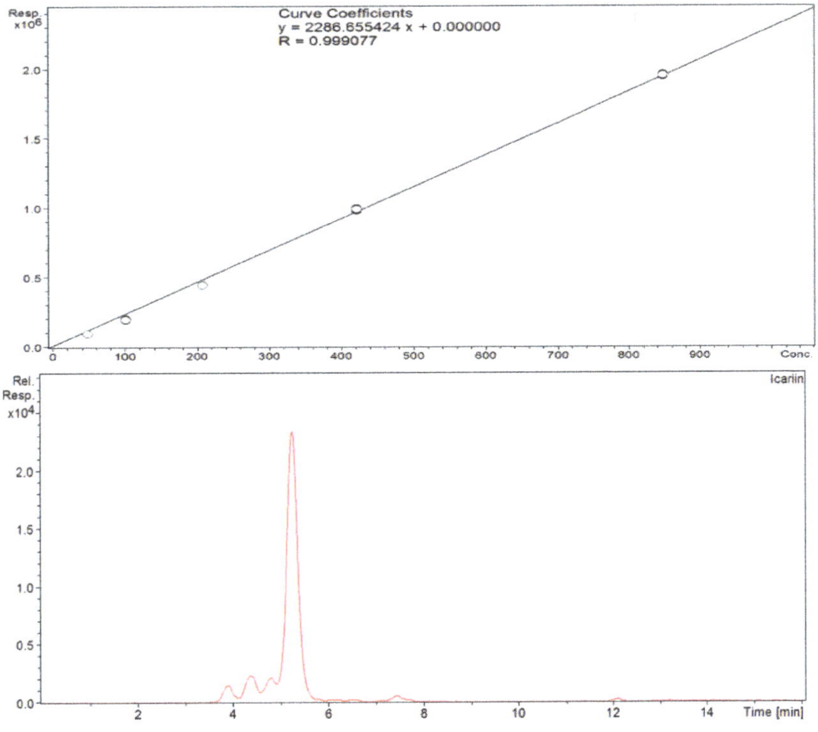

Figure 1. Calibration curve and chromatogram of icariin in *Epimedium* L. (Species: *Epimedium Koreanum*) extract indicating its content of 168.90 ng/mL.

The formulations proposed (cream and gel) for arthritis management were successfully developed so that they met all the acceptance criteria and specifications. All formulations were visually assessed for phase homogeneity, agglomeration, and discoloration. They appeared smooth, homogeneous, free from agglomerates, and initially stable, with no mechanical impurities found in the product masses. This study confirmed that gels and emulsions are effective matrices with which to hold high concentrations of water-soluble molecules such as glucosamine sulfate and hydrophilic extracts and are able to deliver such molecules across the skin via topical application. The physicochemical parameters of the formulations are shown in Table 3. The rheological and sensorial properties, appearance, consistency, dynamic viscosity, density, and pH were reproducible and met the acceptance criteria.

Table 3. Physicochemical properties of all formulations.

Formulation	Color	Homogeneity and Consistency	pH	Viscosity [mPa × s]	Density
cream 1_1 sample	off-white	creamy, excellent homogenous, no visible impurities	6.32	37.800	0.792
cream 1_2 sample	off-white	creamy, excellent homogenous, no visible impurities	6.29	38.200	0794
cream 1_3 sample	off-white	creamy, excellent homogenous, no visible impurities	6.33	37.800	0.791
gel 1_1 sample	slightly brown	nontransparent gel, homogenous, no visible impurities	5.19	33.600	0.987
gel 1_2 sample	slightly brown	nontransparent gel, homogenous, no visible impurities	5.17	34.250	0.984
gel 1_3 sample	slightly brown	nontransparent gel, homogenous, no visible impurities	5.18	33.950	0.989

It has been established that some physicochemical parameters can have a significant impact on the degree of skin penetration by particular substances. Malzfeldt et al. [52] have proved that the viscosity of a medium has an influence on the release rate of the active ingredient and that the rate of release is higher in a solution than in a suspension, cream, or ointment. Therefore, the base components were selected such that the lowest possible viscosity of the medium was ensured while maintaining mass stability. The prepared formulations exhibited sufficient viscosity (33,600–38,200 mPa × s) and there was no appreciable change in viscosity over time. Another parameter that may influence the release rate is the pH of the formulations. As the proposed products are cosmetics, it was necessary to follow general guidelines for cosmetics and adjust the pH so that it would not affect the skin's integrity and would not cause any adverse effects such as redness, rush, swelling, or changes in transepithelial water loss. The pH of the carrier may have a significant effect on the permeation of the substance, as it determines the solubility of active ingredients and the formulations' stability during storage. After application, the pH may impact the partitioning of the active substances between the formulation and the skin. In this study, we did not ascertain the effect of pH changes on the ability to release active substances, so further studies are required in order to improve the release profile.

There is little information about the skin permeability of icariin in the literature. Iin et al. have shown that due to its low water solubility, membrane permeability, and low dissolution rate in biological fluids, icariin has low oral bioavailability of about 12% and poor absorption; thus, its clinical applications are limited [53]. The predicted skin bioavailability was poor, as icariin is a relatively large molecule with a molecular weight of 676.7 Da, poor water solubility, and membrane permeability.

In this study, we used absorption enhancers that, in the initial formulation containing the *Epimedium* L. (Species: *Epimedium Koreanum*) extract, enabled the acquirement of an icariin release value of 15–19%. The use of another formulation containing the same amount of icariin as in the extract, but in the form of a pure substance, enabled the acquirement of an icariin release value of 29–35% (Figure 2). The gel matrix was very effective at holding 70% of the hydrophilic extracts and was able to deliver them across the skin via topical application. The results of the study of the in vitro skin permeation of icariin suggest that the use of pure icariin is more effective than the extract, which indicates that, in the case of a formulation with a herbal extract that contains over 260 other substances, there may be a competitive release of other active ingredients or extract-derived icariin may bond with other ingredients, which would impede permeation [20]. This study also showed potentially higher bioavailability via skin application than oral administration, although this supposition requires proof from further studies as it is based on the accumulated amount of icariin but not the actual level of skin permeation. In view of the fact that Xu et al. have reported a value of released icariin as high as 70%, much higher results may be obtained by the proper modification of the carrier and the use of more appropriate permeation promoters for pure icariin, which is known to display poor solubility both in water and lipid solutions. One group of authors obtained an icariin-loaded nanogel via the reverse microemulsion method, yielding very promising release results [54]. In our study, the challenge was to devise a carrier that is suitable for both substances while accounting for their different natures and that ensures an optimum release profile. This analysis and preliminary characterization of the formulation performed so far have painted a very promising picture of osteoarthritis treatment; however, the skin-penetrating ability, application, and bioavailability enhancement, particularly with respect to icariin, may require further investigation to increase their therapeutic effects. Further studies are underway to improve permeation of the actives, examine the applicational prospects, and confirm the properties of the product.

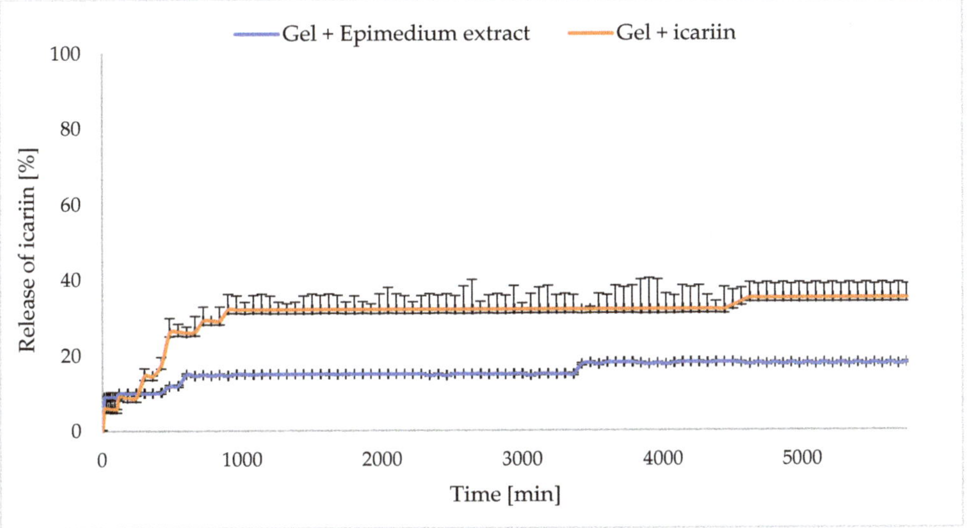

Figure 2. Release of icariin from gel formulation.

Glucosamine is much better known and characterized in the literature than icariin, which facilitated the more precise selection of the analytical conditions. This substance stimulates proteoglycan synthesis via chondrocytes [55] and shows anti-inflammatory activities in different in vivo animal models [56,57]. Clinical studies have demonstrated its superior effect compared to a placebo in the short-term treatment of OA pain and movement limitation [36,58–60]; moreover, it has been shown that long-term (3-years) oral treatment with glucosamine sulfate retards the progression of Knee OA [36]. Not all researchers have been enthusiastic about glucosamine with respect to the treatment of OA, and there are some speculations about its low level of bioavailability [61]. Clinical studies conducted by Persiani et al. have shown increased concentrations of glucosamine after oral administration in both healthy subjects and OA sufferers [62,63]. The concentration was measured in plasma and synovial fluid, which indicates a positive influence on cartilage and relatively high bioavailability, but the levels were still lower than that required to positively affect the cartilage. A great deal of hope was elicited by subsequent studies in which it was shown that much higher concentrations of glucosamine can be absorbed upon topical application compared to the oral route [64], and that it can specifically enter synovial fluid [65]. Lower bioavailability of oral administration of glucosamine is connected to the hepatic first-pass effect [62], which is responsible for drug degradation. The drug's absolute bioavailability is unknown. About 38% of an intravenous dose is excreted in an unchanged form in urine. The absolute bioavailability in humans after a single, oral dose was 25%. In the liver, more than 70% of the administered dose is metabolized. Absorption from the alimentary duct can be of about 90%; approximately 11% of the labeled dose was excreted in the feces [66]. Bypassing the gastrointestinal system is not the only benefit offered by topical administration. It has been widely established that transdermal application offers numerous advantages over oral delivery, especially in terms of targeted drug administration, where its use circumvents most of the undesirable side effects such as nausea, diarrhea, ulcers, and gastric inflammations, and is much safer in long-term treatment [67–69]. Due to its hydrophilic character, glucosamine sulfate is very water soluble, so it can be easily introduced into a medium, which allows for the prediction of its stability under selected conditions and the acquirement of results related to its accumulated amount that are close to the predictions for different carriers. In this study, the best results were obtained for the formulation containing 10 wt.% d-glucosamine sulfate (Table 4). It was also shown that the higher concentration of d-glucosamine sulfate in the formulation led to its higher accumulated amount. The formulations with d-glucosamine salt concentrations above 10 wt.% were not tested because of the difficulty of maintaining their stability. In this study, we used glucosamine sulfate, as it has been reported to exhibit much better effects compared to glucosamine hydrochloride [39]. The findings found in clinical trials with different glucosamine salts indicate the superiority of sulfates over hydrochlorides due to the increase in the sulfur concentration, which is essential for the synthesis of proteoglycans that are important for chondrocyte metabolism [70,71]. In addition, glucosamine sulphate has already shown a synergistic effect when combined with other osteoarthritis treatments [14,40].

Table 4. In vitro cumulative glucosamine release results.

2 g of glucosamine sulfate/100 g of product									
	Cream			Gel			Ointment		
Time [h]	average [μg/cm^2]	SD [μg/cm^2]	[%]	average [μg/cm^2]	SD [μg/cm^2]	[%]	average [μg/cm^2]	SD [μg/cm^2]	[%]
0	0.00	0.00	0.00	0.00	0.00	0.00	0.00	0.00	0.00
1	45	11.70	2.25	132.00	18.50	6.60	0.00	0.00	0.00
2	104	10.40	5.20	152.50	10.90	7.63	12.00	5.50	0.60
3	236	21.30	11.80	213.50	17.70	10.68	64.50	19.70	3.23
4	335	18.50	16.75	229.00	10.20	11.45	84.00	21.70	4.20
5	355	24.90	17.75	294.50	9.70	14.73	107.50	14.70	5.38
5 g of glucosamine sulfate/100 g of product									
	Cream			Gel			Ointment		
Time [h]	average [μg/cm^2]	SD [μg/cm^2]	[%]	average [μg/cm^2]	SD [μg/cm^2]	[%]	average [μg/cm^2]	SD [μg/cm^2]	[%]
0	0.00	0.00	0.00	0.00	0.00	0.00	0.00	0.00	0.00
1	156.00	10.80	3.10	326.10	7.60	6.50	25.70	12.10	0.50
2	385.20	3.10	7.70	472.10	6.10	9.40	39.00	15.90	0.80
3	535.60	61.80	10.70	575.20	15.30	11.50	208.40	54.10	4.20
4	1046.40	34.00	20.90	810.70	10.70	16.20	270.90	80.70	5.40
5	1260.90	49.10	25.20	989.70	11.50	19.80	345.20	51.00	6.90

Table 4. *Cont.*

	10 g of glucosamine sulfate/100 g of product								
	Cream			Gel			Ointment		
Time [h]	average [μg/cm²]	SD [μg/cm²]	[%]	average [μg/cm²]	SD [μg/cm²]	[%]	average [μg/cm²]	SD [μg/cm²]	[%]
0	0.00	0.00	0.00	0.00	0.00	0.00	0.00	0.00	0.00
1	422.20	27.40	4.20	974.30	108.60	9.70	42.40	11.80	0.40
2	1083.10	9.60	10.80	1137.30	29.40	11.40	82.70	4.10	0.80
3	2479.50	20.50	24.80	1589.00	93.20	15.90	299.90	82.60	3.00
4	3179.20	61.20	31.80	1713.90	15.40	17.10	468.80	24.00	4.70
5	3838.10	39.50	38.40	2173.10	36.60	21.70	603.10	36.10	6.00

The in vitro active ingredient cumulative release from the formulations was on average 38.4% for the creams and 21.7% for the gels after 5 h. The formulations with all the concentrations had similar release profiles; initially, glucosamine was released from the gel formulations much faster, but ultimately, the glucosamine was released from the cream formulation in higher concentrations (Figures 3–5).

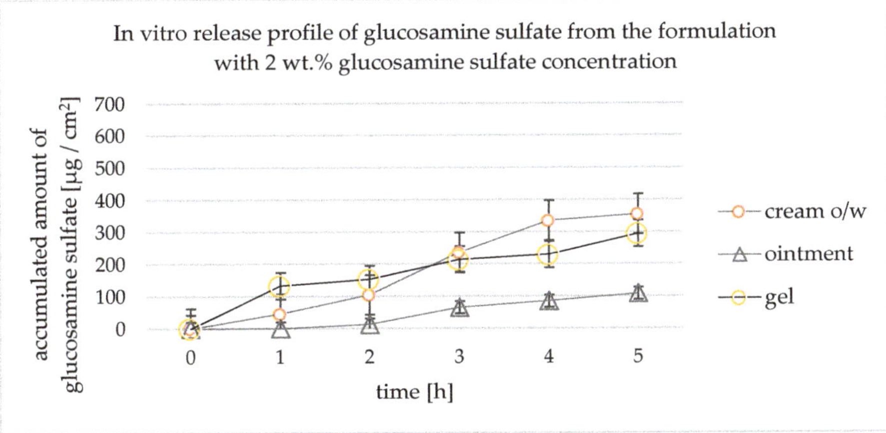

Figure 3. Accumulated amount of glucosamine sulfate—in vitro release study for the formulation with 2 wt.% of glucosamine.

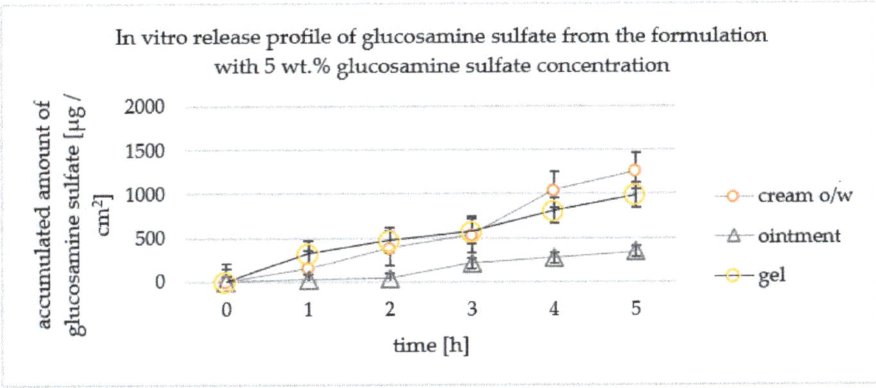

Figure 4. Accumulated amount of glucosamine sulfate—in vitro release study for 5 wt.% formulation.

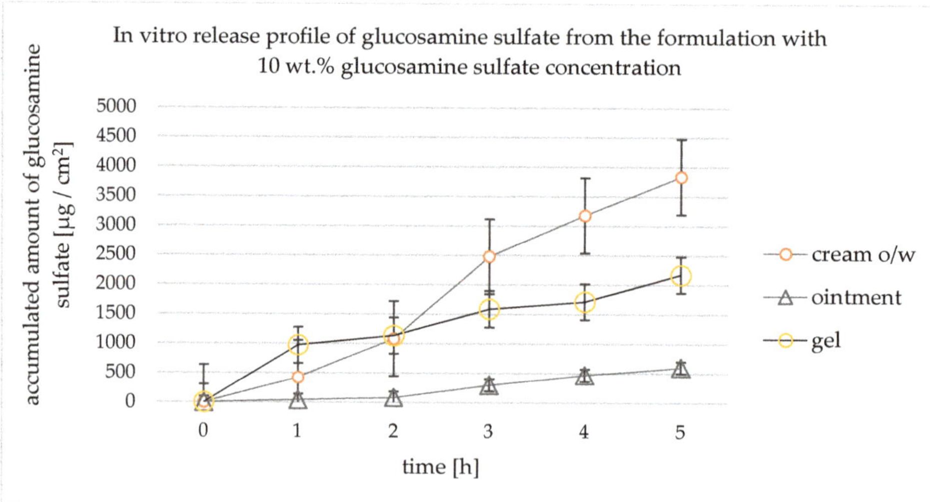

Figure 5. Accumulated amount of glucosamine sulfate—in vitro release profiles of the formulation containing 10 wt.% of glucosamine.

The differences in the in vitro permeation of glucosamine through membranes are correlated with the type of carrier. As the emulsion consists of oil and water phases, it has a greater affinity to the skin than the formulation that consists of only a water phase and a gelling agent. The initial faster release afforded by the gel was probably due to the fact that the active ingredient does not need to partition the two phases and may immediately cross the membrane. The in vitro release study confirmed the potential in an in vitro skin permeation study of both active substances (glucosamine and icariin). Due to the relatively hydrophilic character of glucosamine, the expected in vitro skin permeation was very low especially in the gel formulation. As glucosamine does not contain a chromophore, to make its detection possible via the UV detector used, a derivatization was necessary, which confirmed the results of Kanwischer et al. [72]. A combination of permeation promoters from different groups of chemicals, such as camphor, menthol, propylene glycol, alcohol denat., and essential oils, in the formulations proposed permitted relatively high in vitro release of glucosamine sulfate from the prepared formulation, which was on average 38.4% for the creams after 5 h. A higher permeation result was predicted for the emulsion formulation as it was known that a combination of oil and water phases has greater affinity to the skin and is more compatible with the lipophilic nature of skin, which, eventually, yields a more desirable permeation profile, slower release at the beginning, and much a higher concentration of the active ingredient at the end. A better release profile constitutes a longer duration of action with higher concentrations of the active ingredient.

In permeation studies, the use of different polymeric membranes designed to mimic the human skin, such as polysulfone, polyethersulfone, cellulose, and polydimethylsiloxane, has been suggested by the FDA [73] because they are inert, do not occlude the penetration of the active substances, and provide good permeability [74,75]. Membrane selection is a significant factor in permeation studies as its type should match a given carrier. In this study, synthetic membranes were used to assuage ethical concerns and because of their easier accessibility. Considering the use of semi-liquid carriers in the form of emulsions and gels, it was most advantageous to choose hydrophilic membranes, such as polysulfone and cellulose, which demonstrated significantly better in vitro release results for emulsions and gels than for ointments. These observations confirm that the type and features of the membrane have an impact on the release rate and general accumulation of topical dosage forms. In the case of ointments, the use of a hydrophobic membrane and a carrier that is compatible with it would probably yield much better results. It has been suggested that polydimethylsiloxane membranes could be used to predict the skin permeability of lipophilic substances such as ointments with adequate results [64,76]. Despite their many advantages, such as the ease with which their size and composition can be adjusted; the ease of their use, acquisition, and storage; and the lack of variability, thus matching the intrinsic properties of the skin, unfortunately, the selected membranes (both polysulfone and cellulose) cannot provide an insight into the interaction of the formulation's ingredients with the skin [77]. This is related to the impossibility of accounting for the lipid perturbation effect, cell metabolism, and the heterogeneous nature of the skin, which are difficult to recreate due to the synthetic nature of the membranes. A validated in vitro release test can provide screening information about the potential of a substance to cross the skin barrier, and the way the structural differences between the tested products may affect the rate and amount of the released substance, so long as the analyses are run under the same conditions. The use of artificial membranes instead of human or animal skin provides information on the release characteristics of a given substance instead of the permeation of the active substances, so further research is required to obtain information on the real degree of permeation. Other factors that can have an impact on the active ingredient's release results are the type and characteristics of the prepared carriers. In the case of the formulation composition, the manufacturing process is crucial. The gel formulation was prepared using a cold process to avoid affecting the stabilities of the active ingredients, but the preparation of the emulsion required 75 °C to melt the solid components of the oil phase, which increased the risk of the decomposition of the active substances. Fortunately, the active substances used were stable at such a temperature and the production process did not affect their concentrations in the formulation. The higher total accumulated amount of the active substance released from the emulsion than from the gel formulation can be explained by the fact that the surfactants in the emulsion reduce surface tension and may also improve the wetting of the membrane surface and increase the contact area with the membrane, as confirmed by Ferreira et al. [78]. An initial faster release from the gel was expected, since, in the gel, the active ingredient does not need to partition the two phases and is immediately able to cross the membrane. It has been well known for many years that the skin is a significant barrier to the ingress of many compounds, and protection against any external factors is its most significant function. The penetration of active substances is hindered mostly by the large species diversity and the lipophilic nature of the lipid layer; therefore, to increase the therapeutic effectiveness of topical application, the use of permeation enhancers is needed. In this study, a combination of permeation enhancers that facilitate the absorption of active substances through the skin by temporarily diminishing the skin's impermeability was used. Dimethyl sulfoxide (DMSO) is one of the most important compounds that enhances the transdermal permeation of a variety of drugs; unfortunately, it is forbidden in cosmetic applications, so the search for other enhancers was necessary. Terpenes, such as menthol and camphor and the terpene constituents of essential oils, increase the solubility of the active substances by forming a eutectic mixture with the penetrating compound and altering the barrier properties of the stratum corneum. The lowering of the melting point affects the solubility of many compounds and thus the degree of skin permeation. The use of ethanol offered several potential advantages. Alcohols act as penetration enhancers by leaching lipids from the stratum corneum, and they also release the sulfhydryl groups in keratin proteins of the stratum corneum, thus significantly increasing the possibility of the penetration of hydrophilic substances. They also acts as carriers for the terpenes used to increase the skin penetration and enhance the solubility of icariin [79]. It is important to note that in this study, two main active substances were used: glucosamine sulfate, which is highly hydrophilic, and the aqueous extract of *Epimedium* L. (Species: *Epimedium Koreanum*), which contains hydrophobic icariin. The physiochemical characteristics of the active ingredients' molecules constitute another aspect that is crucial to defining the tendency of their release, and they might have impacted the data obtained in this study. In this research, we considered the use of icariin and glucosamine sulfate in a potential chondrogenic combination. It has been known for many years that the influence of ethanol on the ability to increase the permeation of hydrophilic compounds is usually connected to a simultaneous reduction in the penetration of hydrophobic ones, but icariin displays poor solubility both in water and lipid solutions. Thus, propylene glycol was added as an additional permeation promoter and co-solvent.

4. Conclusions

Demonstrating the possibility of a topical product's therapeutic effects can be challenging. In this study, the potential of active ingredients to permeate the skin was shown. However, the obtained results of the release study do not reflect the actual degree of skin permeation but refer only to the accumulated amount of the active ingredient, as it was impossible to account for the interaction between the skin components and the lipid barrier that naturally occurs on the skin. We obtained the release of icariin at a level of 15.0–19.0% for the substance derived from the plant extract, and at a level of 29.0–35.0% for the pure substance. In both cases, the release was higher than for oral administration (by about 12.0%).

The average release level of glucosamine sulfate was 38.4%, which is sufficient. Both the cream and gel matrices effectively held 70.0% of the hydrophilic extracts, were stable at a 10.0% glucosamine sulfate concentration, and were able to deliver the substance across the synthetic membrane, thus mimicking topical application.

The lowest results were obtained for the ointment carrier, which is most likely due to the strongly hydrophilic nature of the active substances used. Other parameters that also have an impact on the in vitro release of active ingredients and the performance of a topical product were identified, namely, the initial concentration, process parameters, physicochemical properties such as viscosity and pH, the type of selected carrier, and permeation enhancers. This analysis and preliminary research on the formulations created has painted a very promising picture of osteoarthritis treatment, but further investigations are needed to increase their therapeutic effects and verify the actual degree of skin permeation. Due to global demographic changes and gradual lifestyle changes, the number of people suffering from OA is increasing, which, combined with the lack of effective treatments, is mobilizing a constant search for alternative therapies. The idea of using icariin, with its potentially strong ability to activate chondrogenesis and intercellular matrix regeneration by chondrocytes in a hitherto unexplored topical application, provides new insights into the potential use of this active substance and its properties with respect to the treatment of OA. In addition, the novel combination of the two substances (icariin and glucosamine sulfate), which show excellent effects separately due to their completely different mechanisms of action, may exhibit a synergistic effect, which would be worth confirming. Further studies are underway to confirm the properties of the product.

Author Contributions: Conceptualization, K.P. and A.F.-G.; methodology, K.P. and A.F.-G.; investigation, K.P. and A.F.-G.; data curation, K.P. and A.F.-G.; writing—original draft preparation, K.P.; writing—review and editing, A.F.-G. and I.N.; visualization, K.P.; supervision, A.F.-G. All authors have read and agreed to the published version of the manuscript.

Funding: This research received no external funding.

Institutional Review Board Statement: Not applicable.

Informed Consent Statement: Not applicable.

Data Availability Statement: Not applicable.

Conflicts of Interest: The authors declare no conflict of interest. The funders had no role in the design of the study; in the collection, analyses, or interpretation of data; in the writing of the manuscript; or in the decision to publish the results.

References

1. Global Burden of Disease Study 2013 Collaborators. Global, regional, and national incidence, prevalence, and years lived with disability for 301 acute and chronic diseases and injuries in 188 countries, 1990–2013: A systematic analysis for the Global Burden of Disease Study 2013. *Lancet* **2015**, *386*, 743–800. [CrossRef] [PubMed]
2. Long, H.; Liu, Q.; Yin, H.; Wang, K.; Diao, N.; Zhang, Y.; Lin, J.; Guo, A. Prevalence Trends of Site-Specific Osteoarthritis From 1990 to 2019: Findings From the Global Burden of Disease Study 2019. *Arthritis Rheumatol.* **2022**, *74*, 1172–1183. [CrossRef] [PubMed]
3. Fox, S.A.J.; Bedi, A.; Rodeo, S.A. The Basic Science of Articular Cartilage: Structure, Composition, and Function. *Sport. Health A Multidiscip. Approach* **2009**, *1*, 461–468.
4. Frenkel, S.R.; Clancy, R.M.; Ricci, J.L.; Di Cesare, P.E.; Rediske, J.J.; Abramson, S.B. Effects of nitric oxide on chondrocyte migration, adhesion, and cytoskeletal assembly. *Arthritis Rheum.* **1996**, *39*, 1905–1912. [CrossRef]
5. Klein, T.J.; Rizzi, S.C.; Reichert, J.C.; Georgi, N.; Malda, J.; Schuurman, W. Strategies for zonal cartilage repair using hydrogels. *Macromol. Biosci.* **2009**, *9*, 1049–1058. [CrossRef]
6. Grunke, M. Successful treatment of inflammatory knee osteoarthritis with tumour necrosis factor blockade. *Ann. Rheum. Dis.* **2006**, *65*, 555–556. [CrossRef]
7. Deshpande, P.; Patil, K.; Guledgud, M.V.; D'souza, R.S. Diagnostic Imaging in TMJ Osteoarthritis: A Case Report and Overview. *Int. J. Dent. Sci. Res.* **2015**, *3*, 56–59.
8. Kean, W.F.; Kean, R.; Buchanan, W.W. Osteoarthritis: Symptoms, signs and source of pain. *InflammoPharmacology* **2004**, *12*, 3–31. [CrossRef]
9. Hunter, D.J.; McDougall, J.J.; Keefe, F.J. The Symptoms of Osteoarthritis and the Genesis of Pain. *Rheum. Dis. Clin. North Am.* **2008**, *34*, 623–643. [CrossRef]
10. Bannuru, R.R.; Osani, M.C.; Vaysbrot, E.E.; Arden, N.K.; Bennell, K.; Bierma-Zeinstra, S.M.A.; Kraus, V.B.; Lohmander, L.S.; Abbott, J.H.; Bhandari, M.; et al. OARSI guidelines for the non-surgical management of knee, hip, and polyarticular osteoarthritis. *Osteoarthr. Cartil.* **2019**, *27*, 1578–1589. [CrossRef]
11. Pelletier, J.P.; Martel-Pelletier, J.; Rannou, F.; Cooper, C. Efficacy and safety of oral NSAIDs and analgesics in the management of osteoarthritis: Evidence from real-life setting trials and surveys. *Semin. Arthritis Rheum.* **2016**, *45*, S22–S27. [CrossRef] [PubMed]
12. Permuy, M.; Guede, D.; López-Peña, M.; Muñoz, F.; Caeiro, J.R.; González-Cantalapiedra, A. Comparison of various SYSADOA for the osteoarthritis treatment: An experimental study in rabbits. *BMC Musculoskelet Disord.* **2015**, *16*, 120. [CrossRef]
13. Veronese, N.; Ecarnot, F.; Cheleschi, S.; Fioravanti, A.; Maggi, S. Possible synergic action of non-steroidal anti-inflammatory drugs and glucosamine sulfate for the treatment of knee osteoarthritis: A scoping review. *BMC Musculoskelet Disord.* **2022**, *23*, 1084. [CrossRef] [PubMed]
14. Veronese, N.; Cooper, C.; Bruyère, O.; Al-Daghri, N.M.; Branco, J.; Cavalier, E.; Reginster, J.Y. Multimodal Multidisciplinary Management of Patients with Moderate to Severe Pain in Knee Osteoarthritis: A Need to Meet Patient Expectations. *Drugs.* **2022**, *82*, 1347–1355. [CrossRef] [PubMed]
15. Xie, F.; Wu, C.F.; Lai, W.P.; Yang, X.J.; Cheung, P.Y.; Yao, X.S.; Leung, P.C.; Wong, M.S. The osteoprotective effect of Herba epimedii (HEP) extract in vivo and in vitro. *Evid Based Complement Altern. Med.* **2005**, *2*, 353–361. [CrossRef]
16. Pei, L.K.; Guo, B.L. A review on research of raw material and cut crude drug of Herba epimedii in last ten years. *China J. Chin. Mater. Medica.* **2007**, *32*, 466–471.
17. Liu, R.; Li, A.; Sun, A.; Cui, J.; Kong, L. Preparative isolation and purification of three flavonoids from the Chinese medicinal plant *Epimedium* koreanum Nakai by high-speed counter-current chromatography. *J. Chromatogr. A* **2005**, *1064*, 53–57. [CrossRef]
18. Zhang, X.; Li, Y.; Yang, X.; Wang, K.; Ni, J.; Qu, X. Inhibitory effect of *Epimedium* extract on S-adenosyl-L-homocysteine hydrolase and biomethylation. *Life Sci.* **2005**, *78*, 180–186. [CrossRef]
19. Wu, H.; Lien, E.J.; Lien, L.L. Chemical and pharmacological investigations of *Epimedium* species: A survey. *Prog. Drug Res.* **2003**, *60*, 1–57.
20. Ma, H.P.; He, X.R.; Yang, Y.; Li, M.X.; Hao, D.J.; Jia, Z.P. The genus *Epimedium*: An ethnopharmacological and phytochemical review. *J. Ethnopharmacol.* **2011**, *134*, 519–541. [CrossRef]
21. Guo, B.; Xiao, P. Determination of flavonoids in different parts of five *Epimedium* plants. *China J. Chin. Mater. Med.* **1996**, *21*, 523–525.
22. Wang, Y.A.; Guo, Z.M.; Jin, Y.; Zhang, X.L.; Wang, L.; Xue, X.Y.; Liang, X.M. Identification of prenyl flavonoid glycosides and phenolic acids in *Epimedium* koreanum Nakai by Q-TOF-MS combined with selective enrichment on click oligo (ethylene glycol) column. *J. Pharm. Biomed. Anal.* **2010**, *51*, 606–616. [CrossRef] [PubMed]
23. Zhang, H.F.; Zhang, X.; Yang, X.H.; Qiu, N.X.; Wang, Y.; Wang, Z.Z. Microwave assisted extraction of flavonoids from cultivated *Epimedium* sagittatum: Extraction yield and mechanism, antioxidant activity and chemical composition. *Ind. Crop. Prod.* **2013**, *50*, 857–865. [CrossRef]
24. Sze, S.C.; Tong, Y.; Ng, T.B.; Cheng, C.L.; Cheung, H.P. Herba Epimedii: A antioxidative properties and its medical implications. *Molecules* **2010**, *15*, 7861–7870. [CrossRef] [PubMed]
25. Jiang, J.; Song, J.; Jia, X.B. Phytochemistry and ethnopharmacology of *Epimedium* L. species. *Chin. Herb. Med.* **2015**, *7*, 204–222. [CrossRef]

26. Liu, Y.Q.; Han, X.F.; Liu, T.G.; Cheng, M.C.; Xiao, H.B. A cell-based model of bone remodeling for identifying activity of icarrin in the treatment of osteoporosis. *Biotechnol. Lett.* **2015**, *37*, 219–226. [CrossRef]
27. Pan, Y.; Kong, L.D.; Xia, X.; Zhang, W.Y.; Xia, Z.H.; Jiang, F.X. Antidepressant-like effect of icariin and its possible mechanism in rats. *Pharmacol. Biochem. Behav.* **2005**, *82*, 686–694. [CrossRef]
28. Wang, Z.Q.; Lou, Y.J. Proliferation-stimulating effects of icariin and desmethylicaritin in MCF-7 cells. *Eur. J. Pharmacol.* **2004**, *504*, 147–153. [CrossRef]
29. Zhang, X.; Liu, T.; Huang, Y.; Wismeijer, D.; Liu, Y. Icariin: Does It Have An Osteoinductive Potential for Bone Tissue Engineering? *Phytother. Res.* **2013**, *28*, 498–509. [CrossRef]
30. Luo, Y.; Zhang, Y.; Huang, Y. Icariin Reduces Cartilage Degeneration in a Mouse Model of Osteoarthritis and is Associated with the Changes in Expression of Indian Hedgehog and Parathyroid Hormone-Related Protein. *Med Sci Monit.* **2018**, *24*, 6695–6706. [CrossRef]
31. Zhao, J.; Ohba, S.; Komiyama, Y.; Shinkai, M.; Chung, U.; Nagamune, T. Icariin: A Potential Osteoinductive Compound for Bone Tissue Engineering. *Tissue Eng. Part A* **2010**, *16*, 233–243. [CrossRef]
32. Hamerman, D. The biology of osteoarthritis. *N. Engl. J. Med.* **1989**, *320*, 1322–1330. [PubMed]
33. Jerosch, J. Effects of Glucosamine and Chondroitin Sulfate on Cartilage Metabolism in OA: Outlook on Other Nutrient Partners Especially Omega-3 Fatty Acids. *Int. J. Rheumatol.* **2011**, *2011*, 969012. [CrossRef]
34. Serni, U. Profile of glucosamine sulfate as an example of Slow Acting Drug in Osteoarthritis. *Rev. Esp. Reumatol.* **1993**, *20* (Suppl. S1), 222.
35. Avouac, B. Slow Acting Drugs in osteoarthritis: A step towards disease modification. *Rev. Esp. Reumatol.* **1993**, *20* (Suppl. S1), 221–222.
36. Pavelká, K.; Gatterová, J.; Olejarová, M.; Machacek, S.; Giacovelli, G.; Rovati, L.C. Glucosamine Sulfate Use and Delay of Progression of Knee Osteoarthritis. *Arch. Inter. Med.* **2002**, *162*, 2113–2123. [CrossRef] [PubMed]
37. Mullerfasbender, H.; Bach, G.; Haase, W.; Rovati, L.; Sentikar, I. Glucosamine sulfate compared to ibuprofen in osteoarthritis of the knee. *Osteoarthr. Cartil.* **1994**, *2*, 61–69. [CrossRef] [PubMed]
38. Altman, R.D. Glucosamine therapy for knee osteoarthritis: Pharmacokinetic considerations. *Expert Rev. Clin. Pharmacol.* **2009**, *2*, 359–371. [CrossRef] [PubMed]
39. Kucharz, E.J.; Kovalenko, V.; Szántó, S.; Bruyère, O.; Cooper, C.; Reginster, J.Y. A review of glucosamine for knee osteoarthritis: Why patented crystalline glucosamine sulfate should be differentiated from other glucosamines to maximize clinical outcomes. *Curr. Med. Res. Opin.* **2016**, *32*, 997–1004. [CrossRef] [PubMed]
40. Cheleschi, S.; Tenti, S.; Giannotti, S.; Veronese, N.; Reginster, J.Y.; Fioravanti, A. A Combination of Celecoxib and Glucosamine Sulfate Has Anti-Inflammatory and Chondroprotective Effects: Results from an In Vitro Study on Human Osteoarthritic Chondrocytes. *Int. J. Mol. Sci.* **2021**, *22*, 8980. [CrossRef]
41. Rapalli, V.K.; Mahmood, A.; Waghule, T.; Gorantla, S.; Kumar Dubey, S.; Alexander, A.; Singhvi, G. Revisiting techniques to evaluate drug permeation through skin. *Expert Opin. Drug Deliv.* **2021**, *18*, 1829–1842. [CrossRef] [PubMed]
42. Zhang, L.; Zhang, X.; Li, K.F.; Li, D.X.; Xiao, Y.M.; Fan, Y.J.; Zhang, X.D. Icariin Promotes Extracellular Matrix Synthesis and Gene Expression of Chondrocytes In Vitro. *Phytother. Res.* **2012**, *26*, 1385–1392. [CrossRef]
43. Wang, Z.C.; Sun, H.J.; Li, K.H.; Fu, C.; Liu, M.Z. Icariin promotes directed chondrogenic differentiation of bone marrow mesenchymal stem cells but not hypertrophy in vitro. *Exp. Med.* **2014**, *8*, 1528–1534. [CrossRef]
44. Sun, P.; Liu, Y.; Deng, X.; Yu, C.; Dai, N.; Yuan, X. An inhibitor of cathepsin K, icariin suppresses cartilage and bone degradation in mice of collagen induced arthritis. *Phytomedicine* **2013**, *20*, 975–979. [CrossRef]
45. Wei, C.C.; Ping, D.Q.; You, F.T.; Qiang, C.Y.; Tao, C. Icariin Prevents Cartilage and Bone Degradation in Experimental Models of Arthritis. *Mediat. Inflamm.* **2016**, *2016*, 1–10.
46. Liu, Y.J.; Feng, W.; He, D.Y.; Wang, Q.Q. Effect of icariin on bone destruction and serum RANKL/OPG levels in type II collagen-induced arthritis rats. *China J. Chin. Mater. Med.* **2013**, *33*, 1221–1225.
47. Zhang, G.; Qin, L.; Shi, Y. *Epimedium*-derived phytoestrogen flavonoids exert beneficial effect on preventing bone loss in late postmenopausal women: A 24-month randomized, double-blind and placebocontrolled trial. *J. Bone Min. Res.* **2007**, *22*, 1072–1079. [CrossRef] [PubMed]
48. Kang, H.K.; Choi, Y.H.; Kwon, H.; Lee, S.B.; Kim, D.H.; Sung, C.K.; Park, Y.I.; Dong, M.S. Estrogenic/antiestrogenic activities of a *Epimedium* koreanum extract and its major components: In vitro and in vivo studies. *Food Chem. Toxicol.* **2012**, *50*, 2751–2759. [CrossRef]
49. Zanoli, P.; Benelli, A.; Zavatti, M.; Rivasi, M.; Baraldi, C.; Baraldi, M. Improved sexual behavior in male rats treated with a Chinese herbal extract: Hormonal and neuronal implications. *Asian J. Androl.* **2008**, *10*, 937–945. [CrossRef]
50. Yang, A.; Yu, C.; Lu, Q.; Li, H.; Li, Z.; He, C. Mechanism of Action of Icariin in Bone Marrow Mesenchymal Stem Cells. *Stem Cells Int.* **2019**, *4*, 5747298. [CrossRef]
51. Calamia, V.; Mateos, J.; Fernández-Puente, P.; Lourido, L.; Rocha, B.; Fernández-Costa, C.; Montell, E.; Vergés, J.; Ruiz-Romero, C.; Blanco, F.J. A pharmacoproteomic study confirms the synergistic effect of chondroitin sulfate and glucosamine. *Sci. Rep.* **2014**, *4*, 5069. [CrossRef]
52. Malzfeldt, E.; Lehmann, P.; Goerz, G.; Lippold, B.C. Influence of drug solubility in the vehicle on clinical efficacy of ointments. *Arch. Dematol. Res.* **1989**, *281*, 193–197. [CrossRef]

53. Jin, J.; Wang, H.; Hua, X.; Chen, D.; Huang, C.; Chen, Z. An outline for the pharmacological effect of icariin in the nervous system. *Eur. J. Pharmacol.* **2019**, *842*, 20–32. [CrossRef] [PubMed]
54. Xu, D.; Lu, Y.R.; Kou, N.; Hu, M.J.; Wang, Q.S.; Cui, Y.L. Intranasal delivery of icariin via a nanogel-thermoresponsive hydrogel compound system to improve its antidepressant-like activity. *Int. J. Pharm.* **2020**, *586*, 119550. [CrossRef]
55. Bassleer, C.; Henrotin, Y.; Franchimont, P. In-vitro evaluation of drugs proposed as chondroprotective agents. *Int. J. Tissue React.* **1992**, *14*, 231–241.
56. Setnikar, I.; Cereda, R.; Pacini, M.A.; Revel, L. Antireactive properties of glucosamine sulfate. *Arzneimittelforschung* **1991**, *41*, 157–161. [PubMed]
57. Setnikar, I.; Pacini, M.A.; Revel, L. Antiarthritic effects of glucosamine sulfate studied on animal models. *Arzneimittelforschung* **1991**, *41*, 542–545.
58. Noack, W.; Fischer, M.; FSrster, K.K.; Rovati, L.C.; Setnikar, I. Glucosamine sulfate in osteoarthritis of the knee. *Osteoarthr. Cart* **1994**, *2*, 51–59. [CrossRef] [PubMed]
59. Rovati, L.C.; Setnikar, I.; FSrster, K.K.; Reichelt, A.; Noack, W. Glucosamine sulfate in gonarthrosis: Efficacy in placebo controlled studies. *Rev Esp Reum.* **1993**, *20* (Suppl. S1), 72.
60. Giacovelli, G.; Rovati, L.C. Clinical efficacy of glucosamine sulfate in osteoarthritis of the spine. *Rev. Esp. Reum.* **1993**, *20* (Suppl. S1), 96.
61. Silva, J.A.; Apolinário, A.C.; Souza, M.S.R.; Damasceno, B.P.G.L.; Medeiros, A.C.D. Cutaneous Administration of Drugs: Challenges and Strategies for the Development of Trans-dermal Formulations. *Rev. De Ciências Farm. Básica* **2010**, *31*, 125–131.
62. Persiani, S.; Rotini, R.; Trisolino, G.; Rovati, L.C.; Locatelli, M.; Paganini, D.; Antonioli, D.; Roda, A. Synovial and plasma glucosamine concentrations in osteoarthritic patients following oral crystalline glucosamine sulphate at therapeutic dose. *Osteoarthr. Cartil.* **2007**, *15*, 764–772. [CrossRef]
63. Persiani, S.; Roda, E.; Rovati, L.C.; Locatelli, M.; Giacovelli, G.; Roda, A. Glucosamine oral bioavailability and plasma pharmacokinetics after increasing doses of crystalline glucosamine sulfate in man. *Osteoarthr. Cartil.* **2005**, *13*, 1041–1049. [CrossRef]
64. Lee, C.W.; Li, Z.; Patel, K.; Olobo, J.O.; Lee, E.J.D.; Goh, L.B. The Transdermal Profiles of Mediflex™ Glucosamine Cream in Mouse and Man. Available online: https://docplayer.net/31056811-The-transdermal-profiles-of-mediflex-tm-glucosamine-cream-in-mouse-and-man.html (accessed on 10 February 2023).
65. Kong, M.; Hashim, K.B.; Lin, P.; Coestesquis, G.; Xu, A.Y.; Lebes, F.; Ting, C.M. Penetration of Topical Glucosamine Sulfate into the Synovial Fluid of Patients with Knee Osteoarthritis: A Nonrandomized, OpenLabel, Single Dose, Bioavailability Study. *J. Biosci. Med.* **2019**, *7*, 76–90. [CrossRef]
66. Huskisson, E.C. Glucosamine and Chondroitin for Osteoarthritis. *J. Int. Med. Res.* **2008**, *36*, 6. [CrossRef] [PubMed]
67. Leite, C.B.S.; Coehlo, J.M.; Muehlmann, L.A.; de Azevedo, R.B.; Sousa, M.H. Skin Delivery of Glucosamine and Chondroitin Sulphates—A Perspective on the Conservative Treatment for Osteoarthritis of the Knee. *J. Biosci. Med.* **2017**, *5*, 11–20. [CrossRef]
68. Miki, R.; Ichitsuka, Y.; Yamada, T.; Kimura, S.; Egawa, Y.; Seki, T.; Morimoto, Y. Development of a membrane impregnated with a poly(dimethylsiloxane)/poly(ethylene glycol) copolymer for a high-throughput screening of the permeability of drugs, cosmetics, and other chemicals across the human skin. *Eur. J. Pharm. Sci.* **2015**, *66*, 41–49. [CrossRef] [PubMed]
69. Flaten, G.E.; Palac, Z.; Engesland, A.; Filipović-Grčić, J.; Vanić, Ž.; Škalko-Basnet, N. In vitro skin models as a tool in optimization of drug formulation. *Eur. J. Pharm. Sci.* **2015**, *75*, 10–24. [CrossRef] [PubMed]
70. Hoffer, L.J.; Kaplan, L.N.; Hamadeh, M.J.; Grigoriu, A.C.; Baron, M. Sulfate could mediate the therapeutic effect of glucosamine sulfate. *Metabolism.* **2001**, *50*, 767–770. [CrossRef]
71. Cordoba, F.; Nimni, M.E. Chondroitin sulfate and other sulfate containing chondroprotective agents may exhibit their effects by overcoming a deficiency of sulfur amino acids. *Osteoarthr. Cartil.* **2003**, *11*, 228–230. [CrossRef]
72. Kanwischer, M.; Kim, S.Y.; Bian, S.; Kwon, K.A.; Kim, J.S.; Kim, D.D. Evaluation of the Physicochemical Stability and Skin Permeation of Glucosamine Sulfate. *Drug Dev. Ind. Pharm.* **2005**, *31*, 91–97. [CrossRef] [PubMed]
73. FDA. Guidance for Industry. Nonsterile semisolid dosage forms, scale-up and postapproval changes: Chemistry, manufacturing, and controls; in vitro release testing and in vivo bioequivalence documentation. *Cent. Drug Eval. Res.* **1997**, *CMC 7*, 19–25.
74. EMA. *Guideline on Quality of Transdermal Patches*; European Medicines Agency: London, UK, 2014; Volume 24.
75. Argoff, C.E.; Gloth, F.M. Topical Nonsteroidal Anti-Inflammatory Drugs for Management of Osteoarthritis in Long-Term Care Patients. *Ther. Clin. Risk Manag.* **2011**, *7*, 393–399. [CrossRef] [PubMed]
76. Ng, S.F.; Rouse, J.J.; Sanderson, F.D.; Eccleston, G.M. The relevance of polymeric synthetic membranes in topical formulation assessment and drug diffusion study. *Arch. Pharmacal Res.* **2012**, *35*, 579–593. [CrossRef] [PubMed]
77. Ameri, M.; Lewis, H.; Lehman, P. Effect of Skin Model on In Vitro Performance of an Adhesive Dermally Applied Microarray Coated with Zolmitriptan. *J. Pharm.* **2018**, *3*, 7459124–7459130. [CrossRef] [PubMed]

78. Ferreira, L.A.M.; Seiller, M.; Grossiord, J.L.; Marty, J.P.; Wepierre, J. Vehicle influence on in vitro release of metronidazole: Role of w/o/w multiple emulsion. *Int. J. Pharm.* **1994**, *109*, 251–259. [CrossRef]
79. Lin, H.M.; Lin, L.F.; Sun, M.Y.; Liu, J.; Wu, Q. Topical Delivery of Four Neuroprotective Ingredients by Ethosome-Gel: Synergistic Combination for Treatment of Oxaliplatin-Induced Peripheral Neuropathy. *Int. J. Nanomed.* **2020**, *15*, 3251–3266. [CrossRef]

Disclaimer/Publisher's Note: The statements, opinions and data contained in all publications are solely those of the individual author(s) and contributor(s) and not of MDPI and/or the editor(s). MDPI and/or the editor(s) disclaim responsibility for any injury to people or property resulting from any ideas, methods, instructions or products referred to in the content.

cosmetics

MDPI

Article

Analysis of Cosmetic Products Containing *Serratula coronata* Herb Extract

Anna Kroma [1,*], Agnieszka Feliczak-Guzik [2], Mariola Pawlaczyk [1], Tomasz Osmałek [3], Maria Urbańska [1], Iwona Micek [1], Joanna Nawrot [1] and Justyna Gornowicz-Porowska [1]

[1] Department and Division of Practical Cosmetology and Skin Diseases Prophylaxis, Poznan University of Medicinal Sciences, 3 Rokietnicka St., 60-806 Poznań, Poland
[2] Department of Chemistry, Adam Mickiewicz University, 8 Uniwersytetu Poznańskiego St., 61-614 Poznań, Poland
[3] Chair and Department of Pharmaceutical Technology, Faculty of Pharmacy, Poznan University of Medical Sciences, 6 Grunwaldzka Street, 60-780 Poznań, Poland
* Correspondence: anna.kroma@ump.edu.pl

Abstract: Phytoecdysteroids exert significant anti-inflammatory effects, which makes them valuable ingredients in pharmaceutical and cosmetic products. However, data on their use in cosmetics are limited. Here, a new formulation with the extract of the *Serratula coronata* herb containing phytoecdysteroids was developed. The aim of this study was to perform physicochemical characteristics and evaluate the safety of the creams with *S. coronata*. Chromatography was used to detect the dominant phytoecdysteroids in the extract. The chemical and physical description of the creams was performed using the following parameters: viscosity, pH, and stability. The microbiological purity (pharmacopoeial methods) and transdermal permeability (Raman spectroscopy) were assessed to ensure the safety of the plant extracts used in the creams. The study confirmed the presence of phytoecdysteroid fractions of the *S. coronata* herb in the creams (20-hydroxyecdysone, polypodine B, and ajugasterone C). The results indicated that the cosmetics containing the *S. coronata* extract were chemically and microbiologically stable, thereby contributing to their safety. Their effectiveness is the result of transdermal permeability of 20-hydroxyecdysone. In this study, we demonstrated the importance of the *S. coronata* extract as a source of bioactive phytoecdysteroids and proved that the extract's characteristics may make it the key ingredient of safe and stable skincare products that support the treatment of various inflammatory skin diseases. These results were a continuation of those presented in our earlier publication.

Keywords: *Serratula coronata* extract; phytoecdysteroids; chemical analysis; chromatography–electrospray ionization–mass spectrometry; Raman spectroscopy

Citation: Kroma, A.; Feliczak-Guzik, A.; Pawlaczyk, M.; Osmałek, T.; Urbańska, M.; Micek, I.; Nawrot, J.; Gornowicz-Porowska, J. Analysis of Cosmetic Products Containing *Serratula coronata* Herb Extract. *Cosmetics* **2023**, *10*, 18. https://doi.org/10.3390/cosmetics10010018

Academic Editor: Adeyemi Oladapo Aremu

Received: 12 December 2022
Revised: 8 January 2023
Accepted: 10 January 2023
Published: 16 January 2023

Copyright: © 2023 by the authors. Licensee MDPI, Basel, Switzerland. This article is an open access article distributed under the terms and conditions of the Creative Commons Attribution (CC BY) license (https://creativecommons.org/licenses/by/4.0/).

1. Introduction

Recent years have witnessed a noticeable increase in the search for plant ingredients that may be applicable in numerous fields such as the cosmetic industry. However, plant extracts still lack homogeneity and present a considerable challenge in all attempts at standardization and description [1,2]. In addition, plant extract content affects the stability of cosmetic formulations and their transdermal permeability.

Serratula coronata L. (*S. coronata*), genus *Serratula*, is a perennial herb from the Asteraceae family [3]. Its overground parts are rich in ascorbic acid, alkaloids, and flavonoids. The leaves contain numerous amino acids, chief among them aspartic and glutamic acids as well as leucine, whereas the buds are rich in L-arginine. *S. coronata* is also a source of phytoecdysteroids (PEs) such as 20-hydroxyecdysone (20-HE), polypodine B, ajugasterone C, integristerone A, ecdysterone 2,3,20,22-diacetonide, and ecdysterone 20,22-monoacetonide, which belong to the group of polyhydroxylated steroids characterized by low toxicity and broad-spectrum biological activity [4,5]. Total PE content in the overground part of *S. coronata* depends on the vegetative phase of the plant and has been estimated at 0.5–1.5% [4]. So far, various sources have confirmed beneficial effects of preparations containing PEs such as ecdysterone and its esters as well as plant extracts rich in these compounds. Their beneficial effect on the skin is associated with their ability to regulate the process of keratinocyte differentiation, fortify the natural epidermal barrier, reduce transepidermal water loss (TEWL), increase epithelial hydration and strength, and improve exfoliation of the corneal epithelium, thus restoring skin smoothness. Owing to these properties, cosmetics with PEs are suitable for individuals with dry and very dry skin, psoriasis, and seborrheic dermatitis [5,6].

Various analytical techniques may be used for PE characterization [7–9] because their polarity and weak crystallization make separation difficult [10]. The use of combined techniques (HPLC-UV, HPLC-MS, and HPLC-NMR) increases the sensitivity and selectivity of PE analysis [7]; HPLC-NMR is seen as the most reliable source of information on PE structure and content in a sample [5].

Microbiological purity of cosmetics offers safety to their users during application and prevents physicochemical changes in the preparation as well as infections and skin diseases. Testing for stability ensures that a cosmetic product maintains its intended quality. Unfortunately, studies on the microbial quality and stability of cosmetics containing an *S. coronata* herb extract are limited at best. Thus, physicochemical and microbiological assessments are needed to control the quality of the existing cosmetics and to develop new formulations. This study aimed to evaluate the physicochemical and microbiological characteristics of new creams containing the extract of the *S. coronata* herb. A new formulation of the *S. coronata* herb extract containing phytoecdysteroids was developed.

2. Results and Discussion

2.1. Chromatographic Analysis

2.1.1. Analysis of the Chemical Content of the Extract Using Thin-Layer Chromatography

TLC confirmed the presence of the dominant PEs in the *S. coronata* extract. The compounds were identified by means of a comparison between the retention parameters of the extract compounds and those of the model compounds (Figure 1).

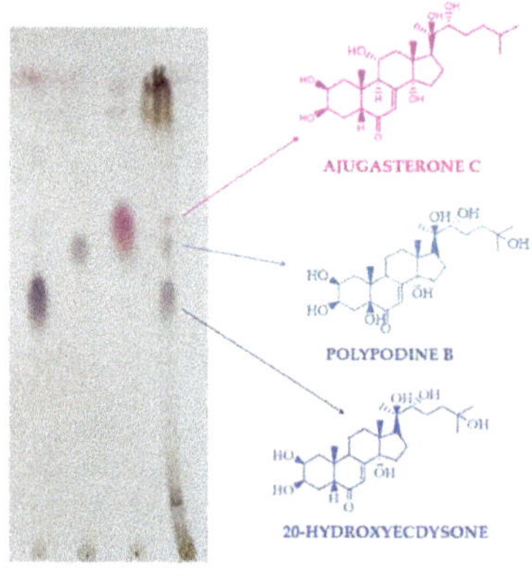

Figure 1. Thin layer-chromatography of the *S. coronata* extract: 1—20-hydroxyecdysone; 2—polypodine B; 3—ajugasterone C; 4—*S. coronata* extract.

2.1.2. Analysis of the Dominant Phytoecdysteroids in the Creams Using HPLC-ESI-MS

The compositions of the prepared cosmetics were as follows: Cream 0 (Lekobaza® and placebo), Cream 1 (Lekobaza® and *S. coronata* extract), Cream 2 (Lekobaza® and salicylic acid), and Cream 3 (Lekobaza®, *S. coronata* extract, and salicylic acid) (Figure 2. Altogether, 80 samples were prepared (20 samples of 30 g for each examined cream).

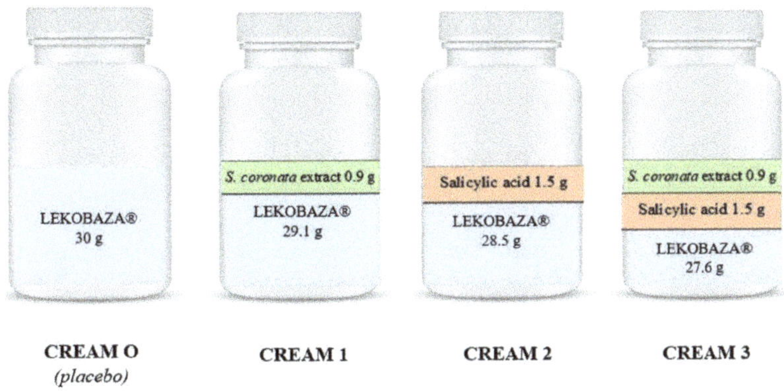

Figure 2. Compositions of the prepared creams.

Chromatograms (Figure 3) and mass spectra (Figure 4) were used to present the results of the qualitative analysis of Creams 1 and 3 using HPLC-ESI-MS. Three phytoecdysteroids (20-HE, ajugasterone C, and polypodine B) were found.

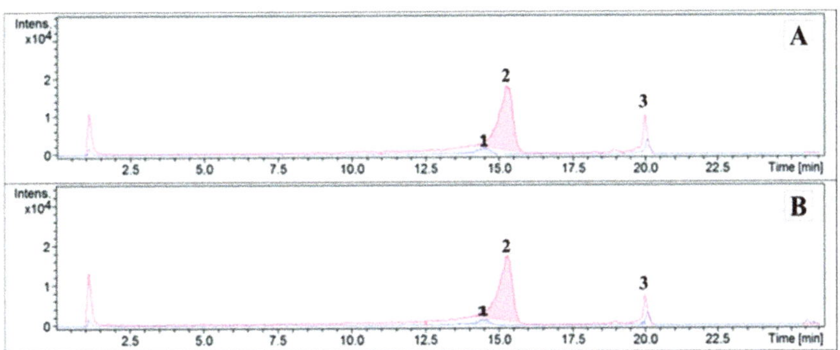

Figure 3. Chromatogram of Cream 1 (**A**) and Cream 3 (**B**) containing 3 wt.% *S. coronata*: (1) polypodine B; (2) 20-hydroxyecdysone; (3) ajugasterone C (red—first analysis of cream 1, blue—second analysis of cream 1).

The following were observed in the ESI-MS (+) spectra in the positive mode: the *m/z* 497.3092 ion (*m/z* represents the mass-to-charge ratio) was identified as the polypodine B molecule (Figure 4A); the *m/z* 481.3136 ion was attributable to 20-HE $[M\text{-}H]^+$ (Figure 4B); and the *m/z* 481,3136 $[M\text{-}H]^+$ ion was identified as the ajugasterone C molecule (Figure 4C). The presence of the subsequently observed ions; e.g., the *m/z* 445.2926 $[M\text{-}H\text{-}2H_2O]^+$ ion, was attributed to the loss of one hydroxyl group, although it may have been the result of a split between C-23 and C-24 at the side chain (481 > 407). The split was not visible in the spectra because it occurred immediately after the loss of two water molecules, thereby resulting in *m/z* 371.2200.

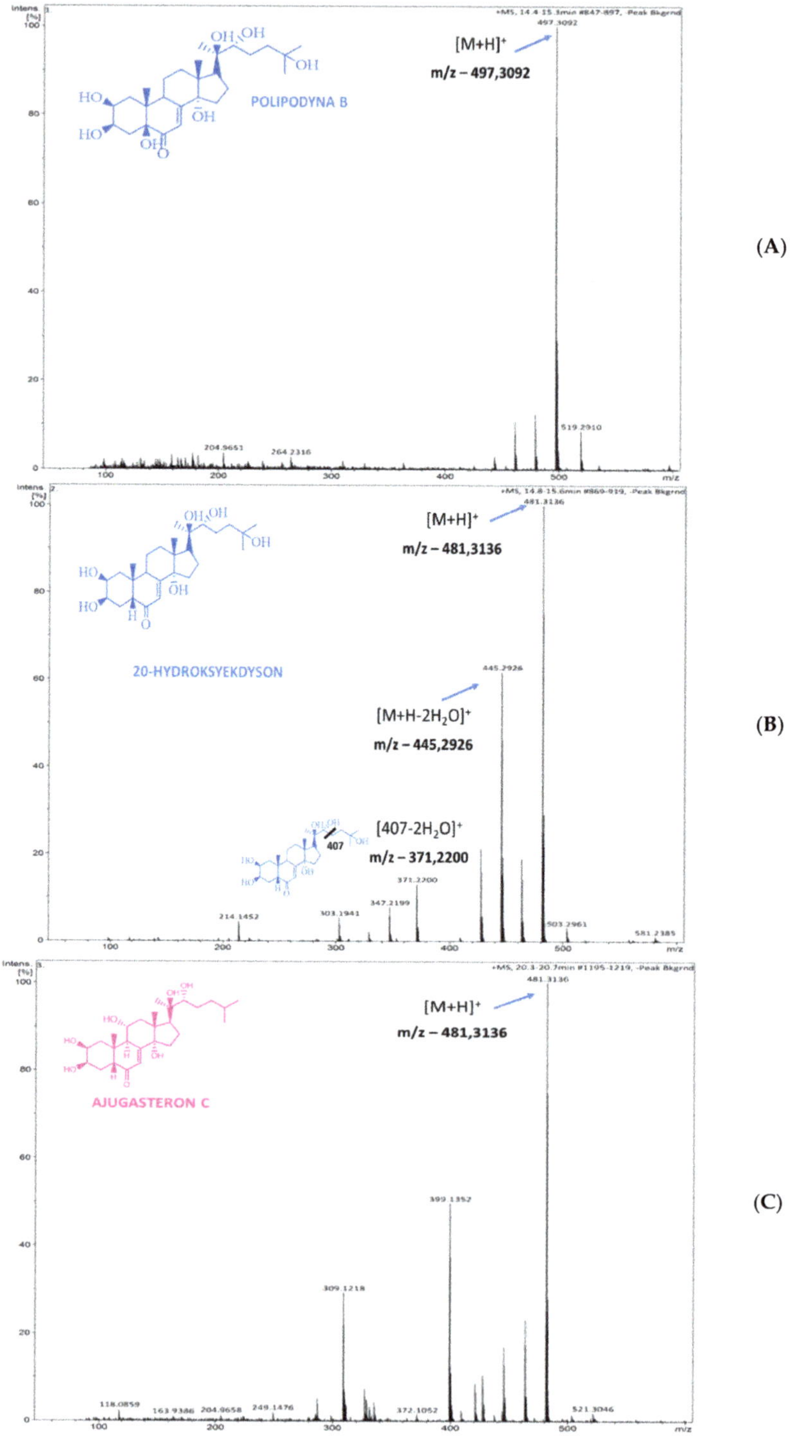

Figure 4. Polypodine B (**A**), 20-hydroxyecdysone (**B**), and ajugasterone C (**C**) in 3 wt.% *S. coronata* creams identified via ESI-MS.

2.2. *Physical and Chemical Analysis of the Creams*

Introduction of a cosmetic preparation to the market requires numerous tests such as physical and chemical analyses. The pH of a cosmetic product significantly affects the barrier function of the skin and its overall condition. The pH values vary depending on the chemical composition of a product and its purpose [11].

2.2.1. Chemical and Microbiological Stability of the Creams

Cosmetic products must remain safe up until their expiry date, which is determined by the manufacturer. Although the composition of the cosmetics is regulated by law, there are no legal requirements regarding stability tests. Most cosmetic preparations are tested for microbiological stability, and chemical stability is only tested only in some cases. Without testing for chemical stability, a manufacturer cannot be certain of a product's safety and stability throughout its entire shelf life. In accordance with the EU directive on cosmetic products, a safety protocol is required for each cosmetic formulation (annex I to the regulation) [12]. Cosmetic preparations are at risk for microbiological contamination during production and usage [13]. Microbes that have been isolated from the contaminated cosmetics include *Staphylococcus* (*S. aureus*, *S. epidermidis*, and *S. warneri*), *Escherichia coli*, and *Pseudomonas aeruginosa* [14]. Both the Food and Drug Administration (FDA) and the EU directive on cosmetic products require the microbial population to be low, stable, and free from harmful microorganisms [15]. The production process should be compliant with the ISO Good Manufacturing Practice (GMP) on the production, monitoring, and transport of cosmetic products to minimize the risk for microbial contamination. The effects of microbial contamination of a cosmetic product may range from moderate (change in color and/or texture) to severe (altered activity/toxicity) [16].

Chemical stability is another key element that ensures product safety. Several testing methods for assessing the chemical stability of cosmetics under predictable transport, storage, and usage conditions have been designed [17]. According to the guidelines of the International Council on Harmonization of Technical Requirements for Registration of Pharmaceuticals for Human Use (ICH) (International Cooperation on Pharmaceuticals), these factors include humidity, temperature, pH, the presence of oxidants, and exposure to light [18]. The so-called "stress tests," during which the product is exposed to the abovementioned factors, are the most important stability tests for active components. Degradation of the active substances in cosmetic formulations is associated with numerous chemical reactions that might occur under the influence of various environmental factors (i.e., hydrolysis, oxidation, isomerization, hydration, dimerization, or decarboxylation). In addition, photostability testing constitutes an integral part of the stability analysis [18].

Chemical Stability

The chemical stability of the creams was determined by exposing the samples to certain temperatures and humidity levels. Chemical stability was confirmed by using the qualitative analysis of the creams (Figures 5 and 6). PEs' presence in the samples after stability testing indicated that the compounds did not undergo degradation processes under the influence of the abovementioned environmental factors.

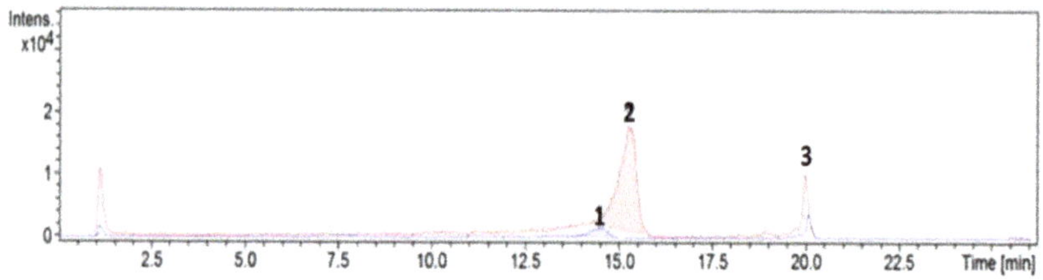

Figure 5. Chromatogram of Cream 1 with the *Serratula coronata* extract after 3 weeks in a climatic chamber: 1—polypodine B; 2—20-hydroxyecdysone; 3—ajugasterone C (red first analysis of Cream 1, blue second analysis of Cream 1).

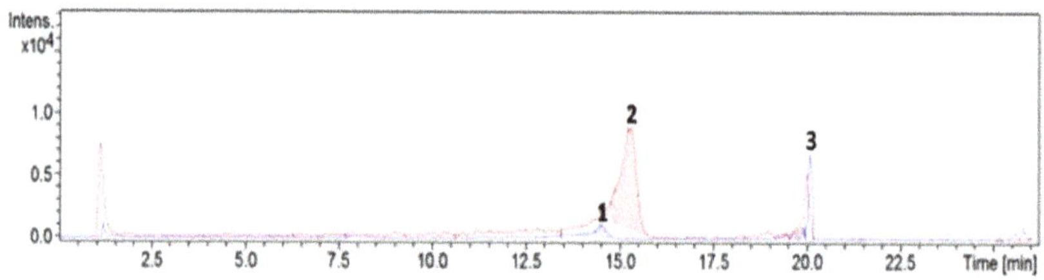

Figure 6. Chromatogram of Cream 3 with the *S. coronata* extract after 3 weeks in a climatic chamber: 1—polypodine B, 2—20-hydroxyecdysone, 3—ajugasterone C (red—first analysis of Cream 3, blue—second analysis of Cream 3).

Microbiological Purity

The microbiological acceptance criteria of non-sterile topicals depend on the microbial count in a sample. Their limits are based on the *Total Aerobic Count* (TAMC) and *Total Yeast/Mold Count* (TYMC). The acceptance criteria for microbiological quality are established by considering TAMC $\leq 1 \times 102$ CFU/g and TYMC $\leq 1 \times 101$ CFU/g (CFU = colony-forming unit) [19]. The results of microbiological testing are presented in Table 1.

Table 1. Results of microbiological testing of the creams.

Creams	Total Aerobic Count Test Result	Total Yeast and Mold Count Test Result	Method
0—Lekobaza®	<10 CFU/g	<10 CFU/g	According to the Polish Pharmacopoeia XII
1—Lekobaza® + *S. coronata* extract	<10 CFU/g	<10 CFU/g	
2—Lekobaza® + salicylic acid	<10 CFU/g	<10 CFU/g	
3—Lekobaza® + salicylic acid + *S. coronata* extract	<10 CFU/g	<10 CFU/g	

According to the Polish Pharmacopoeia XII, the acceptable germ counts for non-sterile topicals are 10^2 CFU for aerobic bacteria and 10^1 CFU for yeast and mold. Therefore, our results were indicative of microbiological stability [20].

Viscosity and pH

The analysis of the pH revealed an acidic reaction of the plant extract (pH 4.13 $\pm$ 0.05). The values of the cream viscosity and pH are presented in Table 2.

Table 2. Viscosities and pHs of the creams.

Creams	pH M ± SD	Viscosity (mPA × S) M ± SD
0—Lekobaza	5.03 ± 0.05	152,900 ± 33,517
1—Lekobaza® + *S. coronata* extract	4.30 ± 0.06	121,737 ± 1744
2—Lekobaza® + salicylic acid	3.07 ± 0.12	194,200 ± 17665
3—Lekobaza® + salicylic acid +*S. coronata* extract	2.47 ± 0.06	134,607 ± 2119

Abbreviations: SD, standard deviation; M, mean; mPA, minipascal; s, second.

The active component lowered the pH of the cosmetic medium (Lekobaza®), and salicylic acid increased the acidic pH of the cosmetic (Table 2). The extract, the pH of which was similar to that of the medium, caused a slight decrease in the pH of the cosmetic formulation. A physiologically slightly acidic pH of the stratum corneum (4.1–5.8) impacts the barrier function of the skin, synthesis and aggregation of the lipids, keratinocyte differentiation, and desquamation, as well as microbiome and antibody activation in the skin [21]. A low skin pH, especially an acidic pH of the stratum corneum, inhibits skin colonization with pathogens such as *Staphylococcus aureus* or *Streptococcus pyogenes* and supports normal microbial growth [22]. An increased skin pH has been observed in various skin dermatoses. In psoriasis, an increase in pH by 0.3–0.4 is associated with a drop in the hydrogen ion (H^+) concentration by half, which affects the biochemical processes of the skin. An elevated pH and increased skin dryness stimulate proteinase-activated receptor-2, which leads to pruritus in psoriatic patients because extra- and intra-cellular concentration of the protons modulates the afferent (pruritus and pain) and the efferent (growth, differentiation, and survival of the cells) functions [21]. Beneficial effects of the creams containing the *S. coronata* extract have been confirmed by applied research [1]. The method of decreasing the skin pH by using acidification has long been applied to wound healing because it increases anti-bacterial activity, changes protease activity and oxygen release, decreases the toxicity of the bacterial end products, and stimulates epithelization and angiogenesis [23].

Rheology plays a vital role in the determination of the stability of cosmetic products. Rheological properties such as viscosity significantly impact the quality of a cosmetic because they determine its consistency and spreadability on the skin [24,25]. The results of the viscosity tests are presented in Table 2. A high relative viscosity of a cosmetic formulation is perceived by the users as the equivalent of high concentration of the active substances and their effectiveness. Rheological properties are often selected to match the type of a cosmetic product and how it affects the skin. Cosmetics that regenerate the skin and soothe skin irritation (e.g., those designed for dry and flaky skin) should remain on the skin's surface for a prolonged time [25,26], and these are most often supplied as creams [27].

2.3. Permeability of 20-Hydroxyecdysone by Raman Spectroscopy in an In Vitro Model

The permeability factor of the active components significantly affects the effectiveness of a cosmetic product. Biologically active substances may be transported through the skin at three levels: topical (absorption), intracutaneous (penetration), and deep (resorption). Active substances need to pass the epithelial barrier and reach the deep layers of the skin to achieve the desired effect [28]. Active components may penetrate the skin via the intercellular and transcellular route as well as through skin appendages. The permeability of the active substances depends on the physical and chemical properties of the diffusion substance, the properties of the medium, and biological (overall condition and thickness of the skin, age and metabolism, and blood flow) and physical (temperature, time of day, and climate) factors [29,30]. The methods that allow the determination of the degree of skin penetration by the active components may be subdivided into two groups: quantitative methods and qualitative or semi-quantitative methods. The former includes diffusion chambers and the Parallel Artificial Membrane Permeability Assay (PAMPA) test; the latter includes microscopic and spectroscopic methods as well as combinations of the two. Quantitative in vitro tests are regularly performed to measure how much of an active component passed through the membrane over time versus the diffusion area associated with the amount of the active component obtained in the acceptor chamber. As far as qualitative tests are concerned, they are used to identify the active substance and will allow the determination of the presence or the relative amount of the active component in various skin layers [31]. A combination of the chemometric methods with spectroscopic imaging on transversal skin sections allows for the precise localization of the molecules in the layers of the skin [32].

In our study, that method was applied to determine the permeability of 20-HE, the dominant PE in the *S. coronata* creams. The test was conducted in an in vitro human skin model using Raman spectroscopy, an efficient spectroscopic technique that detects the vibrational modes of molecules. The obtained Raman spectra constituted highly specific spectroscopic molecule models, thereby allowing to identify them and monitor their passage through the skin. The method offers a non-invasive and dynamic technique to analyze the permeability of active substances through skin layers with a spatial resolution of a few micrometers [32].

According to the available literature, 20-HE remains the dominant ecdysteroid in both the plant and animal kingdoms [33,34]. Thus, a 3% 20-HE cream was composed for our study and investigated via confocal Raman spectroscopy (CRS) to measure the penetration rate of the 20-HE. Based on the CRS spectra, 1570 and 1612 cm^{-1} were identified as the most characteristic bands. Less specific bands were observed at positions 1265, 1369, and 1447 cm^{-1}. The CRS spectra for 20-HE and for the 3 wt.% 20-HE cream are presented in Figure 7A,B, respectively.

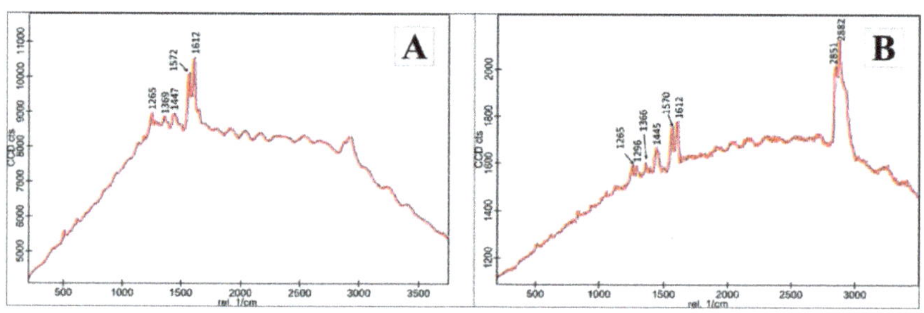

Figure 7. Raman spectra of 20-hydroxyecdysone (**A**) and 3 wt.% 20-hydroxyecdysone cream (**B**).

Raman maps plotting the distribution of the cream with 3 wt.% 20-HE in the cross-section of the skin for a photomicrograph taken in visible light are presented in Figure 8.

The obtained results of the CRS indicted that the transdermal penetration rate was 1200 μm (subcutaneous tissue).

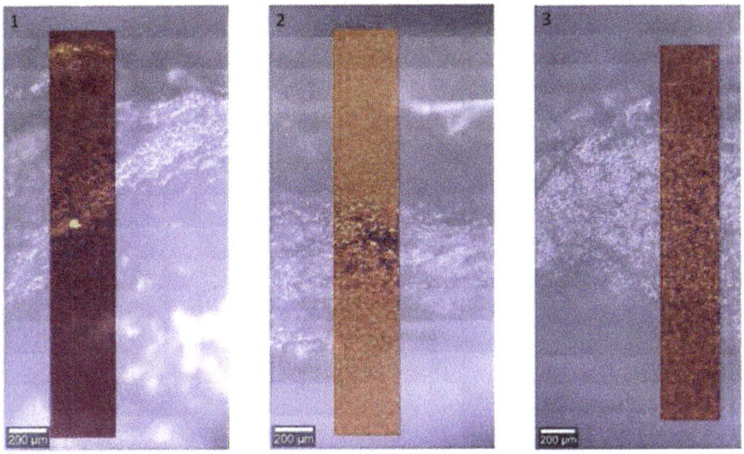

Figure 8. Representative Raman images of the distribution of 3 wt.% 20-hydroxyecdysone cream prepared on the basis of the integration one of the characteristic bands (1612 cm^{-1}) for each of the three biological replicates.

3. Materials and Methods

3.1. Standard and Chemicals

Standard 20-HE, ajugasterone C, and polypodine B were previously purified and characterized for *S. coronata* [35]. All solvents were of HPLC grade and all other chemicals were of analytical reagent grade, and they were used as received.

3.2. Material Extraction [35]

The fragmented and dried sample of *S. coronata* (2000 g) was soaked three times with 10,250 mL of ethanol (96%, POCH). The extract was filtered, and its solvent was evaporated using a rotary evaporator (Rotavapor® R-100, BUCHI, Flawil, Switzerland) at 40 °C. Next, the extract was concentrated to dryness at room temperature. A total of 136.42 g of the extract (extractum siccum) with a dark green color was obtained.

3.2.1. Analysis of the Chemical Composition of the Extract Using Thin-Layer Chromatography

The dominant PEs in the *S. coronata* extract were 20-HE, ajugasterone C, and polypodine B [9,35]. Thin-layer chromatography (TLC) on aluminum slates coated with silica gel (Merck, Germany) was used to confirm the presence of these compounds in the extract. A mixture (5:1) of dichloromethane (p.a., POCH) and methanol (p.a., POCH) was used as the mobile phase. The dishes were sprayed with p-anisaldehyde and heated at 105 °C for visualization [36]. The model compounds used in the TLC analysis (20-HE, ajugasterone C, and polypodine B) were established in our previous study [35].

3.2.2. Analysis of the Dominant Phytoecdysteroids in the S. Coronata Extract Using Liquid Chromatography–Electrospray Ionization–Mass Spectrometry (HPLC-ESI-MS)

HPLC-ESI-MS was used to detect the dominant PEs in the extract. Earlier analyses of the model compounds helped to establish the following retention times (RT): 20-HE—15.2 min, ajugasterone C—20.4 min, and polypodine B—14.8 min. The extract was diluted 1000-fold before testing. The measurements were conducted using a quadrupole time-of-flight mass spectrometer (QTOF, Impact HD, Bruker Daltonics, Bremen, Germany) in positive mode and a liquid chromatograph (LC) Ultimate 3000 (Thermo Scientific/Dionex). The LC conditions are presented in Table 3. We developed and validated a simple, sensitive, and precise method for the isolation and quantification of the dominant ecdysteroids—ajugasterone C, polypodine B, and 20-HE—from the *S. coronata* herb in an earlier study [35].

Table 3. LC conditions using a Kinetex 2.6μ C18 column (100 × 210 mm): temp. = 30 °C, injection volume = 10 μL, and mobile phase flow rate = 0.3 mL/min.

Time (min)	Mobile Phase	
	H_2O + 0.1% FA (%)	ACN + 0.1% FA (%)
0	90.0	10.0
1	90.0	10.0
12	87.5	12.5
14	87.5	12.5
20	55.0	45.0
22	55.0	45.0
24	10.0	90.0
26	10.0	90.0
28	90.0	10.0
35	90.0	10.0

Abbreviations: ACN, acetonitrile; FA, formic acid.

3.3. Cream Preparation

Four creams (0—placebo, 2—cream with salicylic acid, 1 and 3—creams with the extract; 30 g each) were prepared. The composition of the creams is presented in Figure 4. Lekobaza® Pharma Cosmetic (Fagron, Krakow, Poland) was used as the cream base and as the placebo. It is a multi-component medium with a pH of 5.5 containing white Vaseline, glycerol monostearate, and cetyl alcohol, Miglyol® 812, macrogol-20-glycerol-monostearate, propylene glycol, and purified water. It is a universal base that penetrates the skin easily, is less greasy, and creates a protective film on the skin's surface.

Salicylic acid was used as an ingredient in the creams due to its keratolytic properties and to enhance phytoecdysteroid penetration [37].

3.4. Viscosity and pH

The pH of the creams was determined using a VWR ORP15 PEN/PH 10 PEN pH-meter (VWR International, Gdansk, Poland) with a predefined setting range of 4–7 pH. All pH measurements were conducted at 21 °C. The cream viscosity was determined using a RheoTec RCO2 rotational viscometer with an adjustable coagulation threshold (Messtechnik GmbH, Ottendorf-Okrilla, Germany). The creams were placed in narrow glass beakers. Next, the appropriate rotational spindle was mounted and inserted. The measurements were conducted at 25 ± 2 °C using the R5 spindle for the creams with the extract (Creams 1 and 3) and the R7 spindle for Cream 0 at a rotational speed of 3 rpm. The viscosity and pH were measured three times and used to calculate the mean and standard deviation values.

3.5. Chemical Stability

Creams 1 and 3 were placed in the stability chamber (TH-ICH-300, Jeiotech Co., Ltd., Daejeon, Republic of Korea) for 21 days at a temperature of 45 °C and 75% relative humidity (RH) to determine the effects of humidity and temperature on the stability of the dominant PEs in the *S. coronata* extract [38]. Next, all samples were analyzed using the qualitative HPLC-ESI-MS method.

3.6. Microbiological Stability

The microbiological analysis of the creams was conducted in accordance with the method described in the *European Pharmacopoeia* (FP) [39]: 10 g of the preparation was mixed with sterile phosphate buffer (PB, Sigma-Aldrich, Darmstadt, Germany) + 2.5% Tween80 (Sigma-Aldrich, Darmstadt, Germany) to 100 mL (1:10 dilution). The sample was vigorously stirred for approximately 15 min and then further diluted using a 0.2% Tween80 phosphate buffer. Each of the four samples (1:10, 1:100, 1:1000, and 1:10,000 dilutions) was then transferred to a separate Petri dish using 1 ml of the homogenate each (stirred for 5 s before placement). Next, 15–20 mL of liquid medium—Tryptone-Soy-Agar (TSA, 2 dishes) and Sabouraud Dextrose Agar (SDA, 2 dishes)—at 40 °C were added. After mixing and medium solidification, the TSA cultures were incubated at 30–35 °C for 3–5 days and the SDA cultures at 20–25 °C for 5–7 days. Next, the aerobic bacteria, yeast, and mold counts per 1 g of the sample were calculated while taking into the account the dilution factor.

3.7. Analysis of the Dominant Phytoecdysteroids Using HPLC-ESI-MS

A qualitative HPLC-ESI-MS analysis was used to confirm the presence of the dominant PEs in the *S. coronata* creams. We also analyzed the preparations that had been tested in the climate chamber to determine the effects of temperature and humidity on the stability of these components.

3.7.1. Sample Preparation

Before the chromatographic assay, 5.0 g of Creams 1 and 3 were placed in 25 mL glass vials and soaked in 5 mL of 96% ethanol (POCH). The samples were stirred vigorously, placed in an ultrasonic washer (Bandelin Sonorex RK 255 H), and filtered through syringe filters (0.45 µm pore size). Next, 1 mL of the filtrate was measured and placed in a glass vial for the HPLC-ESI-MS analysis.

3.7.2. Conditions of the Analysis

The designed method was subjected to validation in accordance with the ICH guidelines [40].

A qualitative analysis of Creams 1 and 3 was conducted using the same equipment and the same measurement parameters as in the case of the extract analysis (Section 3.2.2).

3.8. Permeability of 20-Hydroxyecdysone as Measured by Raman Spectroscopy in An In Vitro Model

Skin penetration by 20-HE, the dominant compound of the *S. coronata* extract, was examined using confocal Raman spectroscopy: 1 g of 20-HE was dissolved in 5.6 mL of water with 26.6 g of Lekobaza® to prepare the emulsified cream and to analyze its penetration through the skin layers. The study using Raman spectroscopy was performed using a WITec Alpha 300 spectrometer equipped with a confocal microscope (WITec alpha300 R, Ulm, Germany), a TrueSurface attachment, and an electron-multiplying CCD (EMCCD) camera for ultra-fast and sensitive imaging. The measurements were taken with the use of excitation with frequency-doubled Nd:YAG laser line with a wavelength of 532 nm. The radiation power at the focal point was approximately 10 mW. The spectral resolution of the collected spectra was approximately 3 cm^{-1}. The measurements were made using an air objective with a magnification of 20× and a numerical aperture of 0.4. The measurements of the spectra of the samples were recorded for the following parameters: laser power—10 mW, number of accumulations—10, and accumulation time of a single spectrum—0.5 s. In case of mapping, the measurement procedure consisted of the sequential collection of individual spectra from a specific area each time the spectrum was measured over the entire spectral range; i.e., 0–3600 cm^{-1}, using the excitation line with a wavelength of 532 nm. The image measurements were recorded for the following parameters: laser power—10 mW, accumulation time of a single spectrum—0.15 s, and sampling density—10 µm. The analysis of the mapping results consisted of the integration of the observed bands in the spectra. The Raman spectra were collected in the range of 0–3600 cm^{-1} for 532 nm. The permeability analysis was performed on cross-sections through the layers of the skin after incubation with the samples. Skin samples were obtained from excess skin during abdominoplasty from healthy females. Samples were prepared in a 6-well plate with phosphate-buffered saline (PBS) to prevent tissue dehydration. Afterward, the skin was treated with the 3% 20-HE cream and incubated for 6 h at 37 °C (5% CO_2), and then the samples were mounted on slides for the Raman spectroscopy. All experiments were performed in three biological repetitions with each of them in three technical repeats.

4. Conclusions

Based on the results obtained, it was shown that the herb *Serratula coronata* is a source of ecdysone compounds that can be used for the development of formulations for skin care. All of the developed cosmetic formulations showed chemical and microbiological stability. In addition, the biological activity of 20-hydroxyecdysone can be related to its proven transdermal permeation, which was confirmed using Raman spectroscopy in an in vitro model. The regenerative properties of phytoecdysteroids, as already confirmed in our previous work, make cosmetic preparations containing *Serratula coronata* extracts useful for skin problems such as psoriasis.

Author Contributions: Conceptualization, J.G.-P. and M.P.; methodology, J.G.-P., M.P., A.K. and A.F.-G.; investigation, A.K., M.U., A.F.-G. and T.O.; data curation, A.K. and A.F.-G.; writing—original draft preparation, J.G.-P. and M.P.; writing—review and editing, A.K., A.F.-G., M.U., T.O., I.M. and J.N.; supervision, J.G.-P. and M.P.; project administration, A.K. and A.F.-G. All authors have read and agreed to the published version of the manuscript.

Funding: This research received no external funding.

Institutional Review Board Statement: Not applicable.

Informed Consent Statement: Not applicable.

Data Availability Statement: The data presented in this study are available upon request.

Conflicts of Interest: The authors declare no conflict of interest.

References

1. Kroma, A.; Pawlaczyk, M.; Feliczak-Guzik, A.; Urbańska, M.; Jenerowicz, D.; Seraszek-Jaros, A.; Kikowska, M.; Gornowicz-Porowska, J. Phytoecdysteroids from Serratula coronata L. for Psoriatic Skincare. *Molecules* **2022**, *27*, 3471. [CrossRef] [PubMed]
2. Srivastava, P.; Singh, M.; Devi, G.; Chaturvedi, R. Herbal Medicine and Biotechnology for the Benefit of Human Health. In *Animal Biotechnology: Models in Discovery and Translation*; Verma, A.S., Sing, A., Eds.; Elsevier Science Publishing Co. Inc.: San Diego, CA, USA, 2014; pp. 563–575.
3. Olennikov, D.N. Metabolites of *Serratula* L. and *Klasea* Cass. (Asteraceae): Diversity, Separation Methods, and Bioactivity. *Separations* **2022**, *9*, 448. [CrossRef]
4. Ivashchenko, I.; Rakhmetov, D.B. Biomorphological Features of *Serratula coronata* L. (Asteraceae) Introduced in ZHNAEU's Botanical Garden. *Mod. Phytomorphol.* **2016**, *10*, 69–80.
5. Das, N.; Mishra, S.K.; Bishayee, A.; Ali, E.S.; Bishayee, A. The Phytochemical, Biological, and Medicinal Attributes of Phytoecdysteroids: An Updated Review. *Acta Pharm. Sin. B* **2021**, *11*, 1740–1766. [CrossRef] [PubMed]
6. Laekeman, G.; Vlietinck, A. Phytoecdysteroids: Phytochemistry and Pharmacological Activity. In *Natural Products: Phyto-Chemistry, Botany and Metabolism of Alkaloids, Phenolics and Terpenes*; Ramawat, K.G., Mérillon, J.M., Eds.; Springer: Berlin/Heidelberg, Germany, 2013; pp. 3827–3849.
7. Ványolós, A.; Báthori, M. New Perspectives in the Analysis of Ecdysteroids: A Promising Group of Biologically Active Compounds. *Curr. Pharm. Anal.* **2008**, *4*, 162–175. [CrossRef]
8. Garg, A.; Sharma, R.; Dey, P. Analysis of Triterpenes and Triterpenoids. In *Recent Advances in Natural Products Analysis*; Silva, A.S., Nabavi, S.F., Saeedi, M., Nabavi, S.M., Eds.; Elsevier: Amsterdam, The Netherlands, 2020; pp. 393–426.
9. Odinokov, V.N.; Galyautdinov, I.V.; Nedopekin, D.V.; Khalilov, L.M.; Shashkov, A.S.; Kachalab, V.V.; Dinanc, L.; Lafont, R. Phytoecdysteroids from the juice of *Serratula coronata* L. (Asteraceae). *Insect Mol. Biol.* **2002**, *32*, 161–165. [CrossRef]
10. Ghosh, D.; Laddha, K.S. Extraction and Monitoring of Phytoecdysteroids Through HPLC. *J. Chromatogr. Sci.* **2006**, *44*, 22–26. [CrossRef]
11. Lukić, M.; Pantelić, I.; Savić, S.D. Towards Optimal pH of the Skin and Topical Formulations: From the Current State of the Art to Tailored Products. *Cosmetics* **2021**, *8*, 69. [CrossRef]
12. European Union. *Regulation (EC) No 1223/2009 of the European Parliament and of the Council of 30 November 2009 on Cosmetic Products*; European Union: Brussels, Belgium, 2009.
13. Michalek, I.M.; John, S.M.; Caetano Dos Santos, F.L. Microbiological Contamination of Cosmetic Products—Observations from Europe, 2005–2018. *J. Eur. Acad. Dermatol. Venereol.* **2019**, *33*, 2151–2157. [CrossRef]
14. Kim, H.W.; Seok, Y.S.; Cho, T.J.; Rhee, M.S. Risk Factors Influencing Contamination of Customized Cosmetics Made On-the-Spot: Evidence from the National Pilot Project for Public Health. *Sci. Rep.* **2020**, *10*, 1561. [CrossRef]
15. Choubey, S.; Godbole, S. Methods for Evaluation of Microbiological Safety, Guidelines Governing the Quality and Survey on Microbial Contamination of Commercial Cosmetic Products—A Review. *WJPMR* **2017**, *3*, 85–94.

16. Noor, R.; Zerin, N.; Das, K.K.; Nitu, L.N. Safe usage of cosmetics in Bangladesh: A quality perspective based on microbiological attributes. *J. Biol. Res. Thessalon.* **2015**, *22*, 10. [CrossRef] [PubMed]
17. Drechsel, D.; Towle, K.; Fung, E.; Novick, R.; Paustenbach, D.; Monnot, A. Chemical Stability Analysis of Hair Cleansing Conditioners under High-Heat Conditions Experienced during Hair Styling Processes. *Cosmetics* **2018**, *5*, 23. [CrossRef]
18. Jamrógiewicz, M.; Merchel, M. A History of the Physical and Chemical Stability of Pharmaceuticals—A Review. *Acta Pol. Pharm.* **2018**, *75*, 297–304.
19. Wirotesangthong, M. Microbiological Limits in Cosmetics. *Isan J. Pharm. Sci.* **2021**, *17*, 1–22. [CrossRef]
20. Registration of Medicinal Products, Medical Devices and Biocidal Products. *Polish Pharmacopoeia*, 12th ed.; Registration of Medicinal Products, Medical Devices and Biocidal Products: Warsaw, Poland, 2020; Volume I.
21. Proksch, E. pH in Nature, Humans and Skin. *J. Dermatol.* **2018**, *45*, 1044–1052. [CrossRef]
22. Chan, A.; Mauro, T. Acidification in the Epidermis and the Role of Secretory Phospholipases. *Dermato-Endocrinol.* **2011**, *3*, 84–90. [CrossRef]
23. Nagoba, B.S.; Suryawanshi, N.M.; Wadher, B.; Selkar, S. Acidic Environment and Wound Healing: A Review. *Wounds* **2015**, *27*, 5–11.
24. Davies, A.; Amin, S. Rheology of Cosmetic Products: Surfactant Mesophases, Foams and Emulsions. *J. Cosmet. Sci.* **2020**, *71*, 481–496.
25. Gräbner, D.; Hoffmann, H. Rheology of Cosmetic Formulations. In *Cosmetic Science and Technology: Theoretical Principles and Applications*; Sakamoto, K., Lochhead, R.Y., Maibach, H.I., Eds.; Elsevier Inc.: Amsterdam, The Netherlands, 2017; pp. 471–488.
26. Moravkova, T.; Stern, P. Rheological and Textural Properties of Cosmetic Emulsions. *Appl. Rheol.* **2011**, *21*, 1–6. [CrossRef]
27. Sarkar, D.K. Creams and Ointments. In *Pharmaceutical Emulsions: A Drug Developer's Toolbag*; Sarkar, D.K., Ed.; John Wiley & Sons: Hoboken, NJ, USA, 2013; pp. 69–76.
28. Kim, B.; Cho, H.-E.; Moon, S.H.; Ahn, H.-J.; Bae, S.; Cho, H.-D.; An, S. Transdermal Delivery Systems in Cosmetics. *Biomed. Dermatol.* **2020**, *4*, 10. [CrossRef]
29. Shabbir, M.; Ali, S.; Shahid, N.; Rehman, K.; Umair, A.; Moosa, R. Formulation Considerations And Factors Affecting Transdermal Drug Delivery System. *Int. J. Pharm. Integr. Sci.* **2014**, *2*, 20–35.
30. Marwah, H.; Garg, T.; Goyal, A.K.; Rath, G. Permeation enhancer strategies in transdermal drug delivery. *Drug Deliv.* **2016**, *23*, 564–578. [CrossRef] [PubMed]
31. Zsikó, S.; Csányi, E.; Kovács, A.; Budai-Szűcs, M.; Gácsi, A.; Berkó, S. Methods to Evaluate Skin Penetration In Vitro. *Sci. Pharm.* **2019**, *87*, 19. [CrossRef]
32. Essendoubi, M.; Gobinet, C.; Reynaud, R.; Angiboust, J.F.; Manfait, M.; Piot, O. Human Skin Penetration of Hyaluronic Acid of Different Molecular Weights as Probed by Raman Spectroscopy. *Skin. Res. Technol.* **2015**, *22*, 55–62. [CrossRef] [PubMed]
33. Buniam, J.; Chukijrungroat, N.; Rattanavichit, Y.; Surapongchai, J.; Weerachayaphorn, J.; Bupha-Intr, T.; Saengsirisuwan, V. 20-Hydroxyecdysone Ameliorates Metabolic and Cardiovascular Dysfunction in High-Fat-High-Fructose-Fed Ovariectomized Rats. *BMC Complement. Med. Ther.* **2022**, *20*, 140. [CrossRef]
34. Festucci-Buselli, R.A.; Contim, L.A.S.; Barbosa, L.C.A.; Stuart, J.; Otoni, W.C. Biosynthesis and Potential Functions of the Ecdysteroid 20-Hydroxyecdysone—A Review. *Canad. J. Bot.* **2008**, *86*, 978–987. [CrossRef]
35. Napierała, M.; Nawrot, J.; Gornowicz-Porowska, J.; Florek, E.; Moroch, A.; Adamski, Z.; Kroma, A.; Miechowicz, I.; Nowak, G. Separation and HPLC Characterization of Active Natural Steroids in a Standardized Extract from the S. Coronata Herb with Antiseborrheic Dermatitis Activity. *Int. J. Environ. Res. Public Health* **2020**, *17*, 6453. [CrossRef]
36. Nowak, G.; Nawrot, J.; Latowski, K. Arbutin in Serratula quinquefolia M.B. (Asteraceae). *Acta Soc. Bot. Pol.* **2009**, *78*, 137–140. [CrossRef]
37. Arif, T. Salicylic acid as a peeling agent: A comprehensive review. *Clin. Cosmet. Investig. Dermatol.* **2015**, *8*, 455–461. [CrossRef]
38. González-González, O.; Ramirez, I.O.; Ramirez, B.I.; O'Connell, P.; Ballesteros, M.P.; Torrado, J.J.; Serrano, D.R. Drug Stability: ICH versus Accelerated Predictive Stability Studies. *Pharmaceutics* **2022**, *14*, 2324. [CrossRef] [PubMed]
39. European Directorate for the Quality of Medicines EDQM. *European Pharmacopoeia*, 7th ed.; The European Directorate for the Quality of Medicines & HealthCare: Strasbourg, France, 2010; pp. 163–167, 519–520.
40. ICH. Q2 (R2): Validation of Analytical Procedures–Text and Methodology. In Proceedings of the International Conference on Harmonization (ICH), Geneva, Switzerland, 31 March 2022.

Disclaimer/Publisher's Note: The statements, opinions and data contained in all publications are solely those of the individual author(s) and contributor(s) and not of MDPI and/or the editor(s). MDPI and/or the editor(s) disclaim responsibility for any injury to people or property resulting from any ideas, methods, instructions or products referred to in the content.

cosmetics

MDPI

Article

Chemical and Rheological Characterization of a Facial Mask Containing an Olive Pomace Fraction

Raquel Rodrigues [1,†], Joana C. Lobo [1,†], Diana M. Ferreira [1], Ewa Senderowicz [1], M. Antónia Nunes [1], M. Helena Amaral [2,3,*], Rita C. Alves [1,*] and M. Beatriz P. P. Oliveira [1]

[1] REQUIMTE/LAQV, Department of Chemical Sciences, Faculty of Pharmacy, University of Porto, R. J. Viterbo Ferreira, 228, 4050-313 Porto, Portugal

[2] Associate Laboratory i4HB—Institute for Health and Bioeconomy, Faculty of Pharmacy, University of Porto, 4050-313 Porto, Portugal

[3] UCIBIO—Applied Molecular Biosciences Unit, MEDTECH, Laboratory of Pharmaceutical Technology, Department of Drug Sciences, Faculty of Pharmacy, University of Porto, 4050-313 Porto, Portugal

* Correspondence: hamaral@ff.up.pt (M.H.A.); rcalves@ff.up.pt (R.C.A.)

† These authors contributed equally to this work.

Abstract: Cosmetic interest in agro-industrial byproducts is growing. In fact, many studies have shown that these residues present bioactive compounds with several skincare applications. One example is olive byproducts, such as olive pomace, which has a composition rich in phenolic compounds. As the production of olive oil is increasing, the amount of byproducts being generated is escalating, with significant constraints in their safe disposal due to their phytotoxic nature. The present study aimed to, from a zero-waste perspective, characterize and add value to a sub-byproduct, a semi-solid paste (SSP) derived from a patent process of olive pomace extraction. The chemical analysis of this residue revealed high moisture and significant protein, fat, and ash contents. Furthermore, vitamin E total phenolics and flavonoid content were assessed, as well as antioxidant activity, using DPPH• (2,2-diphenyl-1-picrylhydrazyl radical) and FRAP (ferric reducing antioxidant power) methods. Based on this primary assessment, a facial mask with antioxidant properties was developed. Rheological analysis showed that the developed mask presented shear thinning behavior, thixotropy, and texture characteristics desirable for skincare use. The results of this study showed the successful incorporation of SSP into facial masks and provides a preliminary assessment of this byproduct's impact on the appearance and performance of these formulations.

Keywords: olive pomace; cosmetics; sustainability; facial mask; antioxidant; rheology

Citation: Rodrigues, R.; Lobo, J.C.; Ferreira, D.M.; Senderowicz, E.; Nunes, M.A.; Amaral, M.H.; Alves, R.C.; Oliveira, M.B.P.P. Chemical and Rheological Characterization of a Facial Mask Containing an Olive Pomace Fraction. *Cosmetics* **2023**, *10*, 64. https://doi.org/10.3390/cosmetics10020064

Academic Editor: Agnieszka Feliczak-Guzik

Received: 15 February 2023
Revised: 7 April 2023
Accepted: 13 April 2023
Published: 18 April 2023

Copyright: © 2023 by the authors. Licensee MDPI, Basel, Switzerland. This article is an open access article distributed under the terms and conditions of the Creative Commons Attribution (CC BY) license (https://creativecommons.org/licenses/by/4.0/).

1. Introduction

In recent years, sustainability has been a major driver of industry innovation, leading to new procedures and products while bearing in mind waste reduction and eco-friendly processes [1]. Considering the widespread use of skincare products in everyday life, cosmetic industry practices have a significant environmental impact [1]. Designing more sustainable products also complies with newer consumer concerns about environmental footprints. Sustainable development represents a pressing issue as the population increases and resources are depleted. The concept of a circular economy is based on the reduction/elimination of wastes through the upscaling of residues, thus adding value and expanding their life cycle [2]. In that sense, agri-food byproducts have been a recent focus of investigations. In fact, each year, byproducts represent billions of tons of residues that are discarded from various industries, posing a significant disposal problem [1]. Agri-food byproducts are promising sources of compounds with biological properties, representing renewable, low-cost, and sustainable raw materials [3].

One example of a circular economy in the agro-industrial sector is the olive byproduct valorization. The olive oil industry generates high amounts of waste, such as leaves, olive pomace (OP), olive stones, and olive mill wastewater (OMWW), particularly during the agricultural phase and oil production stage [4]. OP, the main residue of olive oil production, is a semi-solid residue that contains olive husk and pulp, crushed olive stones, and water, with a moisture content of more than 60% depending on the cultivation region and the employed extraction method [4–6]. OP also contains a substantial amount of compounds, such as dietary fiber, unsaturated fatty acids, minerals, and phenolic compounds [7]. In fact, this byproduct is particularly rich in phenolic compounds—such as hydroxytyrosol, oleuropein, tyrosol, and others—that, due to their polar character, remain in the residues [8]. Phenolic compounds have demonstrated biological activities with cosmetic potential, such as antioxidant, anti-aging, photoprotector, and anti-inflammatory activities [9]. In turn, olive pomace is a major ecological concern due to its high phenolic and organic load and low pH, contributing to potentially hazardous effects on soil and water [4]. Aside from bringing environmental benefits, byproduct recycling strategies can provide social and economic advantages, thus increasing the sustainability of the olive oil industry [4]. As olive oil has been used as a cosmetic and skin protector since ancient times, olive pomace application in skincare is promising. A literature review suggests there are few cosmetics developed based on other olive byproducts, such as olive leaves [5,10] or OMWW [11], with potential antioxidant, anti-aging, and photoprotector claims; however, to the best of our knowledge, there are no olive pomace-based topic formulations.

The olive pomace fraction used in this study is a secondary byproduct that resulted from the extraction of bioactive components through an eco-friendly physical process (PCT/IB2018/060111). After centrifugation, a semi-solid paste (SSP) was obtained. This extraction method did not involve the use of organic solvents and presented economic advantages when compared with emergent non-conventional extraction processes. This strategy also provided the particular advantage of retaining various functional compounds—namely, fatty acids, polysaccharides, minerals, and phenolics. Achieving synergy between these components can favor multiactivity cosmetic formulations, thus adding functional properties and improving skin benefits. Considering the results of the chemical characterization, a facial mask with moisturizing and antioxidant properties, containing SSP as an ingredient, was developed. Facial masks are increasingly popular as they are accessible, can be easily applied, show instant effects on the skin, and are available in different forms adjusted to consumer needs. Facial masks with clay minerals are frequently used in skincare regimens due to their cleansing properties and skin-toning effects. Other ingredients, such as vitamins, proteins, or substances with exfoliating or moisturizing properties, can also be added to facial masks [12]. In fact, anti-pollution skincare is one of the latest cosmetic trends and is based on ingredients and formulations that can, on one hand, purify and remove chemical pollutants from the skin, and on the other hand, possess biological activities that counteract the negative effects of airborne pollutants, particularly oxidative damage through the formation of reactive oxygen species (ROS) [13]. Air pollution, along with other exogenous factors, such as ultraviolet (UV) radiation and smoking, are recognized factors that lead to premature aging [13]. As a result of an imbalance between endogenous antioxidant production and ROS formation, oxidative damage to the skin cell components escalates [13]. Plant extracts are often rich in bioactive compounds, whose antioxidant and chelating activities can be exploited in anti-pollution formulations by reducing skin oxidative stress [13]. Phenolic compounds act as free radical scavengers preventing cell damage, thus delaying the skin aging process [14]. In that sense, this work presents a preliminary study on the valorization of olive pomace for a specific cosmetic application within a circular economy context.

2. Materials and Methods

2.1. Standards and Reagents

Folin-Ciocalteu's reagent, sodium carbonate, ferrous sulfate, 2,4,6-tri-(2-pyridyl)-S-triazine (TPTZ), ferric chloride, sodium nitrite, aluminum chloride, sodium hydroxide, 2,2-diphenyl-1-picrylhydrazyl radical ($DPPH^\bullet$), 6-hydroxy2,5,7,8-tetramethylchroman-2-

carboxylic acid (Trolox), gallic acid, catechin, and butylated hydroxytoluene, were acquired from Sigma-Aldrich (St. Louis, MO, USA). Absolute ethanol was obtained from Fisher Chemical (Loughborough, England). Kjeldahl catalyst tablets, sulfuric acid, boric acid, potassium hydroxide, anhydrous sodium sulfate, and n-hexane (HPLC grade) were obtained from Merck (Darmstadt, Germany). Tocol (2-methyl-2-(4,8,12-trimethyl-tridecyl) chroman-6-ol) was obtained from Matreya Inc. (State College, PA, USA). Vitamin E standards were received from Calbiochem (La Jolla, CA, USA). For mask formulation, bentonite, glycerin, and panthenol were acquired from Acofarma (Madrid, Spain), and ethanol was purchased from Aga (Prior Velho, Portugal). Water was purified in a Milli-Q system (Millipore, Bedford, MA, USA). All other reagents were of an analytical grade.

2.2. Methods

2.2.1. Sample and Sample Preparation

Fresh olive pomace, collected immediately after olive oil production, was kindly provided by different olive oil producers (two samples from the North of Portugal, Trás-os-Montes, and two from the South, Alentejo). The samples were equally mixed in order to be representative of the Portuguese panorama regarding the main areas of olive oil production. A sample of the final olive pomace (200 g) was subjected to a patented process of extraction (PCT/IB2018/060111). Briefly, the sample was pressed using a hydraulic press (30 min), and the obtained semi-liquid fraction was collected. This fraction was further centrifuged (5000 rpm; 20 min), the supernatant was discarded for other applications, and the obtained pellet (SSP) was freeze-dried (−80 °C, 0.015 mbar; Cryodos, Telstar, Barcelona, Spain) for further characterization and mask development (Figure 1).

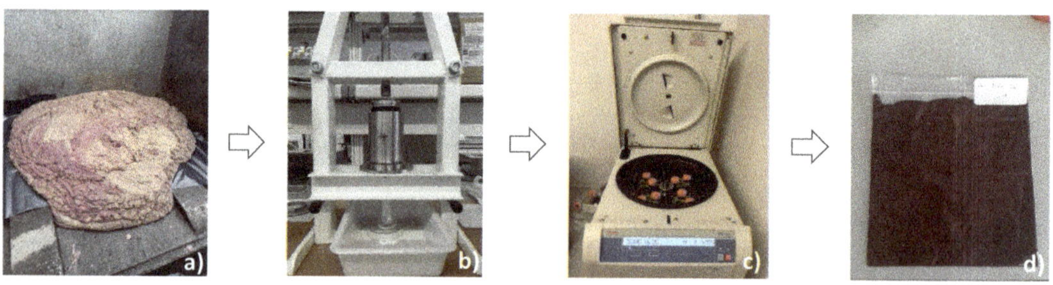

Figure 1. Summary of the steps employed to obtain the semi-solid paste (SSP) from fresh olive pomace. (**a**) fresh olive pomace (**b**) hydraulic press; (**c**) centrifugation; (**d**) SSP.

2.2.2. Proximate Analysis

Sample moisture was determined using an infrared balance (Scaltec model SMO01, Scaltec Instruments, Heiligenstadt, Germany). Total fat and total protein were determined using Soxhlet and Kjeldahl methods, respectively, according to AOAC standard methods [15]. The total protein content was calculated using 6.25 as the nitrogen conversion factor [16]. Ash content was determined through incineration at 500 °C (Thermolyne 48000 Furnace, Barnstead/Thermolyne, Dubuque, IA, USA), following AOAC standard methods [15]. Each analysis was performed in triplicate.

2.2.3. Extracts Preparation

A sample aliquot (150 mg of the SSP or 2 g of the mask) was extracted with 50 mL of ethanol/water (80%/20%). The mixture was continuously stirred at 600 rpm in a heating plate (MS-H-S10 10-Channel Classic Magnetic Hotplate Stirrer, DLAB Instruments Ltd., Beijing, China) at 40 ± 1 °C, for 60 min. Then, it was filtered (Whatman No. 4) and stored at −20 °C until analysis. The extraction was performed in triplicate for the evaluation of antioxidant activity and phenolic and flavonoid content.

2.2.4. Antioxidant Activity

Ferric Reducing Antioxidant Power (FRAP)

The FRAP assay was carried out according to Costa et al. [17]. Briefly, using a microplate, 35 μL of a diluted extract was mixed with 265 μL of the FRAP reagent (0.3 M of acetate buffer, 10 mM of TPTZ solution, and 20 mM of ferric chloride). The mixture was incubated at 37 °C for 30 min and protected from light. Absorbance was measured at 595 nm (Synergy HT Microplate Reader, BioTek Instruments, Inc., Winooski, VT, USA). The analysis was performed in triplicate, and the results were expressed as g of ferrous sulfate equivalents (FSE) per 100 g.

2,2-diphenyl-1-picrylhydrazyl radical ($DPPH^•$) Scavenging Ability

The radical scavenging potential was assessed according to Costa et al. [18]. In brief, 270 μL of an ethanolic $DPPH^•$ solution (6.0×10^{-5} mol/L) was mixed with 30 μL of the extract. The absorbance decrease was measured every 2 min, at 515 nm, in order to observe the kinetic reaction (Synergy HT Microplate Reader, BioTek Instruments, Inc., Winooski, VT, USA). The reaction endpoint was reached at 20 min. The analysis was performed in triplicate, and the results were expressed as g of Trolox equivalents (TE) per 100 g.

2.2.5. Phenolic and Flavonoid Content

Total phenolic content was determined using spectrophotometry (Synergy HT Microplate Reader, BioTek Instruments, Inc., Winooski, VT, USA), according to Costa et al. [18] with minor modifications. Briefly, 30 μL of the extract was mixed with 150 μL of the Folin–Ciocalteu reagent (1:10). Then, 120 μL of sodium carbonate (7.5% *m*/*v*) was added, and the mixture was incubated at 45 °C for 15 min and protected from light. After 30 min at room temperature, absorbance was measured at 765 nm. The analysis was performed in triplicate, and the results were expressed as g of gallic acid equivalents (GAE) per 100 g.

Total flavonoids were estimated using a colorimetric assay based on Costa et al. [17]. In brief, 1 mL of the extract was mixed with 4 mL of deionized water and 300 μL of 5% sodium nitrite. After 5 min at room temperature, 300 μL of aluminum chloride 10% was added to the previous solution. After an additional 1 min at room temperature, 2 mL of sodium hydroxide and 2.5 mL of deionized water were added. The solution was mixed, and the absorbance was read at 510 nm (Microplate reader Synergy HT, BioTek Instruments, Inc., Winooski, VT, USA). The analysis was performed in triplicate, and the results were expressed as g of catechin equivalents (CE) per 100 g.

2.2.6. Vitamin E Profile

For vitamin E analysis, the SSP oil was first extracted using a cold extraction procedure, according to Alves et al. [19]. Briefly, an antioxidant (0.1% BHT, 75 μL), an internal standard (100 μg/mL of tocol; 50 μL), and absolute ethanol (1 mL) were added to a 200 mg sample, being homogenized for 30 min in an orbital vortex mixer (VV3, VWR Int., Lutterworth, UK) with multiple sample supports. Then, 2 mL of n-hexane was added, and the solution was mixed again for 30 min. After that, 1 mL of NaCl 1% (*m*/*v*) was added. The mixture was vortexed and centrifuged (2 min, 5000 rpm; Labofuge Ae, Heraeus Sepatech, Germany), and the organic phase was collected. The residue was re-extracted twice with 2 mL of n-hexane, and the organic phases were combined. Then, a sufficient amount of anhydrous sodium sulfate was added to the final extract. The mixture was vortexed and centrifuged to collect the upper layer. The solvent was evaporated under nitrogen until reaching a volume of 500 μL. The chromatographic analysis was carried out using an HPLC system (Jasco, Tokyo, Japan) equipped with a multiwavelength diode array detector (MD-2015) that was coupled with an FP-2020 fluorescence detector (Jasco, Tokyo, Japan) and programmed for excitation at 290 and emission at 330 nm. The chromatographic separation of the compounds was achieved using a normal phase Supelcosil™ LC-SI column (75 mm × 3.0 mm, 3.0 μm; Supelco, Bellefonte, PA, USA). The vitamin E vitamers were identified based on their UV spectra and by comparing their retention times with those of existing standards. Quantification was obtained using the internal standard method of fluorescence signals.

2.2.7. Facial Mask Formulation

A control base formulation (without SSP) and a base with 5% of SSP as an ingredient were developed. Powders (SSP and bentonite) were mixed in a mortar, and panthenol was added as a moisturizer. The mask vehicle was prepared by mixing water, ethanol, and glycerin at room temperature. Then, the vehicle was slowly added to the solid mixture through manual stirring until reaching complete homogenization.

Clay minerals, such as bentonite, are used in cosmetics—not only as rheology modifiers but also in facial masks—due to their high absorbency capabilities, in order to clean impurities, such as excess sebum or toxins [20,21]. Ethanol, due to its lower vapor pressure compared to water, is frequently used as a drying agent that controls the application time of facial masks, as these should dry quickly after being applied [12]. The correct balance between ethanol and glycerin (glycerin has hygroscopic proprieties, which delay mask drying) contributes to the acceptability and efficiency of the mask formula. The detailed composition of each mask is presented in Table 1.

Table 1. Composition of facial masks (%, *w*/*w*): the control and the 5% SSP-containing mask.

Composition (INCI)	Control	SSP Mask
Bentonite	38.0	33.0
Glycerin	23.1	23.1
Alcohol	23.1	23.1
Panthenol	10.0	10.0
SSP	-	5
Purified water (*Aqua*)	5.8	5.8

2.2.8. pH Measurement

Measurements of the pH were carried out in dispersions (1%, *w/w*) of masks 15 days after preparation, in neutral water, using a potentiometer (Metrohm 827 pH lab, Herisau, Switzerland).

2.2.9. Texture and Rheological Analysis

The evaluation of texture and viscosity was performed 15 days after the preparation of the masks (base and 5% SSP mask) and storage at room temperature while protected from light.

The evaluation of the texture parameters (firmness and adhesiveness) was performed using a texturometer, Stable Micro Systems TA-XT2i (Haslemere, UK). The following test conditions were used: compression mode, cylindrical probe (13 mm, P/0.5), penetration distance of 5 mm, test speed of 3 mm/s, trigger force of 0.049 N, and load cell of 5 kg. The viscosity and the rheological behavior of the facial masks were evaluated using a rotational viscometer, Thermo Haake VT550 (Karlsruhe, Germany). Measurements were obtained at progressively higher shear rates (up to 500 s^{-1}) to obtain the ascending curve, and the procedure was repeated in reverse with progressively slower rates to obtain the descending curve. All measurements were made at room temperature (20.00 $\pm$ 0.04 °C), in triplicate, for each analyzed sample.

2.2.10. Statistical Analysis

Data are reported as mean $\pm$ standard deviation. Statistical analyses were performed using the statistical package IBM SPSS v 26 for Windows (SPSS Inc., Chicago, IL, USA). Student's *t*-test was used to discriminate between any two groups under consideration.

3. Results and Discussion

3.1. Proximate Analysis

Table 2 shows the results for the SSP chemical composition. The proximate analysis demonstrated a high moisture content (68.1%) and significant protein and fat contents (9.9 % and 9.6% in dry weight (dw), respectively). A previous study by Nunes et al. [22], regarding the chemical characterization of olive pomace, revealed lower values for these parameters—namely, 5.8% of fat and 7.4% of protein, expressed in dw. The ash content in SSP was also notably higher, with 5.2% dw versus olive pomace with 1.9% dw being mentioned in the cited work. In fact, olive pomace contains crushed olive kernels, which contributed to the weight of the sample, whereas olive the SSP used in this study was a smooth paste without kernels. This could be a contributing factor to the differences observed. Further, geographical and cultivation differences can account for the variations observed in chemical composition [4]. The samples in this work resulted from a mixture of two different geographical locations, whereas the sample used in the study of Nunes et al. [22] was derived from only one olive oil producer. The observed chemical features of SSP can be of interest in cosmetic applications. As demonstrated by Nunes et al. [22], the olive pomace oil fraction is essentially rich in monosaturated fatty acids, such as oleic acid. These substances provide structural stability to cell membranes and contribute to skin barrier integrity. Further, their emollient properties increase the hydration and softness of the skin. The unsaponifiable fraction also contains squalene, a natural component of human sebum, which has emollient qualities, as well as an effect as a skin barrier against solar radiation and antioxidant proprieties. Minerals have hygroscopic characteristics and are part of the natural moisturizing factor, which is essential to the stratum corneum hydration plasticity and homeostasis [4].

Table 2. Proximate analysis of the semi-solid paste (SSP) obtained from fresh olive pomace.

Moisture	Total Fat	Total Protein	Ash
68.1 ± 0.2	9.6 ± 1.3	9.9 ± 0.9	5.2 ± 0.1

Results expressed in dry weight (dw), g per 100 g of sample, as mean ± standard deviation.

3.2. Vitamin E Profile

In a previous study by Nunes et al. [22], it was evidenced that the vitamin E profile of olive pomace was comprised of the following vitamers: α-tocopherol, β-tocopherol, α-tocotrienol, and γ-tocopherol. α-Tocopherol was the major form (7.64 mg/100 g, dw), while the other vitamers were present in lower amounts. In the SSP, the same vitamers were also observed and, similarly, α-tocopherol was found to be the major vitamin E form (6.83 mg/100 g dw). α-Tocopherol is the most biologically active form of vitamin E, having a major role in the prevention of lipid peroxidation and the scavenging of lipid peroxyl radicals; hence, its use in cosmetics formulations is extensive [22].

3.3. Phytochemicals and Antioxidant Activity

Plants often produce secondary metabolites with antioxidant characteristics as a response to environmental stressors, such as ultraviolet (UV) radiation or high temperatures, in order to preserve physical and metabolic integrity [3]. Olive phenolics have been widely studied for their health-promoting properties, especially oleuropein, hydroxytyrosol (HT), and tyrosol [9]. HT, in particular, is the main phenolic present in olive pomace and has been reported as a powerful antioxidant with anti-aging proprieties, including collagen metabolism influence, radical scavenging activity, and melanogenesis inhibition [6,23,24]. Flavonoids are bioactive polyphenols found in fruit and vegetables with extended nutraceutical, pharmaceutical, and cosmetic applications, due to their antioxidative, anti-inflammatory, anti-mutagenic, and anti-carcinogenic properties [25]. The flavonoids described in olive pomace include quercetin, apigenin, rutin, luteolin glucoside, luteolin, dismetin, taxifolin, and apigenin glucoside [26]. The results for total phenolic content (TPC), total flavonoids (TF), and antioxidant activity are presented in Table 3. The content of phenolic compounds in olive pomace paste was 3.57 g/100 g dw. TPC was also evaluated in the masks, with values of 0.01 g/100 g for the control and 0.17 g/100 g for the 5% SSP-containing mask, showing a significant improvement in this parameter due to SSP incorporation. As phenolic compounds show low stability regarding environmental conditions, such as exposure to light, oxygen, and temperature, chemical assessment through stability testing should be performed in order to guarantee antioxidant efficacy during shelf-life [27].

Table 3. Phytochemical analysis, vitamin E content, and antioxidant activity of the semi-solid paste (SSP) obtained from the fresh olive pomace.

TPC (g GAE/100 g)	TF (g CE/100 g)	Vitamin E (mg/100 g)		DPPH• Inhibition (g TE/100 g)	FRAP (g FSE/100 g)
3.57 ± 0.15	3.18 ± 0.17	Total	7.04 ± 0.10	4.95 ± 0.28	6.79 ± 0.32
		α-tocopherol	6.83 ± 0.11		
		β-tocopherol	0.08 ± 0.01		
		γ-tocopherol	0.12 ± 0.01		
		α-tocotrienol	0.01 ± 0.01		

Results expressed in dry weight, as mean ± standard deviation. GAE, gallic acid equivalents; CE, catechin equivalents; TE, Trolox equivalents; FSE, ferrous sulfate equivalents.

3.4. *Facial Masks' Organoleptic Characterization*

The masks developed in this study presented a homogenous appearance and similar green to grey coloring across both types. The formulations exhibited an opaque appearance with semi-solid consistency and no particular smell. Figure 2 shows the two masks' visual aspects: (A) a base mask (without SSP) as a control, and (B) the mask with 5% of the SSP incorporated.

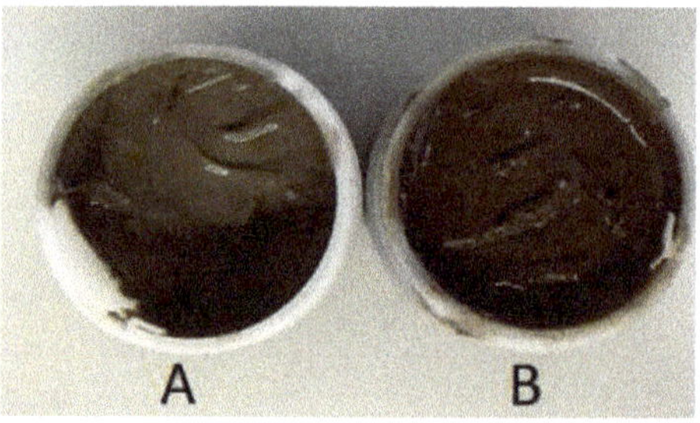

Figure 2. Control mask (**A**) and OPP mask (**B**).

3.4.1. pH Measurement

The pH of the control and SPP-based masks were both considerably high: 9.65 ± 0.01 and 9.39 ± 0.01, respectively. The pH values of cosmetic products are important to maintain optimal skin balance. The acidic skin surface pH (4.0–4.5) is a key factor in its barrier function influencing various factors, such as the composition of stratum corneum lipids and hydration, as well as the resident microflora [28]. Increased skin surface pH may be associated with the pathogenesis or the severity of many skin disorders, including acute eczema, irritant contact dermatitis, atopic dermatitis, ichthyosis, acne vulgaris, and Candida albicans infections [28]. For topical application, suitability formulations should comply with natural skin pH, presenting an acidic to neutral pH [28]. As bentonite has alkaline properties, the addition of citric or lactic acid could enhance skin compatibility by lowering the formulation pH. It is worth mentioning that the SSP presented in the mask slightly lowered the pH, probably due to the acidic characteristics of phenolic and organic acids present in olive pomace.

3.4.2. Textural and Rheological Analysis

Effortless spreading and product adherence to the skin are important characteristics to consider when formulating cosmetic masks [29]. Once applied, the mask layer should remain in place and not drip. Rheology and texture analyses are valid tools to assess the impact of raw materials on the final formulation as well as quality control since these parameters can provide essential information about the structure, which can influence shelf-life as well as sensory and processing attributes [30].

Figure 3 shows the rheograms of the SSP-based mask and the control after 15 days of storage at room temperature. At the highest shear rate (500 s^{-1}), the control presented a lower apparent viscosity compared to the SSP mask (901.7 Pa.s vs. 1164 Pa.s). However, the increase in the shear rate produced a reduction in the viscosity of both masks, therefore displaying a shear thinning behavior. Formulations with this behavior improve application and spreading, thus providing a pleasant sensory feeling and skin coverage [31]. It is also possible to observe the presence of a hysteresis loop in both masks, thus indicating that these formulations are thixotropic materials. Thixotropy corresponds to a decrease in viscosity with time, which is also desirable in topical formulations because it helps to maintain the suspending components' stability and indicates a more cohesive structure [31]. Thixotropy was more pronounced in the base mask, which can indicate that SSP inclusion had an impact on the mask structure. In fact, SSP was incorporated by replacing 5% of bentonite in the base formula; since bentonite composition provides better holding water capacity, this fact should also be accounted for in the changes observed. It should also be noted that, since the SSP-containing mask and the respective control are suspensions, powders tend to deposit over time and separate from the vehicle [32]. Further, powders, such as bentonite, tend to hydrate progressively during storage, which can alter the viscosity and hinder spreadability. Therefore, further studies should include stability testing in order to predict shelf-life and ensure that the formulation's physical characteristics remain constant.

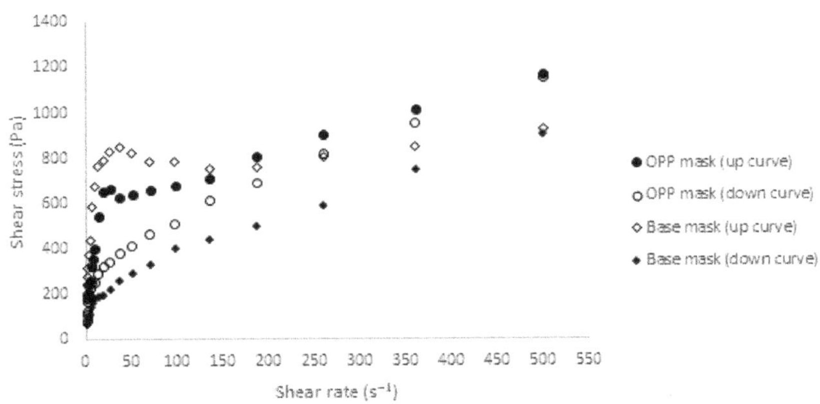

Figure 3. Rheograms, obtained for the control and 5% SSP-containing mask, at 20.00 ± 0.04 °C.

Figures 4 and 5 present the graphs corresponding to the evaluation, respectively, of the maximum force (firmness) and negative area (adhesiveness) of facial masks after 15 days of storage at room temperature. Similar values of adhesiveness were observed in both masks Less adhesive facial masks may slide during application time, reducing their effectiveness and having a negative impact on the consumer experience; thus, these results favor SSP incorporation. Regarding firmness, the SSP-based mask presented higher values (0.329 ± 0.008 N) than the base mask (0.221 ± 0.011 N).

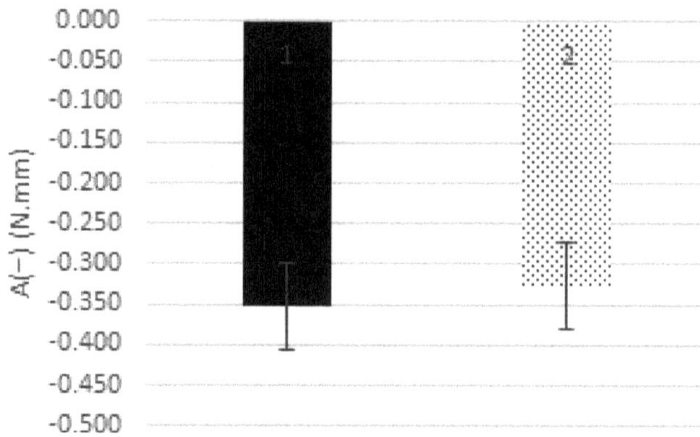

Figure 4. Texture analysis of the facial masks. (**1**) mask containing 5% of SSP; (**2**) control. Adhesiveness (A(−)) is represented in the bars as mean ± standard deviation. No significant differences were found between samples ($p > 0.05$).

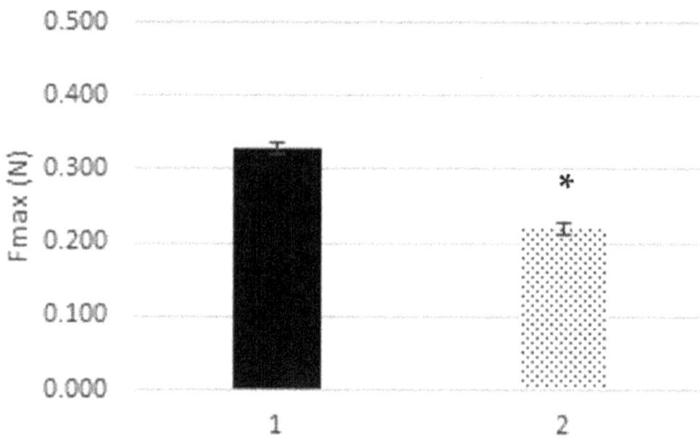

Figure 5. Texture analysis of the facial masks. (**1**) mask containing 5% of SSP; (**2**) control. Firmness (Fmax) is represented in the bars as mean ± standard deviation. *, Significant differences ($p < 0.05$) were found between samples.

4. Conclusions

This work highlights the potential of olive pomace to obtain an innovative ingredient (a semi-solid paste) to be used in facial mask formulations, as a sustainable source of both lipid and polar bioactive compounds. Indeed, the lipid fraction of this new ingredient is a source of antioxidants (such as vitamin E) and hydrosoluble compounds (such as phenolics) that protect the skin against oxidative stress. The 5% SSP-containing mask showed a 17-fold increase in total phenolic content compared to the base mask. This study also demonstrated that the SSP facial mask had a pleasant visual aspect, with no odor and good spreadability. This mask shows a pseudoplastic behavior with thixotropy, which are important characteristics for application and consumer acceptance. However, the SSP inclusion had an impact on the mask formulation, reducing thixotropy, which can be predictive of loss of stability. As this was a preliminary study, some important studies are missing. Further development should be focused on stability testing regarding microbiological, chemical, and physical properties to assess the efficacy, safety, and cosmetic quality of these types of facial masks.

Author Contributions: Conceptualization, J.C.L., M.A.N. and M.B.P.P.O.; methodology, D.M.F., M.A.N., R.C.A. and M.H.A.; validation, M.A.N. and M.H.A.; formal analysis, J.C.L., D.M.F. and E.S.; investigation, J.C.L., E.S., D.M.F. and M.A.N.; resources, M.H.A. and M.B.P.P.O.; writing—original draft preparation, R.R. and E.S.; writing—review and editing, R.C.A., M.H.A. and M.B.P.P.O.; visualization, R.R., E.S. and R.C.A.; supervision, M.H.A. and M.B.P.P.O.; project administration, M.B.P.P.O.; funding acquisition, M.B.P.P.O. All authors have read and agreed to the published version of the manuscript.

Funding: This work was funded by Fundação para a Ciência e Tecnologia and Ministério da Ciência, Tecnologia e Ensino Superior (FCT/MCTES) through projects UIDB/50006/2020 and UIDP/50006/2020. This work was also supported by the AgriFood XXI I&D&I project (NORTE-01-0145-FEDER-000041) co-financed by the European Regional Development Fund (ERDF) through NORTE 2020 (Programa Operacional Regional do Norte 2014/2020).

Institutional Review Board Statement: Not applicable.

Informed Consent Statement: Not applicable.

Data Availability Statement: Not applicable.

Acknowledgments: D.M.F. thanks to FCT/MCTES and European Social Fund (ESF) through NORTE 2020 (Programa Operacional Região Norte) for her PhD grant ref. 2022.13375.BD. R.C.A. extends thanks to FCT for funding through the Scientific Employment Stimulus Individual Call (Ref. CEECIND/01120/2017).

Conflicts of Interest: The authors declare no conflict of interest.

References

1. Mellou, F.; Varvaresou, A.; Papageorgiou, S. Renewable Sources: Applications in Personal Care Formulations. *Int. J. Cosmet. Sci.* **2019**, *41*, 517–525. [CrossRef] [PubMed]
2. Romani, A.; Pinelli, P.; Ieri, F.; Bernini, R. Sustainability, Innovation, and Green Chemistry in the Production and Valorization of Phenolic Extracts from *Olea europaea* L. *Sustainability* **2016**, *8*, 1002. [CrossRef]
3. Hoang, H.T.; Moon, J.-Y.; Lee, Y.-C. Natural Antioxidants from Plant Extracts in Skincare Cosmetics: Recent Applications, Challenges and Perspectives. *Cosmetics* **2021**, *8*, 106. [CrossRef]
4. Rodrigues, F.; Pimentel, F.B.; Oliveira, M.B.P.P. Olive by-products: Challenge Application in Cosmetic Industry. *Ind. Crops Prod.* **2015**, *70*, 116–124. [CrossRef]
5. Nunes, A.; Goncalves, L.; Marto, J.; Martins, A.M.; Silva, A.N.; Pinto, P.; Martins, M.; Fraga, C.; Ribeiro, H.M. Investigations of Olive Oil Industry By-Products Extracts with Potential Skin Benefits in Topical Formulations. *Pharmaceutics* **2021**, *13*, 465. [CrossRef]
6. Nunes, M.A.; Pimentel, F.B.; Costa, A.S.G.; Alves, R.C.; Oliveira, M.B.P.P. Olive By-Products for Functional and Food Applications: Challenging Opportunities to Face Environmental Constraints. *Innov. Food Sci. Emerg. Technol.* **2016**, *35*, 139–148. [CrossRef]
7. Ribeiro, T.B.; Oliveira, A.; Coelho, M.; Veiga, M.; Costa, E.M.; Silva, S.; Nunes, J.; Vicente, A.A.; Pintado, M. Are Olive Pomace Powders a Safe Source of Bioactives and Nutrients? *J. Sci. Food Agric.* **2021**, *101*, 1963–1978. [CrossRef]
8. Granados-Principal, S.; Quiles, J.L.; Ramirez-Tortosa, C.L.; Sanchez-Rovira, P.; Ramirez-Tortosa, M.C. Hydroxytyrosol: From Laboratory Investigations to Future Clinical Trials. *Nutr. Rev.* **2010**, *68*, 191–206. [CrossRef]
9. Nunes, A.; Marto, J.; Gonçalves, L.; Martins, A.M.; Fraga, C.; Ribeiro, H.M. Potential Therapeutic of Olive Oil Industry By-Products in Skin Health: A Review. *Int. J. Food Sci. Technol.* **2021**, *57*, 173–187. [CrossRef]
10. Kesente, M.; Kavetsou, E.; Roussaki, M.; Blidi, S.; Loupassaki, S.; Chanioti, S.; Siamandoura, P.; Stamatogianni, C.; Philippou, E.; Papaspyrides, C.; et al. Encapsulation of Olive Leaves Extracts in Biodegradable PLA Nanoparticles for Use in Cosmetic Formulation. *Bioengineering* **2017**, *4*, 75. [CrossRef]
11. Galanakis, C.M.; Tsatalas, P.; Galanakis, I.M. Implementation of Phenols Recovered from Olive Mill Wastewater as UV Booster in Cosmetics. *Ind. Crops Prod.* **2018**, *111*, 30–37. [CrossRef]
12. Nilforoushzadeh, M.A.; Amirkhani, M.A.; Zarrintaj, P.; Salehi Moghaddam, A.; Mehrabi, T.; Alavi, S.; Mollapour Sisakht, M. Skin Care and Rejuvenation by Cosmeceutical Facial Mask. *J. Cosmet. Dermatol.* **2018**, *17*, 693–702. [CrossRef] [PubMed]
13. Juliano, C.; Magrini, G. Cosmetic Functional Ingredients from Botanical Sources for Anti-Pollution Skincare Products. *Cosmetics* **2018**, *5*, 19. [CrossRef]
14. Liu, J.K. Natural Products in Cosmetics. *Nat. Prod. Bioprospect.* **2022**, *12*, 40. [CrossRef] [PubMed]
15. AOAC. *Official Methods of Analysis of AOAC International*, 19th ed.; AOAC International: Gaithersburg, MD, USA, 2010.
16. Tontisirin, K. Chapter 2: Methods of food analysis. In *Food Energy: Methods of Analysis and Conversion Factors: Report of a Technical Workshop*; Food and Agriculture Organization of the United Nations: Rome, Italy, 2002; pp. 7–17.
17. Costa, C.S.G.; Nunes, M.A.; Almeida, A.A.; Santos-Silva, A.; Oliveira, M. Nutritional, Chemical and Antioxidant/Pro-Oxidant Profiles of Silverskin, a Coffee Roasting By-Product. *Food Chem.* **2018**, *267*, 28–35. [CrossRef]

18. Costa, A.S.G.; Alves, R.C.; Vinha, A.F.; Barreira, S.V.P.; Nunes, M.A.; Cunha, L.M.; Oliveira, M.B.P.P. Optimization of Antioxidants Extraction from Coffee Silverskin, a Roasting By-Product, Having in View a Sustainable Process. *Ind. Crops Prod.* **2014**, *53*, 350–357. [CrossRef]
19. Alves, R.C.; Casal, S.; Oliveira, M.B.P.P. Determination of Vitamin E in Coffee Beans by HPLC Using a Micro-extraction Method. *Food Sci. Technol. Int.* **2009**, *15*, 57–63. [CrossRef]
20. Carretero, M.I. Clay Minerals and Their Beneficial Effects Upon Human Health. A Review. *Appl. Clay Sci.* **2002**, *21*, 155–163. [CrossRef]
21. Benson, H.A.E.; Roberts, M.S.; Leite-Silva, V.R.; Walters, K. *Cosmetic Formulation: Principles and Practice*; CRC Press: Boca Raton, FL, USA, 2019; p. 230.
22. Antónia Nunes, M.; Costa, A.S.G.; Bessada, S.; Santos, J.; Puga, H.; Alves, R.C.; Freitas, V.; Oliveira, M. Olive Pomace as a Valuable Source of Bioactive Compounds: A Study Regarding its Lipid- and Water-Soluble Components. *Sci. Total Environ.* **2018**, *644*, 229–236. [CrossRef]
23. Jeon, S.; Choi, M. Anti-Inflammatory and Anti-Aging Effects of Hydroxytyrosol on Human Dermal Fibroblasts (HDFs). *Biomed. Dermatol.* **2018**, *2*, 21. [CrossRef]
24. Uchida, R.; Ishikawa, S.; Tomoda, H. Inhibition of Tyrosinase Activity and Melanine Pigmentation by 2-Hydroxytyrosol. *Acta Pharm. Sin. B* **2014**, *4*, 141–145. [CrossRef] [PubMed]
25. Panche, A.N.; Diwan, A.D.; Chandra, S.R. Flavonoids: An Overview. *J. Nutr. Sci.* **2016**, *5*, e47. [CrossRef]
26. Nunes, M.A.; Palmeira, J.D.; Melo, D.; Machado, S.; Lobo, J.C.; Costa, A.S.G.; Alves, R.C.; Ferreira, H.; Oliveira, M. Chemical Composition and Antimicrobial Activity of a New Olive Pomace Functional Ingredient. *Pharmaceuticals* **2021**, *14*, 913. [CrossRef] [PubMed]
27. Paini, M.; Aliakbarian, B.; Casazza, A.A.; Lagazzo, A.; Botter, R.; Perego, P. Microencapsulation of Phenolic Compounds From Olive Pomace Using Spray Drying: A study of operative parameters. *LWT Food Sci. Technol.* **2015**, *62*, 177–186. [CrossRef]
28. Barel, A.O.; Paye, M.; Maibach, H.I. *Handbook of Cosmetic Science and Technology*, 3rd ed.; Informa Healthcare Inc.: New York, NY, USA, 2009; pp. 221–232.
29. Klimaszewska, E.; Zieba, M.; Gregorczyk, K.; Markuszewski, L. Application of Blue Honeysuckle Powder Obtained by an Innovative Method of Low-Temperature Drying in Skincare Face Masks. *Molecules* **2021**, *26*, 7184. [CrossRef]
30. Semenzato, A.; Costantini, A.; Meloni, M.; Maramaldi, G.; Meneghin, M.; Baratto, G. Formulating O/W Emulsions with Plant-Based Actives: A Stability Challenge for an Effective Product. *Cosmetics* **2018**, *5*, 59. [CrossRef]
31. Di Mambro, V.M.; Fonseca, M.J. Assays of Physical Stability and Antioxidant Activity of a Topical Formulation Added with Different Plant Extracts. *J. Pharm. Biomed. Anal.* **2005**, *37*, 287–295. [CrossRef]
32. da Silva Favero, J.; dos Santos, V.; Weiss-Angeli, V.; Gomes, L.B.; Veras, D.G.; Dani, N.; Mexias, A.S.; Bergmann, C.P. Evaluation and Characterization of Melo Bentonite Clay for Cosmetic Applications. *Appl. Clay Sci.* **2019**, *175*, 40–46. [CrossRef]

Disclaimer/Publisher's Note: The statements, opinions and data contained in all publications are solely those of the individual author(s) and contributor(s) and not of MDPI and/or the editor(s). MDPI and/or the editor(s) disclaim responsibility for any injury to people or property resulting from any ideas, methods, instructions or products referred to in the content.

cosmetics

MDPI

Article

In Vitro Investigation of the Cytotoxic and Antiproliferative Effects of *Haberlea rhodopensis* Total Extract: A Comparative Study

Martina I. Peeva [1], Maya G. Georgieva [1], Aneliya A. Balacheva [1], Atanas Pavlov [2,3,*] and Nikolay T. Tzvetkov [1]

[1] Department of Biochemical Pharmacology & Drug Design, Institute of Molecular Biology "Roumen Tsanev" Bulgarian Academy of Sciences (BAS), Acad. G. Bonchev Str., bl. 21, 1113 Sofia, Bulgaria; martina.iv.peeva@gmail.com (M.I.P.); mgeorgieva@bio21.bas.bg (M.G.G.); neli_bal@abv.bg (A.A.B.); ntzvetkov@bio21.bas.bg (N.T.T.)
[2] Faculty of Technologies, University of Food Technologies, 26 Maritza Blvd., 4002 Plovdiv, Bulgaria
[3] R&D Department, Innova BM Ltd., 12 B, Stefan Karadja Str., 1000 Sofia, Bulgaria
* Correspondence: at_pavlov@yahoo.com or pavlov@innovabm.com

Abstract: *Haberlea rhodopensis* Friv., known also as Rhodope silivryak and the Orpheus flower, is a Balkan endemic "resurrecting" plant belonging to the *Gesneriaceae* family. In folk medicine, the leaves of *Haberlea rhodopensis* Friv. were widely used to treat wounds and some infectious diseases of stock such as foot-and-mouth disease and hoof rot, while the herb of *Haberlea rhodopensis* Friv. is still used to cleanse the stomach, liver, kidneys, and blood vessels. Because of the content of myconoside, during the last decade, *Haberlea rhodopensis* Friv. extracts have been recognized as valuable cosmetic ingredients. In the present study, we aim to (i) evaluate the cytotoxic and antiproliferative activity of two herb extracts of *Haberlea rhodopensis* Friv. that are commercially used for the preparation of cosmetic ingredients on different cancer cells, with one normal cell line used as a reference, and (ii) compare the investigated effects with those observed for the reference anticancer, non-selective compound doxorubicin. Herein, we observed a decrease in the inhibitory activity of both extracts compared to those of doxorubicin against all tested cell lines. However, the myconoside-enriched *Haberlea rhodopensis* Friv. plant Extract 2 (designated also as M2) showed increased inhibitory activity (cytotoxicity and antiproliferative effects) compared to the *Haberlea rhodopensis* Friv. plant Extract 1 (designated also as E1). Moreover, the *Haberlea rhodopensis* Friv. plant Extract 2 showed a significant increase in cytotoxicity (at 24 h) and antiproliferative activity (at 48 and 72 h post-treatment) at its highest-tested concentration of 100 µg/mL compared to *Haberlea rhodopensis* Friv. plant Extract 1.

Keywords: antiproliferation; cisplatin; cytotoxicity; doxorubicin; flavonoid antioxidants; *Haberlea rhodopensis* Friv.

Citation: Peeva, M.I.; Georgieva, M.G.; Balacheva, A.A.; Pavlov, A.; Tzvetkov, N.T. *In Vitro* Investigation of the Cytotoxic and Antiproliferative Effects of *Haberlea rhodopensis* Total Extract: A Comparative Study. *Cosmetics* **2024**, *11*, 46. https://doi.org/10.3390/cosmetics11020046

Academic Editor: Agnieszka Feliczak-Guzik

Received: 12 January 2024
Revised: 16 February 2024
Accepted: 6 March 2024
Published: 21 March 2024

Copyright: © 2024 by the authors. Licensee MDPI, Basel, Switzerland. This article is an open access article distributed under the terms and conditions of the Creative Commons Attribution (CC BY) license (https://creativecommons.org/licenses/by/4.0/).

1. Introduction

Haberlea rhodopensis Friv. (HR) is a plant of the *Gesneriaceae* family with typical lilac flowers that grows in stony areas and is endemic in the Rhodope Mountains of the Thracian regions of Bulgaria and Greece (Figure S1A). *H. rhodopensis* Friv. belongs to a group of resurrection plants able to withstand prolonged drought periods, tolerating desiccation and quickly resuming growth upon rehydration [1,2]. Due to its growth in limited places, HR is a protected species in Greece and Bulgaria [3].

Primary investigations of Dell'Acqua and Schweikert proved skin benefits of HR extracts that are myconoside-rich [4]. In recent years, HR has been intensively studied in terms of its broad antimigratory and anticancer effects, antioxidant and free radical-scavenging activity, radioprotective and immunostimulatory effects, as well as antibacterial and anti-aging efficacy [5]. Therefore, the attention of researchers is focused on the investigation of the potential applications of HR in phytotherapy, human and veterinary medicine, and cosmetics [5,6]. Extensive investigations showed that different HR extracts may contain high levels of flavonoid antioxidants, including phenolic acids (e.g., ferulic, staric, caffeic, *p*-coumaric, sinapinic acid, and others), flavonoid aglycones, and glycosides (such as luteolin, hesperidin, quercetin, myricitin, rutin, and others) [7–9]. However, spectroscopic analyses using (U)HPLC-MS showed that two glycosides—myconoside and paucifloside—together with three other hispidulin-flavone C-glycopyranosides, represent the main constituents of different ethanolic *H. rhodopensis* Friv. extracts (HREs) [5]. The three new flavone C-glycosides are hispidulin 8-C-(2-O-syringoyl glucopyranoside), hispidulin 8-C-(6-O-acetyl glucopyranoside), and hispidulin 8-C-(6-O-acetyl-2-O-syringoyl gluco-pyranoside), which were also isolated and reported by another scientific group [9]. Noteworthy, alcohol extracts from HR were investigated to show antioxidant, antiviral, antibacterial, and antifungal activities [10]. In addition, the potent antioxidant and hepatoprotective effects of myconoside were recently demonstrated. The chemical structures and IUPAC names of myconoside and paucifloside are presented in Figure S1B.

Recent experiments show that among the components with biological activity found in HR, myconoside has potent antioxidant and hepatoprotective effects [11]. Together with the phenolic content, these compounds significantly contribute to the antioxidant properties of HREs [7]. In particular, ethanolic extracts of HR leaves have been reported to exhibit vital antimicrobial and antioxidant activities, reduce the clastogenic effect of γ-irradiation, and exert in vivo anticlastogenic and antimutagenic potential against the anticancer drug cyclophosphamide [12–15].

Phenolic acids accumulated in high amounts in the resurrection plants possess therapeutic properties due to their ability to capture free radicals and decrease oxidative stress [9]. Berkov and colleagues reported their study on the polar and apolar fractions of methanol HREs by gas chromatography-mass spectrometry and they have identified five free phenolic acids, namely syringic, vanillic, caffeic, dihydrocaffeic, and *p*-coumaric [16]. Furthermore, another study demonstrated that in alcohol HRE, the most abundant phenolic acids were the sinapic, ferulic, caffeic, and p-coumaric acids, as well as at least five other phenolic acids, which, although not so abundant, were still present in the extracts [7].

HR is a rare and threatened plant and, therefore, its harvesting from nature is prohibited. In vitro cultivation of plants is a promising technology for the sustainable production of bioactive plant metabolites. On this basis, Bulgarian company Innova BM Ltd. developed in vitro cultures of HR with different degrees of differentiation that are used for industrial sources of water–alcoholic extracts [17]. These extracts are the subject of the current investigation.

In the present study, we aim to (i) evaluate the cytotoxic and antiproliferative effects of two herb extracts of HR on different cancer cell lines, with one normal cell line used as a reference, and (ii) compare the investigated effects with those observed for the reference anticancer, non-selective compound doxorubicin.

2. Materials and Methods

2.1. *Haberlea rhodopensis Friv. In Vitro Culture Extracts*

Haberlea rhodopensis Friv. in vitro culture extracts were provided by Innova BM Ltd., Bulgaria https://innovabm.com/ (accessed on 30 January 2020). The extracts were produced by the company's proprietary technology by ethanol extraction of *Haberlea rhodopensis* Friv. in vitro culture biomass [17]. The myconoside content was 115 mg/g dry extract and 215 mg/g dry extract for E1 and M2 extracts, respectively.

2.2. *Preparation of Stock Solutions*

Two extracts (indicated throughout as Extract 1 or E1 at 5.0 mg/mL in ddH_2O stock, and myconoside-enriched Extract 2 or M2 at 10.0 mg/mL in ddH_2O stock) were tested for their cytotoxic and antiproliferative effects, respectively, after 24-, 48- and 72-h treatment of three cancer and one non-cancer (reference) cell lines. The stock solutions of both HR extracts were stored at −20 °C in the dark and tempered before use. The experiments were carried out on triple-negative, epithelial human breast adenocarcinoma (MDA-MB-231; ATCC; Manassas, VA, USA), human colorectal adenocarcinoma (HT-29; ATCC, USA), human hepatocellular carcinoma (HepG2; ATCC, USA), and mouse embryonic fibroblasts (3T3/L1; ATCC, USA) cells. Cells were incubated with serial dilutions (0.01–100 μg/mL) of both extracts using a modified MTT assay [18,19]. The anticancer, nonselective agent doxorubicin (Key Organics Ltd., Camelford, UK) (100 mM stock in DMSO) was tested in the concentration range of 0.001–100 μM (DMSO concentration $\leq$ 0.05% v/v) and was included in each experiment as a reference substance. For the dissolution of the insoluble formazan, a lysis solution containing N,N-dimethylsulfoxide (DMSO, Sigma-Aldrich, St. Louis, MO, USA) was used. The tests were performed in triplicate (eight replications per test) and the results presented as the mean % of the untreated controls ± SD from three independent experiments (n = 3).

2.3. Preparation of Cell Cultures

Human breast adenocarcinoma (MDA-MB-231), human colorectal adenocarcinoma (HT-29), human hepatocellular carcinoma (HepG2), and mouse embryonic fibroblasts (3T3/L1, reference non-cancer cells) cells were cultured in Dulbecco's modified Eagle's medium (DMEM, Gibco, Vienna, Austria) in the presence of 10% fetal bovine serum (FBS; Gibco, Austria), penicillin (100 U/mL) and streptomycin (0.1 mg/mL) solution (Gibco, USA). All cells were cultivated under an atmosphere of CO_2 (5.0%) at 37 °C and passaged by trypsinization with trypsin-EDTA (Greiner, Pleidelsheim, Germany) at a confluence of approximately 80%. The experiments were performed with cells in the exponential phase of growth (at a density of 5000 cells/well) using 96-well flat-bottom plates at a final volume of 100 μL/well. Cells were incubated overnight before the addition of doxorubicin and/or tested extracts.

2.4. Determination of Cell Viability

The cytotoxicity and antiproliferation of doxorubicin (1.0 mM stock solution in DMSO), Extract 1 (E1, 10.0 mg/mL stock solution in ddH_2O), and Extract 2 (M2, 5.0 mg/mL, stock solution in ddH_2O) were assessed in HepG2, MDA-MB-231, and HT-29 cancer cell lines, as well as in 3T3 mouse embryonic fibroblasts cells used as a reference cell line by colorimetric assay, applying 3-(4,5-dimethylthiazol-2-yl)-2,5-diphenyltetrazolium bromide (MTT) (VWR, Darmstadt, Germany), as previously reported [20].

The MTT test allows for determining the linear dependency between metabolically active cells (viable cells) and the measured color intensity of the purple-colored formazan solution, which can be quantified by spectrophotometric measuring at a certain wavelength (usually 550 or 570 nm). The obtained information is then used to assess the change in cells, e.g., death and/or proliferation. The loss of intensity of the purple color is directly associated with the loss of viable cells in the presence of a cytotoxic compound (agent). The quantity of in situ-formed formazan product can be determined spectrophotometrically by the measurement of the absorption/optical density (Amax/DO), after solubilization with a lysis solution (usually with DMSO, Sigma-Aldrich, St. Louis, MO, USA). The measured OD/Amax corresponds to the number (in %) of viable cells after a certain incubation period (24, 48 or 72 h) with the tested substance.

2.5. MTT Assay

The MTT assay was performed with slight modifications according to the literature [18,20]. Cells were seeded in 96-well culture plates with a density of 5×10^3 cells per well. Following adherence, the cells were treated with test extracts (0.01, 0.1, 1.0, 5.0, 10, 25, 50, and 100 μg/mL) and the reference compound doxorubicin (0.01, 0.1, 1.0, 2.5, 5.0, 10, 25, 50, and 100 μM), and further incubated for 24, 48 or 72 h at 37 °C (under 5.0% CO_2 atmosphere). After the respective incubation period, 10 μL MTT solution (5.0 mg/mL) per well was added and cells were incubated for a further 180 min. Then, the medium was removed and a lysing solution containing DMSO was added to each well. The plates were then shaken at room temperature until the completed dissolution of the purple crystalline product (formazan). The quantification of the produced formazans after the reduction of MTT was monitored using a microplate ELISA reader VarioscanTM LUX (Thermo Fisher Scientific Inc., Waltham, MA, USA) at 550 and 570 nm. The determined cytotoxicity/antiproliferation is expressed as percentage cell viability using the following equation:

$$\%\text{Cell viability} = (\text{ODsample} - \text{ODblank})/(\text{ODcontrol} - \text{ODblank}) \times 100 \quad (1)$$

In this equation, OD sample, OD blank and OD control are the measured absorption of the test, blank, and control sample. The results were presented as the mean % of the untreated controls ± SD from three independent experiments ($n = 3$).

2.6. Statistical Analysis

The statistical analysis and the representative graphs thereof were carried out using GraphPad Prism 6.0 (GraphPad Software, La Jolla, CA, USA). The respective half maximal inhibitory concentration (IC_{50}) values were obtained by non-linear regression analysis. The inhibitory curves were built using the log[inhibitor] vs. normalized response—Variable slope equation with least squares fit. Since the measured inhibitory concentration (growth inhibitory activity) of the tested Extract 1 and 2 is given in μg/mL, the IC_{50} values of doxorubicin (initially obtained as μM) were further converted into their respective μg/mL units using the following equation:

$$\mu g/mL = \mu M \times \text{MWdoxo } (g/mol \text{ or } \mu g/\mu mol)/1000 \quad (2)$$

For further simplification, the inhibitory activity (IC_{50} values) of doxorubicin is given in μg/mL in order to better compare the respective inhibitory activity for both tested extracts (see Table S1). The experimentally estimated IC_{50} values are obtained from at least three independent experiments and given with their standard deviation ($IC_{50} \pm SD$). For the one-way ANOVA test, p values ≤ 0.05 were considered as statistically significant.

3. Results

In order to evaluate cell viability, a modified MTT assay was used [19]. This method is one of the most used to evaluate the viability of different cancer and non-cancer cells in the presence of tested (screened) new compounds (naturally occurring and synthetic molecules). The MTT assay is a colorimetric test based on an enzyme–catalytic reduction of 3-(4,5-dimethylthiazol-2-yl)-2,5-diphenyltetrazolium bromide (MTT) (a yellow tetrazolium salt dye) by Nicotinamide adenine dinucleotide phosphate (NAD(P)H) to its insoluble purple crystalline form formazan. The assay is used for indirectly assessing cell viability based on the capacity of viable cells to reduce the tetrazolium dye MTT in the presence of mitochondrial NAD(P)H-dependent cellular oxidoreductase enzymes [19]. Therefore, the MTT test can also be used to measure cytotoxicity (loss of viable cells), usually after 24 h treatment with an investigated substance, or cytostatic activity (antiproliferation), usually after 48 and/or 72 h treatment with a compound of interest.

3.1. Growth Inhibitory Activity of E1 and M2 against MDA-MB-231 Cancer Cell Line

Both extracts were first tested for their ability to inhibit the growth of triple-negative, epithelial human breast adenocarcinoma (MDA-MB-231) cell line. The MTT tests were performed in triplicate and, in each experiment, doxorubicin was used as a reference compound. The cells were incubated for 24, 48 and 72 h with the tested extracts and doxorubicin in order to evaluate their cytotoxicity (after 24 h) and antiproliferative effects (after 48 and 72 h). The results are summarized in the form of bar diagrams in Figure 1.

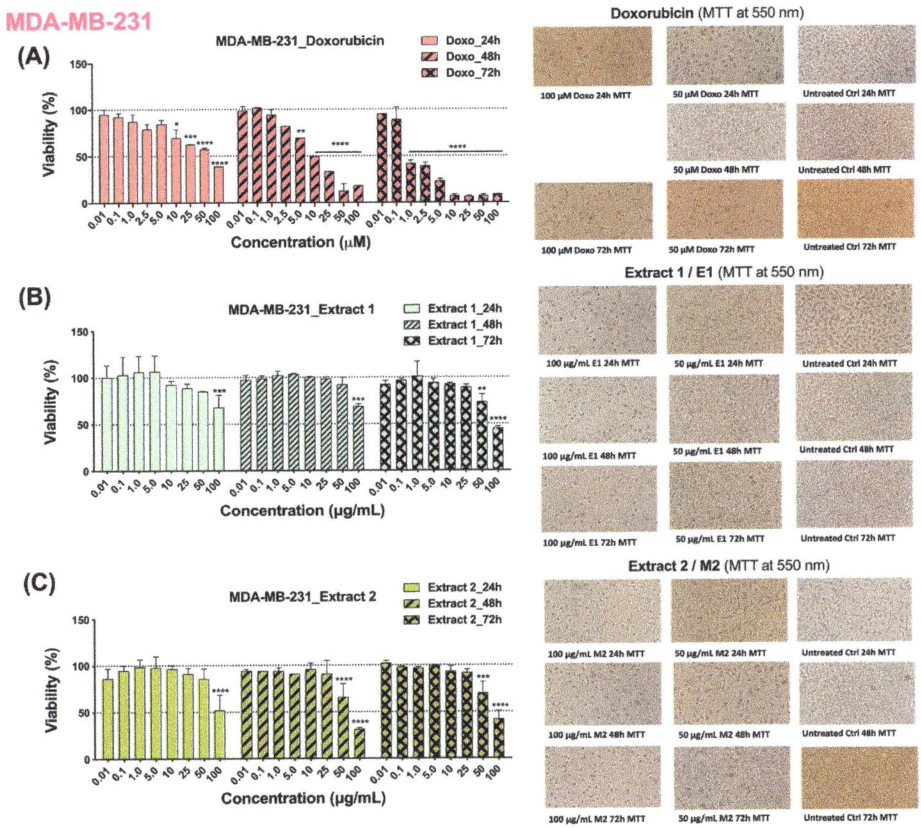

Figure 1. Cytotoxicity profile of doxorubicin (**A**), Extract 1/E1 (**B**), and Extract 2/M2 (**C**) measured on MDA-MB-231 cells after 24, 48, and 72 h incubation in the presence of different concentrations of doxorubicin and tested extracts (0.01 to 100 µM for doxorubicin; 0.01 to 100 µg/mL for E1 and M2). Right: Selected pictures of MDA-MB-231 cell culture after 24, 48 and 72 h incubation with doxorubicin and tested extracts (100 µm with 20× magnification). The respective tested concentrations are indicated. The results are the mean % of untreated controls (Ctrl.) ± SD (n = 3). One-way ANOVA and Dunnett's multiple comparison test: *, $p < 0.05$; **, $p < 0.01$; ***, $p < 0.001$; ****, $p < 0.0001$ vs. control.

As expected, the reference doxorubicin exhibits moderate (after 24 h treatment at a concentration of 100 µM), significantly improved (after 48 h in the concentration range of 25–100 µM, <50% cell viability) and much higher inhibitory activity (after 72 h treatment in the concentration range of 1.0–100 µM, <50% cell viability) against MDA-MB-231 cell line (Figure 1A). The obtained bar diagrams for the inhibitory activity of both HR extracts on MDA-MD-231 cell growth showed that Extract 1 (E1) exhibits a slight effect after 72 h incubation only at its highest-tested concentration of 100 µg/mL (approx. 48% cell viability, Figure 1B), while Extract 2 (M2) moderately inhibits MDA-MB-231 cells at the same highest-tested concentration of 100 µg/mL after 48 h (about 25% cell viability) and 72 h (about 30–40% cell viability) incubation time (Figure 1C). However, both extracts did not show any cytotoxic effects after 24 h incubation with MDA-MB-231 cells. In this first series of tests, we found that the tested extracts possibly interfere with the emission spectra of the MTT reagent used throughout the experiments. This is probably due to the antioxidative properties of both extracts, which may cause in situ reduction of MTT (550 and 570 nm emission), and, therefore, may lead to a huge deviation of the measured OD values. With respect to this, it is also recommended to use an additional assay in order to validate these initial data for the cytotoxic and antiproliferative effects of the tested extracts.

3.2. Growth Inhibitory Activity of E1 and M2 against HT-29 Cancer Cell

Next, both extracts were tested for their inhibitory activity on human colon adenocarcinoma (HT-29) cell line. The MTT tests were performed in triplicate and, in each experiment, doxorubicin was used as a reference compound. The cells were incubated for 24, 48 and 72 h with the tested extracts and doxorubicin in order to evaluate their cytotoxicity (after 24 h) and antiproliferative effects (after 48 and 72 h). The results are summarized in the form of bar diagrams in Figure 2.

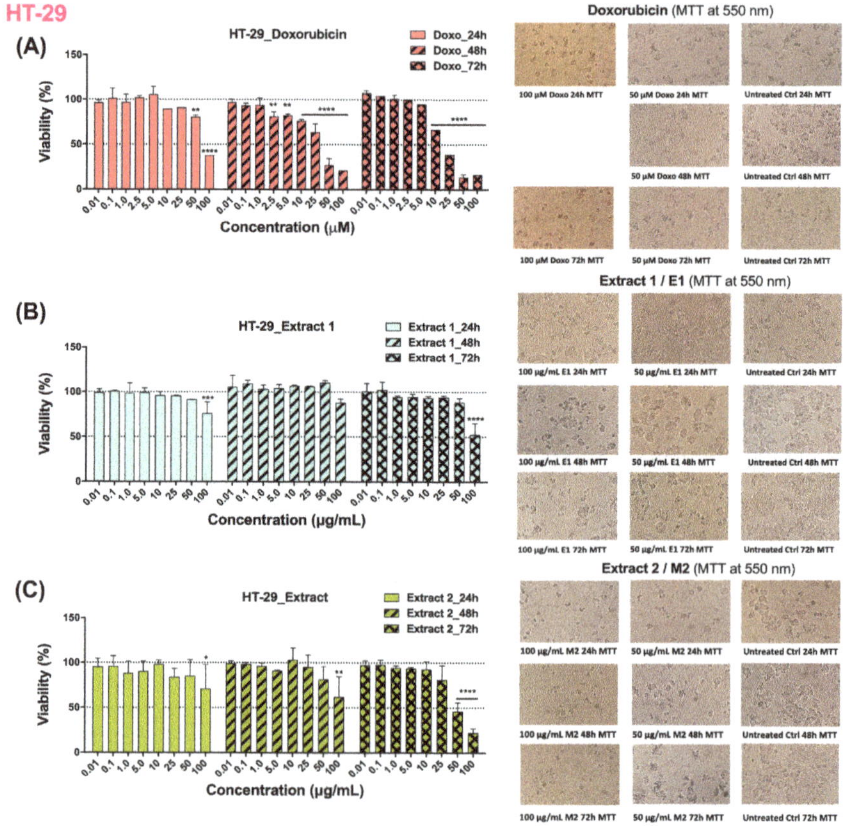

Figure 2. Cytotoxicity profile of doxorubicin (**A**), Extract 1/E1 (**B**), and Extract 2/M2 (**C**) measured on human colon adenocarcinoma HT-29 cells after 24, 48, and 72 h incubation in the presence of different concentrations of doxorubicin and tested extracts (0.01 to 100 μM for doxorubicin; 0.01 to 100 μg/mL for Extract 1 and 2). Right: Selected pictures of HT-29 cell culture after 24, 48 and 72 h incubation with doxorubicin and tested extracts (100 μm with 20× magnification). The respective tested concentrations are indicated. The results are the mean % of untreated controls (Ctrl.) ± SD (n = 3). One-way ANOVA and Dunnett's multiple comparison test: *, $p < 0.05$; **, $p < 0.01$; ***, $p < 0.001$; ****, $p < 0.0001$ vs. control.

Compared to its inhibitory activity against MDA-MB-231 cancer cells, doxorubicin showed similar cytotoxic effects after 24 h incubation with human colon adenocarcinoma cells (cell viability about 40% at 100 μM), but a slight decrease in antiproliferative effects after 48 h (cell viability about 20–25% at 50 and 100 μM) and 72 h exposure (cell viability about 10–40% at 25, 50 and 100 μM) (Figure 2A). Similar to the effects against MDA-MB-231 cells, Extract 1 (E1) did not exhibit the growth of HT-29 adenocarcinoma cells (Figure 2B), while Extract 2 (M2) decreased its inhibitory activity after 48 h treatment (>50% cell viability at 100 µg/mL) with a slight increase in activity after 72 h exposure with HT-29 cells (<50% cell viability at 50 and 100 µg/mL) (Figure 2C). However, the antiproliferative activity of Extract 2 is comparable to that of doxorubicin at the highest-tested concentration of 100 µM. The cell viability of the HT-29 cells was measured to be 16.7 vs. 22.6% after 72 h treatment with doxorubicin and M2, respectively.

3.3. Growth Inhibitory Activity of E1 and M2 against HepG2 Cancer Cell Line

Following our experimental design, both extracts were tested for their ability to inhibit the growth of human hepatocellular carcinoma (HepG2) cell line. The MTT tests were performed in triplicate and, in each experiment, doxorubicin was used as a reference compound. The cells were incubated for 24, 48 and 72 h with the tested extracts and doxorubicin in order to evaluate their cytotoxicity (after 24 h) and antiproliferative effects (after 48 and 72 h). The results are summarized in the form of bar diagrams in Figure 3.

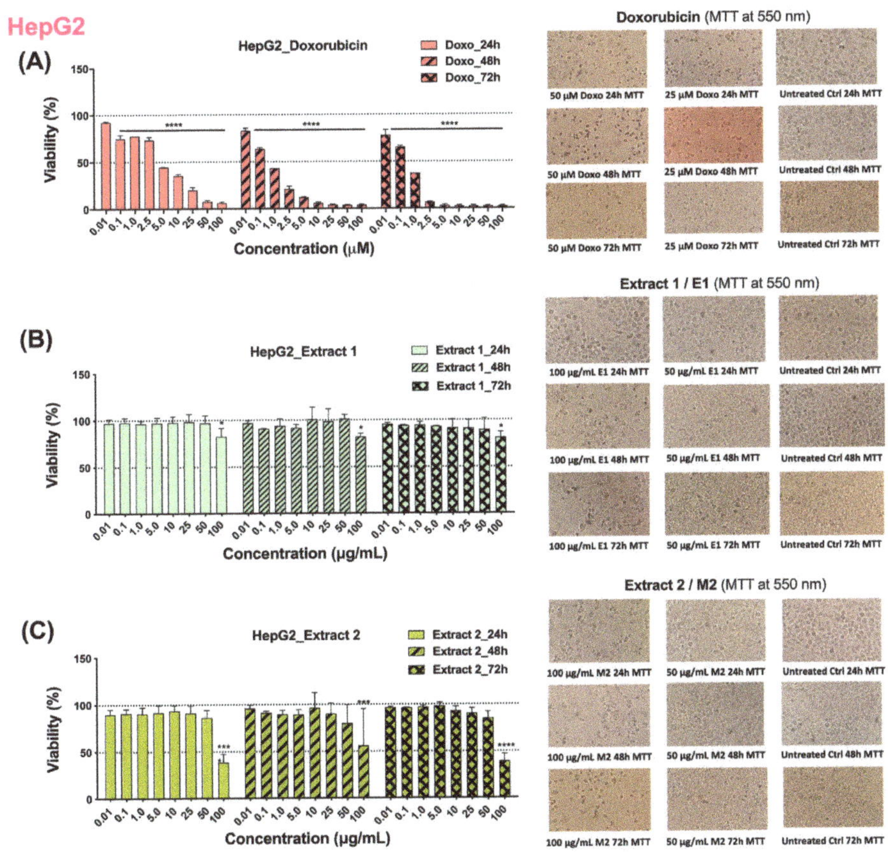

Figure 3. Cytotoxicity profile of doxorubicin (**A**), Extract 1/E1 (**B**), and Extract 2/M2 (**C**) measured on human hepatocellular carcinoma HepG2 cells after 24, 48, and 72 h incubation in the presence of different concentrations of doxorubicin and tested extracts (0.01 to 100 µM for doxorubicin; 0.01 to 100 µg/mL for Extract 1 and 2). Right: Selected pictures of HepG2 cell line after 24, 48 and 72 h incubation with doxorubicin and tested extracts (100 µm with 20× magnification). The respective tested concentrations are indicated. The results are the mean % of untreated controls (Ctrl.) ± SD (n = 3). One-way ANOVA and Dunnett's multiple comparison test: *, $p < 0.05$; ***, $p < 0.001$; ****, $p < 0.0001$ vs. control.

As illustrated in Figure 3A, doxorubicin showed significant improvement in inhibitory activity against HepG2 when compared to its effects on MDA-MB-231 and HT-29 cancer cell lines. The compound inhibits HepG2 cells after 24 h incubation in the concentration range 5.0–100 µM, and the proliferation of HepG2 cells at concentration ranges 1.0–100 µM (after 48 and 72 h incubation period). The antiproliferative effect of doxorubicin at both time periods is almost equal (Figure 3A). The results of both extracts in terms of their cytotoxicity (after 24 h) and antiproliferation (after 48 and 72 h) showed that Extract 1 (E1) did not inhibit the growth of HepG2, neither after 24 h, nor after 48 or 72 h incubation with HepG2 cells (Figure 3B). The inhibitory activity of Extract 2 (M2) against HepG2 cells is similar to that observed against the other two cancer cell lines (MDA-MB-231 and HT-29) after 48 and 72 h treatment of the respective cancer cells. Compared to its cytotoxic activity against MDA-MB-231 and HT-29 cells, Extract 2 inhibits the growth of HepG2 cells at its highest-tested concentration of 100 µg/mL by more than 60% (Figure 3C). Extract 2 showed similar inhibitory activity after 72 h treatment of MDA-MB-231 at a concentration of 100 µg/mL. Interestingly, inhibitory effects for Extract 2 against HepG2 cells were not observed after a 48 h incubation period, but, as seen from the bar diagram, the measured SD is higher than 20%. Both tested extracts did not show significant improvement in inhibitory activity against HepG2 cancer cells when compared to the strongest cytotoxic and antiproliferative effects of doxorubicin.

3.4. Growth Inhibitory Activity of E1 and M2 against 3T3 Cancer Cell Line

In order to evaluate the cytotoxic and antiproliferative effects, both extracts were tested for their ability to inhibit the growth of a non-cancer mouse embryonic fibroblasts 3T3/L1 cell line. The MTT tests were performed in triplicate and, in each experiment, doxorubicin was used as a reference compound. The cells were incubated for 24, 48 and 72 h with the tested extracts and doxorubicin. The results are summarized in the form of bar diagrams in Figure 4.

As indicated in the figure, the reference doxorubicin showed comparable cytotoxic (after 24 h treatment) and antiproliferative effects (after 48 and 72 h treatment) against the non-cancer 3T3 cell line (Figure 4A). After 24 h treatment of 3T3 cells, doxorubicin inhibits the cell growth in the concentration range of 2.5–100 µM; however, the estimated cell viability was between 40 and 50%. In contrast, the inhibitory activity of doxorubicin after 48 and 72 h treatment of 3T3 cells was found to be in the ranges 2.5–100 µM (after 48 h) and 0.5–100 µM (after 72 h). The observed cytotoxic and antiproliferative effects for doxorubicin are comparable to its effects against HepG2. Interestingly, the non-cancer cell line 3T3 was the only one that showed to be sensitive to both tested extracts. In comparison to the determined effects against the tested cancer cell lines (e.g., MDA-MB-231, HT-29, and HepG2), E1 showed slight inhibitory effect on cell growth at its highest-tested concentration of 100 µg/mL after 24, 48 and 72 h incubation with 3T3 cells (Figure 4B). M2 showed similar inhibitory activity against 3T3 cells. It inhibited 3T3 cells at its highest-tested concentration of 100 µg/mL (Figure 4C).

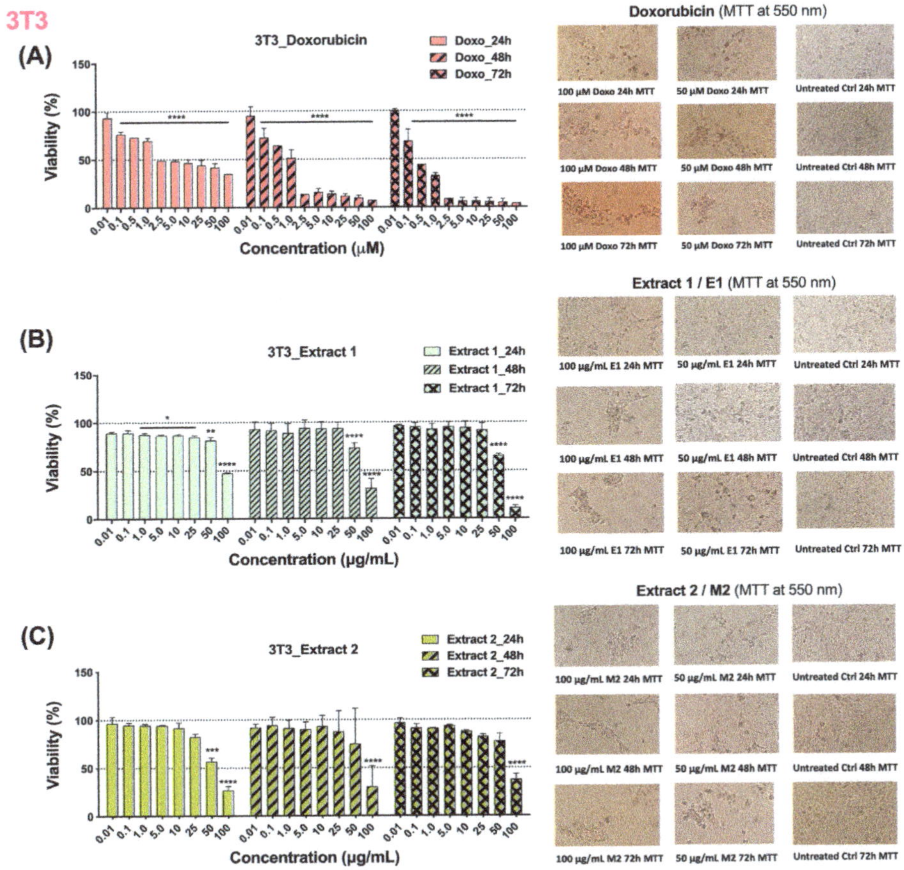

Figure 4. Cytotoxicity profile of doxorubicin (**A**), Extract 1 (**B**), and Extract 2 (**C**) measured on mouse embryonic fibroblasts 3T3 cells after 24, 48, and 72 h incubation in the presence of different concentrations of doxorubicin and tested extracts (0.01 to 100 μM for doxorubicin; 0.01 to 100 μg/mL for Extract 1 and 2). Right: Selected pictures of 3T3 cell cells after 24, 48 and 72 h incubation with doxorubicin and tested extracts (100 μm with 20× magnification). The respective tested concentrations are indicated. The results are the mean % of untreated controls (Ctrl.) ± SD (n = 3). One-way ANOVA and Dunnett's multiple comparison test: *, $p < 0.05$; **, $p < 0.01$; ***, $p < 0.001$; ****, $p < 0.0001$ vs. control.

4. Discussion

The obtained results from all performed MTT tests are summarized in Figure S2. The reference doxorubicin showed inhibitory activity against all investigated cell lines, the cancer cells MDA-MB-231, HT-29, and HT-29, as well as against reference 3T3 cells. The highest inhibitory activity of doxorubicin was measured against HepG2 and 3T3 cells after 48 and 72 h post-treatment (Figure S2, left). The strongest antiproliferative effect for doxorubicin was observed after 72 h incubation with the non-cancer cell line 3T3 (at concentrations between 0.5 and 100 μM). Similar antiproliferative effect for doxorubicin is measured in the concentration range of 1.0–100 μm against HepG2 cancer cells. The lowest inhibitory activity of doxorubicin was assessed against HT-29 cancer cells, independently of the incubation time (at 24, 48, or 72 h).

Compared to doxorubicin, both tested HR extracts showed an overall decrease in inhibitory activity against all cancer (MDA-MB-231, HT-29, and HepG2) and non-cancer (3T3) cell lines. The MTT tests indicated that both extracts did not show comparable dose- and time-dependent (after 24, 48 or 72 h post-treatment) inhibition of cell growth to those of doxorubicin. However, it can be seen that Extract 1 is not able to inhibit the growth of cancer cell lines (e.g., MDA-MB-231, HT-29, or HepG2), while its inhibitory activity against the non-cancer 3T3 cells is comparable to that observed for Extract 2. Extract 1 was able to inhibit the growth of 3T3 cells after 48 and 72 h post-treatment at a concentration of 100 µg/mL (Figure S2, middle). From both extracts, the HR Extract 2 is the most active one. Extract 2 is able to inhibit the growth of all examined cancer cells (e.g., MDA-MB-231, HT-29, or HepG2) after 72 h incubation of the respective cells with the highest-tested concentration of 100 µg/mL (*cf.* the red arrow in Figure S2, right panel). The strongest effect for Extract 2 is determined after 24, 48 and 72 h against the non-cancer 3T3 cells (at the highest-tested concentration of 100 µg/mL). The cytotoxicity and antiproliferative effects of HR Extract 2 against 3T3 cells are almost similar to the effects of Extract 1. In order to obtain the respective IC_{50} values for doxorubicin (in µM), HR Extract 1 and 2 (in µg/mL), dose-dependent non-linear curves were built (Figure 5).

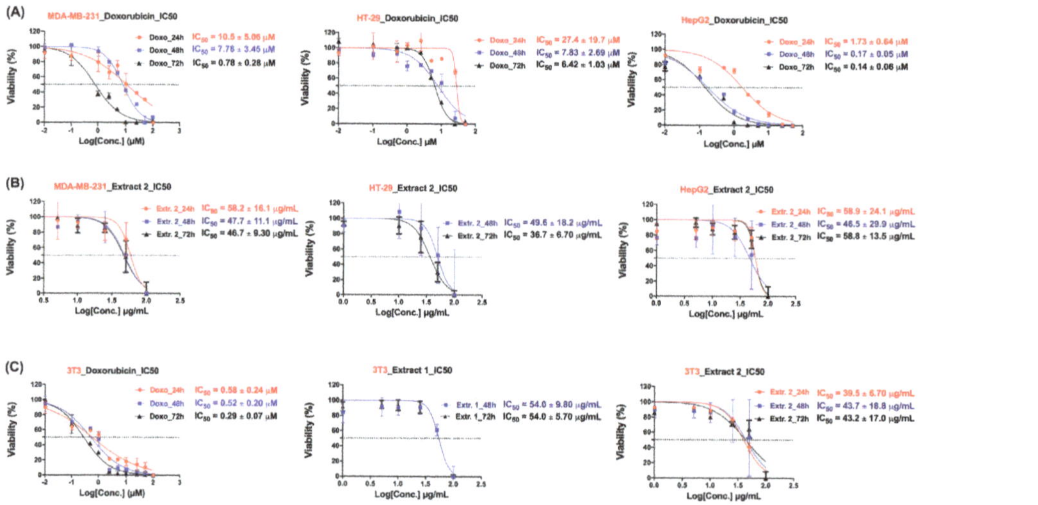

Figure 5. Comparison of non-linear inhibitory curves (log[inhibitor] vs. normalized response/viability in % by variable slope) for doxorubicin (**A**), Extract 2 (**B**), measured on MDA-MB-231, HT-29, and HepG2 cancer cells as well as for doxorubicin, Extract 1 and Extract 2 on 3T3 non-cancer cells (**C**) after 24, 48, and 72 h incubation with different concentrations of doxorubicin and tested extracts (0.01 to 100 µM for doxorubicin; 0.01 to 100 µg/mL for Extract 1 and 2). The respective IC_{50} values (in µM for doxorubicin or µg/mL for Extract 1 and 2) are provided (right). The results are the mean % of untreated controls (Ctrl.) ± SD (*n* = 3).

The dose-response curves were built after the normalization of transformed values, so that the curves run from 100% (vs. control) down to 0%, depending on cell viability (in % vs. control group) that is measured in MTT assay for each cell line and experiment (e.g., after 24, 48 or 72 h treatment). It can be seen that only one curve was built for Extract 1 since it was evaluated to antiproliferative activity (after 48 and 72 h) only against 3T3 cells. Based on the obtained dose-dependent curves, the inhibitory activity of both tested HR extracts and the reference doxorubicin against three cancer (e.g., MDA-MB-231, HT-29, and HepG2) and one non-cancer cell line (3T3) was determined with higher IC_{50} values for 24 h (cytotoxic effects) compared to 48 and 72 h post-treatment, respectively. The IC_{50} values (expressed in µg/mL) for both investigated HR extracts and doxorubicin are summarized in Table 1.

Table 1. Inhibitory potencies (expressed as IC_{50} values) of the reference doxorubicin (Doxo), Extract 1/E1, and Extract 2/M2 after 24, 48 or 72 h treatment of cancer cell lines (MDA-MB-231, HT-29, and HepG2), and the reference non-cancer cell line 3T3.

Extract	$IC_{50} \pm SD$ (μg/mL) [(a)]											
	MDA-MB-231			HT-29			HepG2			3T3		
	24 h	48 h	72 h	24 h	48 h	72 h	24 h	48 h	72 h	24 h	48 h	72 h
Doxo [(b)]	10.5 ± 5.06	7.78 ± 3.45	0.78 ± 0.28	27.4 ± 19.7	7.83 ± 2.69	6.42 ± 1.03	1.73 ± 0.64	0.17 ± 0.05	0.14 ± 0.06	0.58 ± 0.24	0.52 ± 0.20	0.29 ± 0.07
E1	>100	>100	>100	>100	>100	>100	>100	>100	>100	>100	54.0 ± 9.8	54.0 ± 5.7
M2	58.2 ± 16.1	47.7 ± 11.1	46.7 ± 9.3	>100	49.6 ± 18.2	36.7 ± 6.7	58.9 ± 24.1	46.5 ± 29.9	58.8 ± 13.5	39.5 ± 6.7	43.7 ± 18.8	43.2 ± 17.0

(*a*) Data are from three independent experiments ($n = 3$). (*b*) Reference compound doxorubicin (Doxo).

In regard to HR extract E1, the IC_{50} values at 48 and 72 h were determined to be equal to 54.0 μg/mL (50% inhibition of cell growth) for 3T3 cells, whereas, for all other cell lines, they could not be estimated. Therefore, the IC_{50} values for E1 for all cancer lines are higher than the highest-tested concentration of 100 μg/mL ($IC_{50} > 100$ μg/mL, cf. Table 1). In contrast, the HR extract M2 inhibits the cell growth of all cancer cell lines by about 50–60% (with the exception of HT-29 cells at 24 h), and by about 40% of 3T3 non-cancer cells. Therefore, the IC_{50} values for M2 at 24 h were determined to be higher (approx. 58 μg/mL for MDA-MB-231, approx. 59 μg/mL for HepG2, and approx. 40 μg/mL for 3T3) than those measured at 48 or 72 h post-treatment, showing that M2 exhibits stronger antiproliferative effects than cytotoxic activity (increasing by about 10–15%, on average). Moreover, M2 showed increased cytotoxicity and antiproliferative affects compared to E1, which was not able to inhibit the growth of all cancer cell lines.

The highest cytotoxic effect at 24 h for M2 was determined against 3T3 cells (approx. 40 μg/mL), while the strongest antiproliferative effect was estimated at 72 h post-treatment of HT-29 cancer cells (approx. 37 μg/mL). Overall, it can be concluded that M2 moderately inhibits MDA-MB-231, HT-29, HepG2, and 3T3 growth with IC_{50} values in the range of 40–59 μg/mL, whereas Extract 1 (with the exception of 3T3 cells) did not exhibit cytotoxic (at 24 h) or antiproliferative effects (at 48 and 72 h) against all investigated cancer cells (Table 1). However, the cytotoxicity and antiproliferative activity of M2 are not comparable to the effects observed for the reference doxorubicin. The strongest IC_{50} values for doxorubicin were determined to be 0.17 and 0.14 μg/mL against HepG2 cancer cells at 48 or 72 h, respectively (*cf.* Table 1). In general, depending on the tested cells, doxorubicin was evaluated to be between 10- and 300-fold more active than M2 at the different evaluated time points (e.g., 24, 48, or 72 h post-treatment).

5. Conclusions

In conclusion, the myconoside-enriched HR plant Extract 2 (M2) showed increased inhibitory activity (cytotoxicity and antiproliferative effects) compared to the HR plant Extract 1. Moreover, the plant Extract 2 showed a significant increase in cytotoxicity (at 24 h) and antiproliferative activity (at 48 and 72 h post-treatment) at its highest-tested concentration of 100 μg/mL compared to plant Extract 1. As obtained herein, both plant Extracts 1 and 2 inhibited the growth of the non-cancer cell line 3T3 at their highest-tested concentration of 100 μg/mL. The reference compound doxorubicin, used in this study as a reference compound, showed strong cytotoxic (at 24 h) and antiproliferative effects (at 48 and 72 h post-treatment) against all tested cell lines (cancer and non-cancer cells) being between 10- and 300-fold more active than the most active plant Extract 2. Since it is considered a skin-protecting cosmetic product, the plant Extract 2 can also be considered for further time- and dose-dependent experiments in order to evaluate its cytotoxicity and antiproliferative effects against a panel of metastatic melanoma cancer cell lines, e.g., SK-MEL-3, SH-4, SK-MEL-24, and other cells.

Supplementary Materials: The following supporting information can be downloaded at: https://www.mdpi.com/article/10.3390/cosmetics11020046/s1, Figure S1: (A) The plant *Habelea rhodopensis* Friv. (B) Chemical structures and IUPAC names of the main constituents of *H. rhodopensis* Friv. myconoside and paucifloside.; Table S1: Transformation of doxorubicin concentrations (from μM to μg/mL); Figure S2: Comparison of the cytotoxicity profile for doxorubicin (left), Extract 1 (middle), and Extract 2 (right) measured on MDA-MB-231, HT-29, and HepG2 cancer cells, as well as on 3T3 non-cancer cells after 24, 48, and 72 h exposure to different concentrations of compounds (0.01 to 100 μM for doxorubicin; 0.01 to 100 μg/mL for Extract 1 and 2). The respective substance concentrations are indicated. For simplification, the red arrow (right) showed the determined trend in inhibitory activity for extract 2, determined at its highest tested concentration of 100 μg/mL. Untreated (control, Ctrl.) cells were used as a positive control. The results are expressed as the mean % of untreated controls ± SD ($n = 3$). Statistical analysis was performed by one-way ANOVA and Dunnett's multiple comparison test. *, $p < 0.1$; **, $p < 0.01$; ***, $p < 0.001$; ****, $p < 0.0001$ *vs.* control.

Author Contributions: Conceptualization, collection and assembly of data, data analysis and interpretation, N.T.T.; methodology, collection and assembly of data, data analysis and interpretation M.I.P., M.G.G. and A.A.B.; software, N.T.T. and M.G.G.; validation of data, M.I.P., M.G.G. and A.A.B.; investigation, M.I.P., M.G.G. and A.A.B.; resources, N.T.T. and A.P.; data curation, M.I.P.; writing—original draft preparation, N.T.T.; writing—review and editing, M.I.P., A.P. and N.T.T.; visualization, M.I.P. and M.G.G.; supervision, N.T.T.; project administration, A.P.; funding acquisition, N.T.T. and A.P. All authors have read and agreed to the published version of the manuscript.

Funding: This research received no external funding.

Institutional Review Board Statement: Not applicable.

Informed Consent Statement: Not applicable.

Data Availability Statement: Data are contained within the article and the Supplementary Materials.

Conflicts of Interest: The authors declare no conflicts of interest. Atanas Pavlov is employed by the company Innova BM Ltd. The remaining authors declare that the research was conducted in the absence of any commercial or financial relationships that could be construed as potential conflicts of interest. The authors declare that the company Innova BM Ltd. provided the materials. They were not involved in the study design, collection, analysis, interpretation of data, the writing of this article or the decision to submit it for publication.

Abbreviations

ANOVA	analysis of variance
DMSO	*N,N*-dimethyl sulfoxide
Doxo	doxorubicin
HR	*Habelea rhodopensis* Friv.
HREs	*H. rhodopensis* Friv. extracts
IUPAC	International Union of Pure and Applied Chemistry
MTT	3-(4,5-dimethylthiazol-2-yl)-2,5-diphenyltetrazolium bromide
SD	Standard Deviation
(U)HPLC-MS	(ultra-)High Performance Liquid Chromatography-Mass Spectrometry.

References

1. Gechev, T.S.; Benina, M.; Obata, T.; Tohge, T.; Sujeeth, N.; Minkov, I.; Hille, J.; Temanni, M.-R.; Marriott, A.S.; Bergstrom, E.; et al. Molecular mechanisms of desiccation tolerance in the resurrection glacial relic *Haberlea rhodopensis*. *Cell. Mol. Life Sci.* **2013**, *70*, 689–709. [CrossRef]
2. Curtis, W. *Botanical Magazine*, 3rd ed.; L. Reeve & Co.: London, UK, 1882; Volume XXXVIII, p. 108.
3. Georgiev, Y.N.; Ognyanov, M.H.; Denev, P.N. The ancient Thracian endemic plant *Haberlea rhodopensis* Friv. and related species: A review. *J. Ethnopharmacol.* **2019**, *249*, 112359. [CrossRef] [PubMed]
4. Dell'acqua, G.; Schweikert, K. Skin benefits of a myconoside-rich extract from resurrection plant *Haberlea rhodopensis*. *Int. J. Cosmet. Sci.* **2011**, *34*, 132–139. [CrossRef] [PubMed]
5. Bankova, R. *Haberlea rodopensis*—Effects and potential applications. *Tradit. Mod. Vet. Med.* **2022**, *7*, 128–138.

6. Spyridopoulou, K.; Kyriakou, S.; Nomikou, A.; Roupas, A.; Ermogenous, A.; Karamanoli, K.; Moyankova, D.; Djilianov, D.; Galanis, A.; Panayiotidis, M.I.; et al. Chemical profiling, antiproliferative and antimigratory capacity of *Haberlea rhodopensis* extracts in an in vitro platform of various human cancer cell lines. *Antioxidants* **2022**, *11*, 2305. [CrossRef] [PubMed]
7. Mihaylova, D.S.; Bahchevanska, S.T.; Toneva, V.T. Microwave-assisted extraction of flavonoid antioxidants from leaves of *Haberlea rhodopensis*. *J. Int. Sci. Publ. Mater. Methods Technol.* **2011**, *5*, 104–114.
8. Djilianov, D.; Ivanov, S.; Georgieva, T.; Moyankova, D.; Berkov, S.; Petrova, G.; Mladenov, P.; Christov, N.; Hristova, N.; Peshev, D.; et al. A holistic approach to resurrection plants. *Haberlea rhodopensis—A case study. Biotechnol. Biotechnol. Equip.* **2009**, *23*, 1414–1416. [CrossRef]
9. Staneva, D.; Dimitrova, N.; Popov, B.; Alexandrova, A.; Georgieva, M.; Miloshev, G. *Haberlea rhodopensis* extract tunes the cellular response to stress by modulating DNA damage, redox components, and gene expression. *Int. J. Mol. Sci.* **2023**, *24*, 15964. [CrossRef] [PubMed]
10. Mihaylova, D.; Banchevanska, S.; Toneva, V. Examination of the antioxidant activity of *Haberlea rhodopensis* leaf extracts and their phenolic constituents. *J. Food Biochem.* **2013**, *37*, 255–261. [CrossRef]
11. Kondeva-Burdina, M.; Zheleva-Dimitrova, D.; Nedialkov, P.; Girreser, U.; Mitcheva, M. Cytoprotective and antioxidant effects of phenolic compounds *from Haberlea rhodopensis* Friv. (Gesneriaceae). *Pharmacogn. Mag.* **2013**, *9*, 294–301. [PubMed]
12. Dimitrova, N.; Staneva, D.; Popov, B.; Alexandrova, A.; Georgieva, M.; Miloshev, G. *Haberlea rhodopensis* alcohol extract normalizes stress-responsive transcription of the human TP53 gene. *J. Exp. Biol. Agric. Sci.* **2023**, *11*, 405–415. [CrossRef]
13. Popov, B.; Dobreva, Z.I.; Georgieva, S.; Stanilova, S. Enhancement of anti-KLH IgG antibody production in rabbits after treatment with *Haberlea rhodopensis* extract. *Trakia J. Sci.* **2010**, *8* (Suppl. 2), 92–97.
14. Ebrahimi, N.S.; Gafner, F.; Dell'Acqua, G.; Schweikert, K.; Hamburger, M. Flavone 8-C-glycosides from *Haberlea rhodopensis* Friv. (Gesneriaceae). *Helv. Chim. Acta* **2011**, *94*, 38–45. [CrossRef]
15. Popov, B.; Georgieva, S.; Gadjeva, V.; Petrov, V. Radioprotective, anticlastogenic and antioxidant effects of total extract of *Haberlea Rhodopensis* on rabbit blood samples exposed to gamma radiation in vitro. *Rev. Med. Vet.* **2011**, *162*, 34–39.
16. Berkov, S.H.; Nikolova, M.T.; Hristozova, N.I.; Momekov, G.Z.; Ionkova, I.I.; Djilianov, D.L. GC–MS profiling of bioactive extracts from *Haberlea rhodopensis*: An endemic resurrection plant. *J. Serbian Chem. Soc.* **2011**, *76*, 211–220. [CrossRef]
17. Georgiev, V.G.; Pavlov, A.I.; INNOVA BM Ltd. Standardized Plant Extract from Biomass of In Vitro Cultures, Method for Preparation and Use Thereof. WO2021184086A1. U.S. Patent Application 17/912,237, 22 June 2023.
18. Kühl, T.; Georgieva, M.G.; Hübner, H.; Lazarova, M.; Vogel, M.; Haas, B.; Peeva, M.I.; Balacheva, A.A.; Bogdanov, I.P.; Milella, L.; et al. Neurotensin(8–13) analogs as dual NTS1 and NTS2 receptor ligands with enhanced effects on a mouse model of Parkinson's disease. *Eur. J. Med. Chem.* **2023**, *254*, 115386. [CrossRef] [PubMed]
19. Mossman, T. Rapid colorimetric assay for cellular growth and survival: Application to proliferation and cytotoxicity assays. *J. Immunol. Methods* **1983**, *65*, 55–63. [CrossRef] [PubMed]
20. Tzvetkov, N.T.; Stammler, H.-G.; Georgieva, M.G.; Russo, D.; Faraone, I.; Balacheva, A.A.; Hristova, S.; Atanasov, A.G.; Milella, L.; Antonov, L.; et al. Carboxamides vs. methanimines: Crystal structures, binding interactions, photophysical studies, and biological evaluation of (indazole-5-yl)methanimines as monoamine oxidase B and acetylcholinesterase inhibitors. *Eur. J. Med. Chem.* **2019**, *179*, 404–422. [CrossRef] [PubMed]

Disclaimer/Publisher's Note: The statements, opinions and data contained in all publications are solely those of the individual author(s) and contributor(s) and not of MDPI and/or the editor(s). MDPI and/or the editor(s) disclaim responsibility for any injury to people or property resulting from any ideas, methods, instructions or products referred to in the content.

cosmetics

MDPI

Review

A Recent Update on the Potential Use of Catechins in Cosmeceuticals

Soraya Ratnawulan Mita [1,*], Patihul Husni [1], Norisca Aliza Putriana [1], Rani Maharani [2], Ryan Proxy Hendrawan [1] and Dian Anggraeni Dewi [1]

1 Department of Pharmaceutics and Pharmaceutical Technology, Faculty of Pharmacy, Universitas Padjadjaran, Jatinangor 45363, Indonesia; patihul.husni@unpad.ac.id (P.H.); norisca@unpad.ac.id (N.A.P.); ryan20001@mail.unpad.ac.id (R.P.H.); dian19004@mail.unpad.ac.id (D.A.D.)

2 Departement of Chemistry, Faculty of Mathematics and Natural Sciences, Universitas Padjadjaran, Jatinangor 45363, Indonesia; r.maharani@unpad.ac.id

* Correspondence: soraya@unpad.ac.id

Abstract: Catechins are a type of flavonoid known for their beneficial functions as antioxidants and antibacterials. Recent research indicates the antioxidant potential of catechins on the skin. Catechin and epigallocatechin are reported to have significant potential in preventing ageing. Epigallocatechin gallate, gallocatechin gallate, and epigallocatechin can inhibit hyperpigmentation processes. Additionally, catechins exhibit potential in UV protection and inflammation inhibition in acne. Consequently, catechins are now being used in the cosmetics industry, with formulations containing catechins as the active ingredient developed to produce various products such as soap, sunscreen, creams, etc. Herein, this paper reviews the antioxidant potential of catechins for use in cosmetic formulations and the current status of clinical trials of catechins in cosmetics.

Keywords: catechins; flavonoid; antioxidant; cosmeceutical

Citation: Mita, S.R.; Husni, P.; Putriana, N.A.; Maharani, R.; Hendrawan, R.P.; Dewi, D.A. A Recent Update on the Potential Use of Catechins in Cosmeceuticals. *Cosmetics* **2024**, *11*, 23. https://doi.org/10.3390/cosmetics11010023

Academic Editor: Agnieszka Feliczak-Guzik

Received: 26 December 2023
Revised: 22 January 2024
Accepted: 22 January 2024
Published: 6 February 2024

Copyright: © 2024 by the authors. Licensee MDPI, Basel, Switzerland. This article is an open access article distributed under the terms and conditions of the Creative Commons Attribution (CC BY) license (https://creativecommons.org/licenses/by/4.0/).

1. Introduction

Cosmetics are widely used by everyone and can be classified as decorative or skincare cosmetics. They are defined by BPOM Indonesia as materials or preparations intended for use on the external parts of the human body such as the epidermis, hair, nails, lips, and external genital organs, or on the teeth and oral mucous membranes, especially for cleaning, perfuming, changing the appearance and/or improving body odour, or to protect or maintain the body in good condition. Based on the literature, cosmetics are often used for aesthetic and self-care benefits, and that they also intersect with skin health. The skin is the largest human organ and protects the body from external insults, therefore, it is essential to care for the skin according to the unique needs of the individual. The term 'cosmeceutical' refers to a cosmetic product that provides medical effects and contains certain active compounds. These cosmeceutical products are not only for whitening, antiageing, and sunscreen purposes, but also for hyperpigmentation, photoageing, wrinkles, and hair loss [1]. It is important to note that cosmeceutical preparations are not intended to be given systematically, rather, they are applied locally/topically, that is, the 'dermal delivery' as the site of action is usually the stratum corneum (SC), the viable epidermis and/or dermis [2].

Skin ageing typically starts to manifest when an individual reaches their late 20s to early 30s due to two distinct sources [3]: intrinsic ageing, which results from issues within the network of elastin fibres and collagen, or extrinsic ageing due to exposure to environmental factors such as sun radiation. Oxidative stress triggers inflammation, constraining epidermal cell renewal and ultimately leads to a reduction in epidermal thickness and a weakening of the protective barrier [4]. Sun radiation triggers the creation of reactive oxygen radicals (ROS) that cause keratinocytes to generate pro-inflammatory cytokines, including tumor necrosis factor-α (TNF-α) and interleukin-8 (IL-8), thereby producing more ROS [5]. Excessive ROS leads to the development of wrinkles by causing the breakdown and abnormal interlinking of structural proteins such as glycosaminoglycans, collagen, and elastin fibres in the skin's extracellular matrix. Hence, antioxidants isolated from natural products can be used to suppress ROS production to slow down skin ageing [6].

Catechins are natural flavan-3-ols (or flavonols), a type of polyphenolic compound belonging to the flavonoid family. They are present in a variety of fruits, vegetables, and plant-based beverages [7] and are particularly concentrated in tea leaves, red wine, broad beans, rock-rose leaves, apricots, black grapes, and strawberries. Epicatechin is abundant in chocolate, apples, broad beans, pears, black grapes, cherries, and certain types of berries including blackberries and rasp6berries [8].

Catechins offer numerous health benefits by effectively eliminating free radicals and slowing down the breakdown of the extracellular matrix caused by exposure to ultraviolet (UV) radiation and pollution. They stimulate collagen production while preventing the generation of matrix metalloproteinase enzymes. Due to the presence of hydroxyl in the gallate group, epigallocatechin gallate (EGCG) and epigallocatechin (ECG) can neutralise free radicals, surpassing several antioxidants like trolox, ascorbic acid, and tocopherol [9]. Thus, catechins have the potential to be used in cosmetic and dermatological products [10] and are now commonly included in pharmaceutical, medical, and cosmetic products. For example, a transethosomal gel form of catechins can reduce total cholesterol in mice [11] and *Uncaria gambir* is used to treat diarrhoea, sore throat, spongy gums, dysentery, arteriosclerosis, and obesity [12].

The anti-inflammatory and antioxidant properties of EGCG have been extensively examined for their impact on apoptosis, proliferation, and differentiation. EGCG is also used as a skincare ingredient due to its potential for skin hydration and as an anti-pigmentation agent, although further study is needed [13,14].

2. Physicochemical Properties of Catechins

Catechins are bioactive polyphenols and are typically isolated from green tea (*Camellia sinensis* L.) and gambir leaves (*Uncaria gambir* Roxb). There are a few types of catechins which possess the flavan-3-ol structure consisting of two benzene rings, a heterocycle dihydropyran, and a hydroxyl [15,16]. The structures are shown in Figure 1.

Catechins have been utilised in many pharmacological formulations [17,18]. The use of catechins is based on its physicochemical characteristics such as polarisability, dipole moment, molecular weight, surface area, and van der Waals volume, as well as macroscopic traits including solubility, octanol/water partition coefficient, acidity or basicity in solution, etc. [19]. In the Indonesian standard guide, namely the *Indonesian Herbal Pharmacopoeia*, there are several physicochemical profiles of catechins obtained from gambir plants (*Uncaria gambir* (Hunter) Roxb.) as shown in Table 1.

Table 1. Physicochemical parameters of catechins based on the *Pharmacopeia* handbook.

Physicochemical Parameters	Accepted Value	Reference
Organoleptic	Solid form, it appears as a light brown to dark reddish-brown substance with a distinctive odour. It possesses a chelate taste that is slightly bitter at first but ends with a sweet aftertaste.	[20]
Water Content	Quantity should not exceed 14%	[20]
Ash Content	Quantity should not exceed 0.5%	[20]
Ash, not soluble in acid	Quantity should not exceed 0.1%	[20]

Table 1. *Cont.*

Physicochemical Parameters	Accepted Value	Reference
Purity	Contains no less than 90% tannins counted as catechins.	[20]
Identification	The assay was carried out using spectrophotometry with a wavelength of 294 nm.	[20]
Molecular weight	290.27 g/mol	[20]
Solubility	Soluble in water and polar organic solvents; soluble in pressurised hot water between 298.75 to 415.85 K; soluble in mixtures of supercritical carbon dioxide (SC-CO_2) and ethanol at 313 K and pressures ranging from 80 to 120 bar; soluble in SC-CO_2 between 313.15 and 343.15 K and pressures ranging from 12 to 26 MPa using ethanol as the co-solvent.	[21]

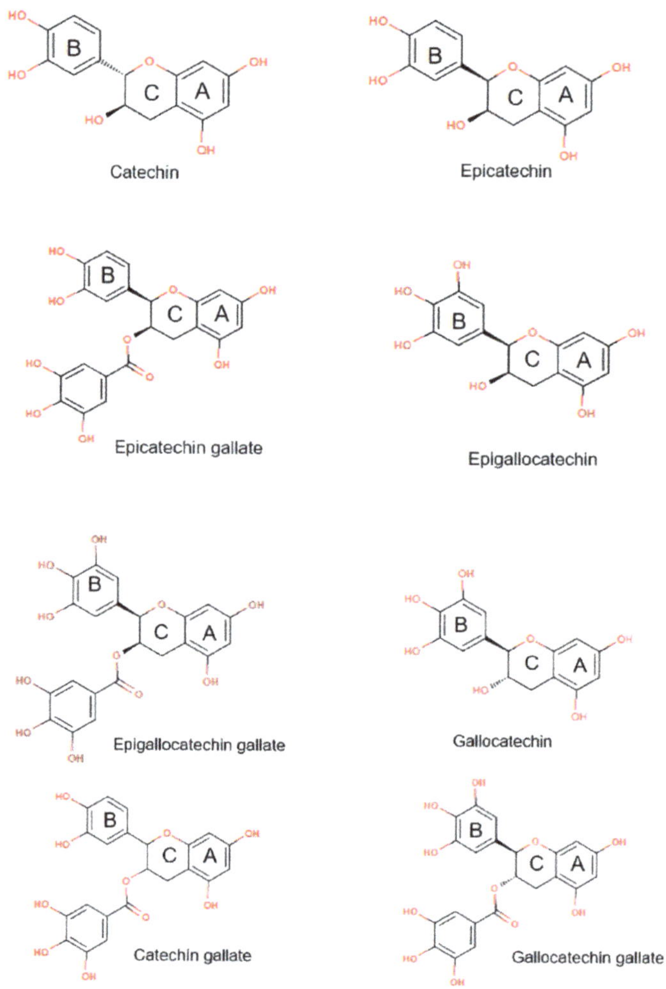

Figure 1. Various type of catechin structures.

3. Activities of Catechin

3.1. Antioxidant

Antioxidants scavenge oxygen free radicals (singlet and triplet), ROS, peroxide decomposers, and enzyme inhibitors to protect vital molecules from harm [22]. The use of antioxidants derived from natural ingredients is increasing. Natural antioxidants can be single pure compounds/isolates, combinations of compounds, or plant extracts. The secondary metabolites, polyphenols, are the most common phyto-antioxidants [23]. Polyphenols have benzene rings with attached -OH groups which determine the antioxidant activity based on their number and position [24]. Protein phosphorylation is influenced by phenol groups by inhibiting lipid peroxidation. Flavonoids are the main source of polyphenols, while carotenoids are the most abundant sources of terpenes [25].

Catechin produces and discards free radicals [26] through several key direct and indirect antioxidant mechanisms. The direct mechanism involves the scavenging of ROS, whereas the indirect mechanism occurs through increased antioxidant enzymes and the inhibition of the pro-enzyme that participates in oxidant stress [8,27]. The phenolic hydroxyl group in catechin is involved in the scavenging of ROS, therefore more hydroxyl groups will improve the antioxidant activity. According to the structure, the hierarchy of antioxidant activity of catechins is EGCG, EGG, EGC, EC, and, lastly, catechin [8].

The structural characteristics of flavan-3-ols, specifically their resorcinol and catechol components, which consist of A and B rings connected by the Pyron ring (C ring), are responsible for their antioxidant properties (Figure 1) [28]. The ability of flavan-3-ols to scavenge radicals primarily relies on the arrangement of hydroxyl groups and their capacity to donate hydrogen atoms [29]. The stability of the phenoxy radical produced after hydrogen atom transfer (HAT) also plays a role in their ability to counteract reactive oxygen radicals [30]. Catechins can exist in four different diastereoisomers, which arise from two chiral centers (2^n) at C2 and C3. These diastereoisomers are referred to as (+) catechin (2R, 3S), (−) epicatechin (2R, 3R), (−) catechin (2S, 3R), and (+) epicatechin (2S, 3S) [31]. Overall, the stereoisomerisms are determined by the positioning of the B ring connected to the C ring at the C2 atom and also the chirality of R1 and R2 attached to the C ring at the C3.

The structures are indicative of a site of a projected bond—R stands for dashed wedge bonds that extend away from the viewer, and S represents solid wedge bonds that project out of the paper towards the viewer, as seen in Figure 2. The degree of polymerization also plays a role in determining the antioxidant properties [32]. The capability of catechin and epicatechin to scavenge radicals is attributed to the dihydroxyl group at C-3′ and C-4′ on the B ring, a C3-hydroxyl group on the C ring without a 2,3 double bond, and hydroxyl groups at C5 and C7 on ring A. In this context, the catechol ring (B) exhibits greater electron-donating capacity than the other rings because it contains an ortho-dihydroxyl group (detail see Figures 2 and 3) [33]. The scavenging of free radicals by the hydroxyl groups occurs via hydrogen atom transfer and single-electron atom transfer, with catechin undergoing oxidation to form the relatively reactive quinone [34] as could be observed from Figure 4.

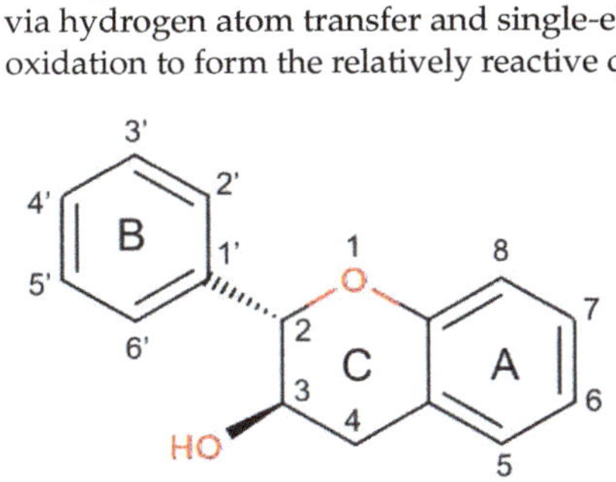

Figure 2. Structure of Catechin in Detail.

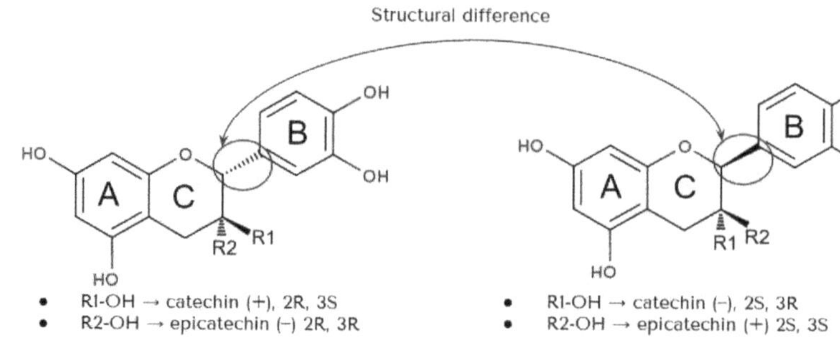

Figure 3. Structure of stereoisomers of catechin (+) and epicatechin (−), also catechin (−) and epicatechin (+).

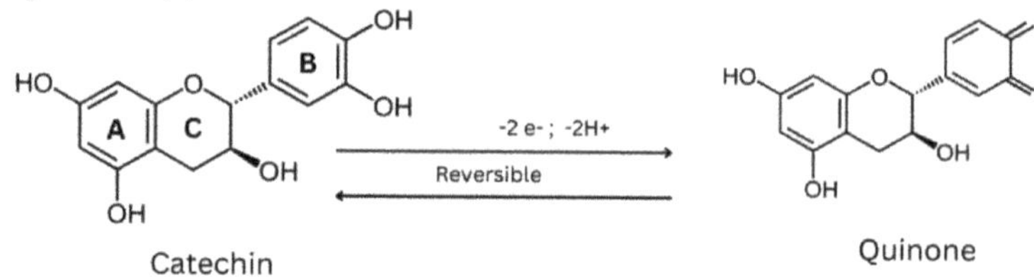

Figure 4. Scavenging of free radicals by catechin.

Numerous studies have reported that the primary source of antioxidant activity in CT and ECT is their elevated redox characteristics. The antioxidant properties of catechin and epicatechin, which are found in different plant extracts, were assessed using various experimental antioxidant tests, including ferric reducing antioxidant power (FRAP), 1,1-diphenyl-1-picrylhydrazyl (DPPH), nitric oxide (NO), and 2,2-azino-bis (3-ethylbenzothiazoline-6-sulfonic acid) ($ABTS^+$), demonstrating that CT and ECT exhibit strong antioxidant capabilities [35]. CT and ECT share a similar molecular structure, but they exhibit distinct chemical reactivity properties because CT's antioxidant capability relies on its planar geometry, whereas ECT's antioxidant activity is due to the interaction of hydrogen bonds at the catechol moiety. The lack of a double bond at C2 = C3 in the C ring significantly affects ECT's antioxidant capacity. Additionally, it has been demonstrated that the reactivity of flavan-3-ols is thermodynamically modified depending on the solvents used [36].

There are many uses of catechins. Aside from being used as pure antioxidant compounds, catechins can be formulated into many products including sunscreen, lip balm, anti-dandruff shampoo, cosmetic cleansers, and cosmetic creams. They have also been extracted from many natural sources, as detailed in Table 2.

Table 2. Antioxidant activity of catechins from various sources.

Methods	Sample	Result	Ref.
DPPH Assay	*Camellia sinensis*	• Solvent Used: Methanol • Green Tea = 67.3% • White Tea = 47.9% • Black Tea = 28.9%	[37]
	Lepisanthes alata (Blume) Leenh	• Solvent Used: Water • Rind: 61.61% • Flesh: 47.93% • Seeds: 48.66% • Whole Fruit: 69.30% • Leaves: 59.35% • Bark: 49.91%	[38]

Table 2. *Cont.*

Methods	Sample	Result	Ref.
		• Solvent Used: Methanol • Rind: 86.17% • Flesh: 27.47% • Seeds: 89.58% • Whole Fruit: 78.34% • Leaves: 61.71% • Bark: 87.03%	
		• Solvent Used: Ethanol • Rind: 85.81% • Flesh: 21.23% • Seeds: 90.12% • Whole Fruit: 46.20% • Leaves: 79.61% • Bark: 87.03%	
	Sterculia quadrifida R.	Bark: 51.5 µg/mL (50%)	[22]
Malondialdehyde	*Uncaria Gambir* Roxb	• Dose of 5 mg/kg: 0.19% • Dose of 10 mg/kg: 31.28% • Dose of 20 mg/kg: 57.63% • Control + (Vit E): 5.55% • Control-: −77.79%	[39]
	Uncaria Gambir Roxb.	Varies from 2.732% to 3.792%	[40]
	Combination of Isolated Catechin and Quercetin	• Insignificant antioxidant activity • Dose of catechin used was 100 µg/mL ($p < 0.05$)	[41]

In the context of the skin, several studies indicate that antioxidants play a crucial role in inhibiting the effects of radiation [42]. The various roles of antioxidants on the skin are presented in Table 3.

Table 3. The various roles of antioxidants on skin.

Antioxidant Function	Mechanism of Action	Reference
Anti-ageing	Antioxidants inhibit the action of superoxide dismutase (SOD) enzymes, which play a role in degrading collagen.	[43]
Skin Brightening	Antioxidants have whitening effects by inhibiting tyrosinase and act as anti-inflammatory agents for hyperpigmentation caused by UV exposure (commonly known as melasma).	[44]
UV Filters	Topically administered antioxidants can enhance the photoprotective capabilities of UV filters by reducing erythema, inhibiting the development of sunburned skin cells, and causing immunosuppression.	[45]
Skin Hydration and Anti-Hyperpigmentation	Antioxidants suppress the production and secretion of melanin in melanoma cells to enhance skin hydration and improve hyperpigmentation.	[46]

3.2. Anti-Ageing

Elastase, a protein kinase enzyme, cleaves specific polypeptide bonds to reduce elastin levels. Preventing elastase functions within the dermis layer can be utilised to maintain the skin's flexibility, therefore, elastase activity inhibitors can serve as cosmetic ingredients to counteract the signs of skin ageing [47]. Polyphenols found in isolated white tea inhibit collagenase and elastase enzymes, specifically catechin and EGCG (Figure 5). Furthermore, given that collagenase is a zinc-containing metalloproteinase, these catechins could potentially attach to the Zn^{2+} ion present in the enzyme, thereby obstructing its ability to bind to the substrate [48].

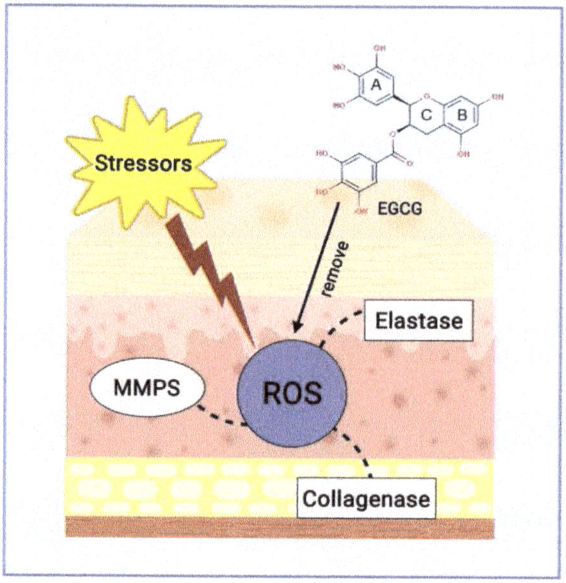

Figure 5. Catechin mechanism of action as anti-ageing agent.

3.3. Skin Brightener

Catechin directly inhibits tyrosinase activity and reduces the expression of tyrosinase [49]. EGCG, GCG, and EGC demonstrate great potential as tyrosinase activity inhibitors [50]. The catechins have a substantial inhibitory effect on tyrosinase activity and melanin production by downregulating the cAMP/CREB/MITF signalling pathway in B16F10 cells (Figure 6), with ECG demonstrating the strongest effect, followed by EGCG and GCG [51]. EGCG also suppresses the production of melanin induced by α-MSH in B16 melanoma cells [49].

3.4. Anti-Hyperpigmentation

UV radiation results in melanogenesis. The mass production of melanin in the skin minimises UV radiation but causes the skin to slightly darken to a brownish color. Catechin as a depigmentation agent will inhibit melanin formation by inhibiting melanin synthesis through tyrosinase (TYR) and microphthalmia-associated transcription factor (MITF). MITF is a transcription factor that is responsible for melanocyte development in melanogenesis. Catechin will inhibit MILTF, thereby preventing melanocytes from producing melanin and hyperpigmentation [52].

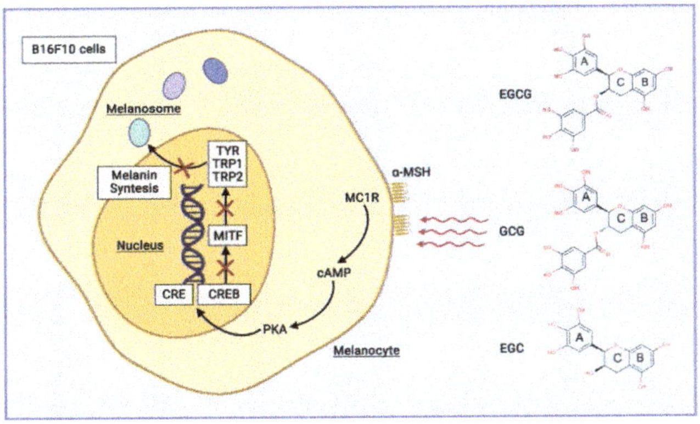

Figure 6. Catechin mechanism of action as anti-hyperpigmentation and brightening agent.

3.5. UV-Reduction and Sunscreen

UV radiation is categorised into UV-A (315–400 nm), UV-B (280–315 nm), and UV-C (280–100 nm). Continuous exposure to UV-B radiation can lead to disruptions in the skin caused by free radicals and ROS, which stimulate melanin production and melanocyte proliferation. Tyrosinase facilitates melanin synthesis and represents a pivotal point in melanogenesis. High melanin levels can disrupt pigmentation in human skin, leading to conditions such as age spots, melasma, malignant melanomas, and freckles. EGCG is acknowledged as a natural antioxidant to neutralise free radicals, modulate the activity of antioxidant enzymes, diminish the effects of oxidative stress, and inhibit tyrosinase activity [53].

Sunscreen is a cosmetic that functions as a skin protector [54], filtering UV to reduce the radiation emitted from the sun [55,56]. Aromatic compounds conjugated to carbonyl groups convert UV energy into minimised UV energy [51], preventing the chemical properties of UV-absorbing potency without the need for significant photodegradation (Figure 7). The usage of sunscreen is determined based on the Sun Protection Factor (SPF), which is defined as the ability to supply a minimal erythema dose (MED) on the skin divided by the variable of UV energy that is needed to supply MED on unprotected skin [57].

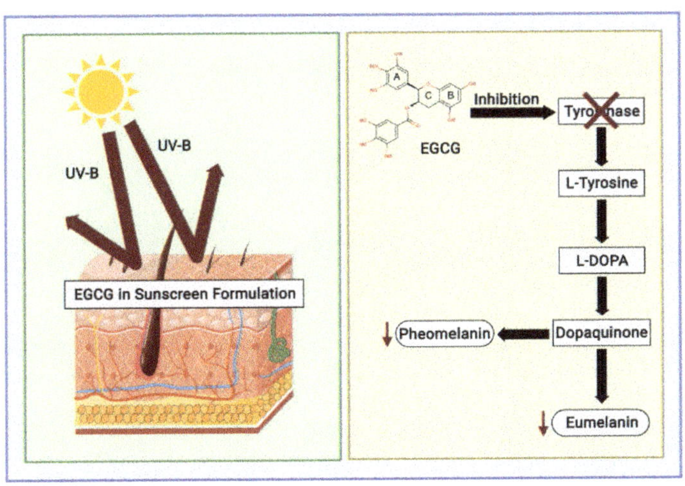

Figure 7. Catechin mechanism of action as a sunscreen.

The higher the SPF, the more protective the sunscreen [58] to prevent sunburn as well as skin cancer in the long term. The SPF can be determined with in vivo or in vitro methods. The in vivo test involves human volunteers while the in vitro is differentiated through two categories: measuring the UV absorption or transmission of the test product or spectrophotometric analysis. Effective sunscreen products at least need to have an absorbance of 290 to 400 mm [59].

3.6. Anti-Acne

Acne is a skin disease caused by the overgrowth of Staphylococcus epidermidis and Propionibacterium acnes. Some recent research showed that catechin from gambir and green tea could inhibit the growth of acne pathogens. Recent research showed the MIC of catechin as anti-acne is 0.1%. Another study showed that the inhibition zone of catechin is 18.45 mm for P. acnes and 15.68 mm for S. epidermidis [60]. The antibacterial mechanism of catechin involves disrupting the cell membrane and inhibiting intracellular enzymes (Figure 8). The mechanism of catechin to counter acne inflammation is through the inhibition of the pro-inflammatory cytokines IL-8 and TNF-α [61].

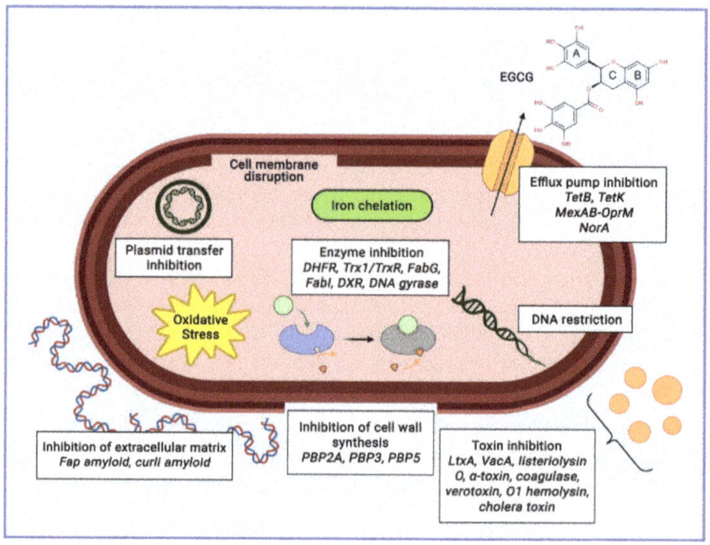

Figure 8. Catechin mechanism of action as anti-acne agent.

4. Application and Development of Catechin Formulations in Cosmetic Preparations

The skin serves as the protective outer covering and comprises three primary structural layers: The outermost part is the epidermis, the middle layer is the dermis, and the innermost part is the subcutaneous layer [62]. Over time, the skin experiences alterations influenced by various external and internal factors. These skin changes represent some of the most apparent ageing indicators and encompass features such as loose skin, fine lines, dryness, and age-related spots, often accompanied by a reduction in fat content and natural skin smoothness [63]. Since catechin is a potential antioxidant agent [34], numerous cosmetic and skincare products have been developed using catechin extracts as their main ingredient including soaps, sunscreens, shampoos, and lip balms (Table 4).

Table 4. The formulation of catechin as cosmetics modified from various formulations.

Dosage Form	Formulation	Evaluation Available	Reference
Solid Soap	Catechin (obtained from extract), Aquadest, EDTA, Olive oil, Palm oil, Stearic acid	• Organoleptic: solid, brown, smelly • Foam Test: varies from 50.94% to 64.78% • pH Test: around 8–9 • Moisture content: 15%	[64]
Sunscreen	Catechin, Cera alba, Tween 80, Ceryl Alcohol, Stearyl Alcohol	• Organoleptic: stable in colour • pH: 4.2–7.4 • SPF: 16	[65]
	Gambir Leaf Extract, Glycerin, Triethanolamine, Propylene Glycol, Aquadest	• Organoleptic: physically stable during storage • pH: 5.55–6.93 • SPF: 7 to 26.55	[66]
Solid Shampoo	Sodium Cocoyl Isethionate, Coco Glucoside, Beeswax, Shea Butter, Panthenol, Essential Oil, Lactic Acid, Tocopherol, BHT, Mango Peel Extract (containing catechin)	• Organoleptic: physically stable with surface tension of water to at least 40 mN/m • pH: 6.0–7.0 • Accelerated thermal stability test: no change regarding the texture, smell, and colour • Oxidative stability: low oxidation state	[67]
Lip Balm	Catechin Extract, Ethyl Alcohol, Lanolin, Cera alba, Propylene Glycol, Oleum rosae, Nipagin, Dye, Liquid Paraffin	• Organoleptic: physically stable during storage • Irritation Test: does not show irritation	[65]
Cream	Gambir Leaf Extract, Stearic Acid, Cetyl Alcohol, Paraffin, Isopropyl Myristic, Methylparaben, Triethanolamine, Glycerine, Perfume, Aquadest	• Organoleptic: yellow or green cream that is homogeneous and non-greasy • pH: 7.24–7.80	[68]
Pell-Off Gel Mask	Catechins, PVA, PVP K-30, Propylene Glycol, Methylparaben, Propylparaben, Ethanol 70%, Citrus Essential Oil, Distilled Water	• Organoleptic: pale yellow to light brown and featuring a scent reminiscent of oranges • pH: 5.39–5.92 • Irritation Test: does not show irritation	[69]

Zeng, 2017, found that tea polyphenols at pH 3–6 remained stable during storage at 4 and 25 °C. The more the pH decreases, the more stable the solution is. The color of the tea polyphenol solution changed from green to dark yellow with increasing temperature. The total catechin content decreased significantly when heating reached 100 °C, in addition to epimerization [70].

The pH influences the stability of catechins, with catechins being more stable at a low pH, providing more stable antioxidant activity. Temperature and light also affect stability, thus it is necessary to make appropriate formulations to maintain their stability in preparations [15]. Active compounds for topical antioxidant purposes must penetrate through the stratum corneum and enter the deeper layers of the skin but should not enter the blood vessels so that they circulate in the body. It reported that on catechin permeation, catechins are retained in the stratum corneum [15].

In another research study, Yamamoto et al. developed a novel methylated catechin produced (Figure 9) by using, as a substrate, epigallocatechin-3-O-gallate, epicatechin-3-O-gallate, or an isomer thereof. These researchers found that the novel methylated catechins can be efficiently manufactured using catechins such as EGCG as substrates. Furthermore, the novel methylated catechins obtained have excellent effects such as antiallergic, anti-cancer, anti-obesity, anti-arteriosclerosis, antihypertensive, and antimicrobial effects and can be applied to various products such as food and beverage products, pharmaceutical drugs, quasi-drugs, and cosmetics [71]:

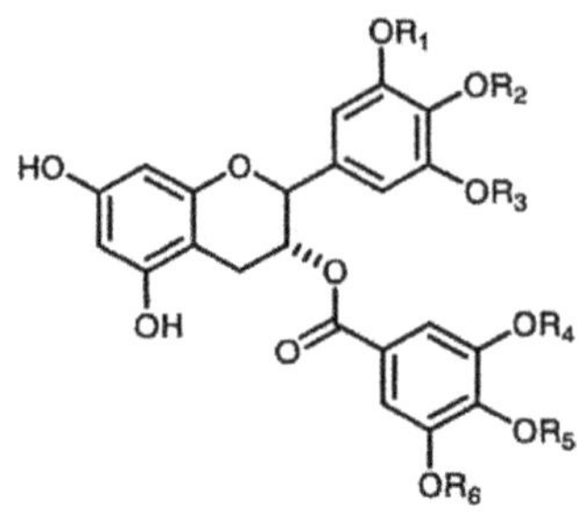

Figure 9. Novel methylated catechin [71].

The liposomal form is considered to be the best protection against possible oxidation of polyphenolic compounds. This structure is able to encapsulate both hydrophilic and lipophilic compounds, protecting them from degradation and dilution in systemic circulation. The interaction between phosphatidylcholine as a component of liposomes and polyphenolic compounds is due to the formation of hydrophobic and covalent bonds between them. This phenomenon causes a decrease in lipid absorption, while the lipids trap and protect the polyphenols. Phenolic compounds such as anthocyanins, ferulic acid, resveratrol, quercetin, and EGCG are already available in the cosmetic market in liposomal form. The liposomes have a particle size range of 50–500 nm. This form not only protects against polyphenol degradation but also enhances skin absorption [72].

5. Pre-Clinical Dan Clinical Trials of Catechins in Cosmetics and Skincare

Table 5 details the clinical trials of catechin-containing cosmetics. As of November 2023, there is an ongoing trial to determine the effect and safety of tea catechins performed by the H. Lee Moffitt Cancer Center & Research Institute. The latest trials of catechin were performed by Farrar in 2015 with the team of the Centre for Dermatology and Institute of Inflammation and Repair in Manchester, showing little to no effect of catechin from green tea on sunburn. Even with the currently available information, it is worth mentioning that catechin has an abundance of pharmacological activities that could be used as potential treatments for many diseases. The trial of catechin as an antioxidant as part of cosmetic function would increase the understanding of catechin as a multifunctional compound. Although the 2015 trial shows little potential for catechin to reduce UV, a modification and revision of the method in the future is worthy of investigation [73].

In a study on skin whitening in vivo, the catechin's transfersome form was found to have better permeation compared to the solution form of the catechin. It was also effective in inhibiting thyrokinase and was well-tolerated in guinea pig test animals. These results suggest that catechin-containing transferosomes could be a potential treatment strategy for UV-induced oxidative damage to the skin through topical administration [74].

Table 5. Clinicals trials of catechins.

Study Title	Catechin	Result/Conclusion	Ref.
Double-blinded, placebo-controlled trial of green tea extracts in the clinical and histologic appearance of photoageing skin	Green Tea, EGCG	Skin elasticity No significant differences, although histologic grading showed improvement in elastic tissue Experiment: supplement of green tea and addition of 10% green tea cream in an 8-week trial	[75]
The green tea polyphenol (-) epigallocatechin gallate and green tea can protect human cellular DNA from UV and visible radiation-induced damage	Green Tea, EGCG	Photoprotective UV radiation inhibition, prevention, and minimal DNA cell damage Experiment: study of 540 mL of green tea in 10 subjects	[76]
A randomised controlled trial of green tea beverages on the in vivo radical scavenging activity in human skin	Green Tea, EGCG	Antioxidant Increasing radical scavenging of skin by 28–29% compared to the control group Experiment: 3 cups of tea (Benifuuki tea and Yabukita tea) for 3 weeks	[77]
UV radiation-induced degradation of the dermal extracellular matrix and protection by green tea catechins	Green Tea, EGCG	Photoprotective Specific UVR protection and significant changes in acute UVR Experiment: 50 subjects were randomised to green tea catechin and vitamin C for 12 weeks with twice-daily consumption	[78]
Treatment of atopic dermatitis associated with *Malassezia sympodialis* by green tea extracts bath therapy: a pilot study	Green Tea, EGCG	Atopic dermatitis Significant improvement in 1 of 3 patients with a total reduction of 50.3% Experiment: 3 subjects bathed in a combination of green tea extract and tap water for 1 month	[79]
A randomised controlled trial of green tea catechins in protection against UV radiation-induced cutaneous inflammation	Green Tea, EGCG	Photoprotective No significant difference between the test and placebo groups Experiment: 50 subjects were randomly placed in 2 groups; group 1 was given encapsulated green tea extract with the addition of vitamins and group 2 was given a placebo twice daily for 3 months	[73]
Formulation of Gambir (*Uncaria gambir* Roxb.) ethanol extract as acne powder	Gambir Extract, Catechin	Anti-acne Inhibition diameters of 3%, 6%, and 9% produced a diameter of 3.6 mm, 4.2 mm, and 6.8 mm respectively Experiment: catechin from an ethanol extract of gambir (3%, 6%, and 9%) tested on *Staphylococcus epidermidis*	[60]
The use of green tea extract in cosmetic formulation: not only antioxidant active ingredient	Green Tea Leaf Extract	Moisturiser Significant increase in skin moisture and improved skin texture Experiment: cosmetic formulation with 6% *Camellia sinensis* extract with 24 volunteers	[80]

6. Conclusions

Catechins are flavonoids found in various plants and particularly abundant in certain foods and beverages, particularly in green tea, as well as in some fruits, such as apples and berries. These compounds have gained attention in the field of cosmeceuticals due to their potential skincare benefits. Cosmeceuticals are cosmetic products that contain biologically active ingredients with potential pharmaceutical properties designed to improve skin health and appearance. Catechins are commonly incorporated into various cosmeceutical products such as creams, serums, toners, and masks, and are often paired with other skincare ingredients for enhanced benefits. When considering products containing catechins, it is essential to consider factors like product formulation, concentration, and the specific type of catechin used. As with any skincare product, it is a good idea to consult with a dermatologist or skincare professional to determine which catechin-containing products are most suitable for specific skin concerns and needs. Additionally, it is important to perform a patch test before using any new cosmeceutical product to ensure that it does not cause adverse reactions.

Author Contributions: Conceptualisation, S.R.M., P.H. and R.M.; methodology, S.R.M. and P.H.; software, R.P.H. and D.A.D.; validation, S.R.M., P.H. and R.M.; formal analysis, N.A.P.; investigation, N.A.P., R.M. and P.H.; resources, S.R.M., R.P.H. and D.A.D.; data curation, P.H; N.A.P. and R.M.; writing—original draft preparation, R.P.H. and D.A.D.; writing—review and editing, S.R.M., P.H. and N.A.P.; visualisation, R.P.H. and D.A.D.; supervision, S.R.M., P.H. and R.M.; project administration, N.A.P.; funding acquisition, R.M. All authors have read and agreed to the published version of the manuscript.

Funding: This review was funded by UNIVERSITAS PADJADJARAN through Review Article Grant (Grant number 1549/UN6.3.1/PT.00/2023) and The APC was funded by by Academic Leadership Grant Universitas Padjadjaran of Prof. Rani Maharani (Grant number 1549/UN6.3.1/PT.00/2023.

Institutional Review Board Statement: Not applicable.

Conflicts of Interest: The authors declare no conflicts of interest.

References

1. Bellad, K.A.; Nanjwade, B.K.; Kamble, M.S.; Srichana, T.; Idris, N.F. Development of cosmeceuticals. *World J. Pharm. Pharm. Sci.* **2017**, *6*, 643–691.
2. Yamada, M.; Mohammed, Y.; Prow, T.W. Advances and controversies in studying sunscreen delivery and toxicity. *Adv. Drug Deliv. Rev.* **2020**, *153*, 72–86. [CrossRef] [PubMed]
3. Cavinato, M.; Jansen-Durr, P. Molecular mechanisms of UVB-induced senescence of dermal fibroblasts and its relevance for photoaging of the human skin. *Exp. Gerontol.* **2017**, *94*, 78–82. [CrossRef] [PubMed]
4. Rinnerthaler, M.; Bischof, J.; Streubel, M.K.; Trost, A.; Richter, K. Oxidative stress in aging human skin. *Biomolecules* **2015**, *5*, 545–589. [CrossRef] [PubMed]
5. Chang, H.H.; Chien, C.Y.; Chen, K.H.; Huang, S.C.; Chien, C.T. Catechins Blunt the Effects of oxLDL and its Primary Metabolite Phosphatidylcholine Hydroperoxide on Endothelial Dysfunction Through Inhibition of Oxidative Stress and Restoration of eNOS in Rats. *Kidney Blood Press. Res.* **2017**, *42*, 919–932. [CrossRef] [PubMed]
6. Lee, S.; Yu, J.S.; Phung, H.M.; Lee, J.G.; Kim, K.H.; Kang, K.S. Potential Anti-Skin Aging Effect of (-)-Catechin Isolated from the Root Bark of Ulmus davidiana var. japonica in Tumor Necrosis Factor-alpha-Stimulated Normal Human Dermal Fibroblasts. *Antioxidants* **2020**, *9*, 981. [CrossRef]
7. Braicu, C.; Ladomery, M.R.; Chedea, V.S.; Irimie, A.; Berindan-Neagoe, I. The relationship between the structure and biological actions of green tea catechins. *Food Chem.* **2013**, *141*, 3282–3289. [CrossRef] [PubMed]
8. Bernatoniene, J.; Kopustinskiene, D.M. The Role of Catechins in Cellular Responses to Oxidative Stress. *Molecules* **2018**, *23*, 965. [CrossRef]
9. Matsubara, T.; Wataoka, I.; Urakawa, H.; Yasunaga, H. Effect of reaction pH and CuSO4 addition on the formation of catechinone due to oxidation of (+)-catechin. *Int. J. Cosmet. Sci.* **2013**, *35*, 362–367. [CrossRef]
10. Ferreira-Nunes, R.; da Silva, S.M.M.; de Souza, P.E.N.; de Oliveira Magalhães, P.; Cunha-Filho, M.; Gratieri, T.; Gelfuso, G.M. Incorporation of *Eugenia dysenterica* extract in microemulsions preserves stability, antioxidant effect and provides enhanced cutaneous permeation. *J. Mol. Liq.* **2018**, *265*, 408–415. [CrossRef]
11. Mita, S.R.; Abdassah, M.; Supratman, U.; Shiono, Y.; Rahayu, D.; Sopyan, I.; Wilar, G. Nanoparticulate System for the Transdermal Delivery of Catechin as an Antihypercholesterol: In Vitro and In Vivo Evaluations. *Pharmaceuticals* **2022**, *15*, 1142. [CrossRef] [PubMed]
12. Munggari, I.P.; Kurnia, D.; Deawati, Y.; Julaeha, E. Current Research of Phytochemical, Medicinal and Non-Medicinal Uses of Uncaria gambir Roxb.: A Review. *Molecules* **2022**, *27*, 6551. [CrossRef] [PubMed]
13. Kim, J.M.; Heo, H.J. The roles of catechins in regulation of systemic inflammation. *Food Sci. Biotechnol.* **2022**, *31*, 957–970. [CrossRef] [PubMed]
14. Kim, E.; Hwang, K.; Lee, J.; Han, S.Y.; Kim, E.-M.; Park, J.; Cho, J.Y. Skin protective effect of epigallocatechin gallate. *Int. J. Mol. Sci.* **2018**, *19*, 173. [CrossRef] [PubMed]

15. Bae, J.; Kim, N.; Shin, Y.; Kim, S.-Y.; Kim, Y.-J. Activity of catechins and their applications. *Biomed. Dermatol.* **2020**, *4*, 1–10. [CrossRef]
16. Isemura, M. Catechin in Human Health and Disease. *Molecules* **2019**, *24*, 528. [CrossRef] [PubMed]
17. Syukri, D.; Azima, F.; Aprialdho, R. Study on the Utilization of Catechins from Gambir (*Uncaria Gambir* Roxb) Leaves as Antioxidants Cooking Oil. *Andalasian Int. J. Agric. Nat. Sci.* **2022**, *3*, 12–25. [CrossRef]
18. Zillich, O.; Schweiggert-Weisz, U.; Eisner, P.; Kerscher, M. Polyphenols as active ingredients for cosmetic products. *Int. J. Cosmet. Sci.* **2015**, *37*, 455–464. [CrossRef]
19. Moldoveanu, S.C.; David, V. *Essentials in Modern HPLC Separations*; Elsevier: Amsterdam, The Netherlands, 2022.
20. Hariyati, N. *Ministry of Health of the Republic of Indonesia Farmakope Herbal Indonesia*, 2nd ed.; Ministry of Health Indonesia: Kota Jakarta, Indonesia, 2017.
21. Cuevas-Valenzuela, J.; González-Rojas, Á.; Wisniak, J.; Apelblat, A.; Pérez-Correa, J.R. Solubility of (+)-catechin in water and water-ethanol mixtures within the temperature range 277.6–331.2 K: Fundamental data to design polyphenol extraction processes. *Fluid Phase Equilibria* **2014**, *382*, 279–285. [CrossRef]
22. Riwu, A.G.; Nugraha, J.; Purwanto, D.A.; Triyono, E.A. Determination of (+)-Catechin and Antioxidant Activity in Faloak (*Sterculia quadrifida* R. Br) Stem Bark Infusion. *Sci. Technol. Indones.* **2023**, *8*, 59–65. [CrossRef]
23. Delarosa, A.; Hendrawan, R.P.; Halimah, E. Screening of Costus speciosus and Determination of Antioxidant Potential Using DPPH Method: A Review. *Eur. J. Med. Plants* **2023**, *34*, 17–28. [CrossRef]
24. Chen, Z.; Liu, Q.; Zhao, Z.; Bai, B.; Sun, Z.; Cai, L.; Fu, Y.; Ma, Y.; Wang, Q.; Xi, G. Effect of hydroxyl on antioxidant properties of 2, 3-dihydro-3, 5-dihydroxy-6-methyl-4 H-pyran-4-one to scavenge free radicals. *RSC Adv.* **2021**, *11*, 34456–34461. [CrossRef] [PubMed]
25. Hoang, H.T.; Moon, J.-Y.; Lee, Y.-C.J.C. Natural antioxidants from plant extracts in skincare cosmetics: Recent applications, challenges and perspectives. *Cosmetics* **2021**, *8*, 106. [CrossRef]
26. Grzesik, M.; Naparło, K.; Bartosz, G.; Sadowska-Bartosz, I. Antioxidant properties of catechins: Comparison with other antioxidants. *Food Chem.* **2018**, *241*, 480–492. [CrossRef] [PubMed]
27. Sheng, Y.; Sun, Y.; Tang, Y.; Yu, Y.; Wang, J.; Zheng, F.; Li, Y.; Sun, Y. Catechins: Protective mechanism of antioxidant stress in atherosclerosis. *Front. Pharmacol.* **2023**, *14*, 1144878. [CrossRef] [PubMed]
28. Šeruga, M.; Tomac, I. Influence of chemical structure of some flavonols on their electrochemical behaviour. *Int. J. Electrochem. Sci.* **2017**, *12*, 7616–7637. [CrossRef]
29. Hassanpour, S.H.; Doroudi, A. Review of the antioxidant potential of flavonoids as a subgroup of polyphenols and partial substitute for synthetic antioxidants. *Avicenna J. Phytomed.* **2023**, *13*, 354. [PubMed]
30. Platzer, M.; Kiese, S.; Tybussek, T.; Herfellner, T.; Schneider, F.; Schweiggert-Weisz, U.; Eisner, P. Radical scavenging mechanisms of phenolic compounds: A quantitative structure-property relationship (QSPR) study. *Front. Nutr.* **2022**, *9*, 882458. [CrossRef]
31. Anitha, S.; Krishnan, S.; Senthilkumar, K.; Sasirekha, V. Theoretical investigation on the structure and antioxidant activity of (+) catechin and (−) epicatechin–a comparative study. *Mol. Phys.* **2020**, *118*, e1745917. [CrossRef]
32. Zhou, H.-C.; Tam, N.F.-y.; Lin, Y.-M.; Ding, Z.-H.; Chai, W.-M.; Wei, S.-D. Relationships between degree of polymerization and antioxidant activities: A study on proanthocyanidins from the leaves of a medicinal mangrove plant *Ceriops tagal*. *PLoS ONE* **2014**, *9*, e107606. [CrossRef]
33. Spiegel, M.; Andruniów, T.; Sroka, Z. Flavones' and Flavonols' Antiradical Structure–Activity Relationship—A Quantum Chemical Study. *Antioxidants* **2020**, *9*, 461. [CrossRef] [PubMed]
34. Munteanu, I.G.; Apetrei, C. Assessment of the Antioxidant Activity of Catechin in Nutraceuticals: Comparison between a Newly Developed Electrochemical Method and Spectrophotometric Methods. *Int. J. Mol. Sci.* **2022**, *23*, 8110. [CrossRef] [PubMed]
35. Doshi, P.; Adsule, P.; Banerjee, K.; Oulkar, D. Technology. Phenolic compounds, antioxidant activity and insulinotropic effect of extracts prepared from grape (*Vitis vinifera* L) byproducts. *J. Food Sci. Technol.* **2015**, *52*, 181–190. [CrossRef] [PubMed]
36. Dias, M.C.; Pinto, D.C.; Silva, A.M.S. Plant flavonoids: Chemical characteristics and biological activity. *Molecules* **2021**, *26*, 5377. [CrossRef] [PubMed]
37. Nuryana, I.; Ratnakomala, S.; Fahrurrozi, A.B.J.; Andriani, A.; Putra, F.J.N.; Rezamela, E.; Wulansari, R.; Prawira-Atmaja, M.I.; Lisdiyanti, P. Catechin Contents, Antioxidant and Antibacterial Activities of Different Types of Indonesian Tea (*Camellia Sinensis*). *Ann. Bogor.* **2020**, *24*, 107. [CrossRef]
38. Anggraini, T.; Wilma, S.; Syukri, D.; Azima, F.J.I.J.o.F.S. Total phenolic, anthocyanin, Catechins, DPPH radical scavenging activity, and toxicity of *Lepisanthes alata* (Blume) Leenh. *Int. J. Food Sci.* **2019**, *2019*, 9703176. [CrossRef]
39. Musdja, M.Y.; Rahman, H.A.; Hasan, D.J.L.I.J.H.L.-S. Antioxidant activity of catechins isolate of Uncaria gambier Roxb in male rats. *LIFE Int. J. Health Life-Sci.* **2018**, *4*, 34–46. [CrossRef]
40. Rahmi, M.; Rita, R.S.; Yetti, H. Gambir Catechins (*Uncaria gambir* Roxb) Prevent Oxidative Stress in Wistar Male Rats Fed a High-Fat Diet. *Maj. Kedokt. Andalas* **2021**, *44*, 436–441.
41. Yetuk, G.; Pandir, D.; Bas, H. Protective role of catechin and quercetin in sodium benzoate-induced lipid peroxidation and the antioxidant system in human erythrocytes in vitro. *Sci. World J.* **2014**, *2014*, 874824. [CrossRef]
42. Amber, K.T.; Shiman, M.I.; Badiavas, E.V. The use of antioxidants in radiotherapy-induced skin toxicity. *Integr. Cancer Ther.* **2014**, *13*, 38–45. [CrossRef]

43. Zheng, M.; Liu, Y.; Zhang, G.; Yang, Z.; Xu, W.; Chen, Q. The Applications and Mechanisms of Superoxide Dismutase in Medicine, Food, and Cosmetics. *Antioxidants* **2023**, *12*, 1675. [CrossRef] [PubMed]
44. Boo, Y.C. Arbutin as a Skin Depigmenting Agent with Antimelanogenic and Antioxidant Properties. *Antioxidants* **2021**, *10*, 1129. [CrossRef] [PubMed]
45. Jesus, A.; Mota, S.; Torres, A.; Cruz, M.T.; Sousa, E.; Almeida, I.F.; Cidade, H. Antioxidants in Sunscreens: Which and What For? *Antioxidants* **2023**, *12*, 138. [CrossRef] [PubMed]
46. Nahhas, A.F.; Abdel-Malek, Z.A.; Kohli, I.; Braunberger, T.L.; Lim, H.W.; Hamzavi, I.H. The potential role of antioxidants in mitigating skin hyperpigmentation resulting from ultraviolet and visible light-induced oxidative stress. *Photodermatol. Photoimmunol. Photomed.* **2019**, *35*, 420–428. [CrossRef] [PubMed]
47. Andrade, J.M.; Domínguez-Martín, E.M.; Nicolai, M.; Faustino, C.; Rodrigues, L.M.; Rijo, P. Screening the dermatological potential of plectranthus species components: Antioxidant and inhibitory capacities over elastase, collagenase and tyrosinase. *J. Enzyme Inhib. Med. Chem.* **2021**, *36*, 258–270. [CrossRef] [PubMed]
48. Sonawane, G.B.; Jadhav, S.P.; Patil, C.D.; Kamble, P.R.; Somavanshi, D.B. A Review on the Antioxidant and Antiaging Properties of White Tea. *J. Pharm. Res. Int.* **2021**, *33*, 129–136.
49. Wang, W.; Di, T.; Wang, W.; Jiang, H. EGCG, GCG, TFDG, or TSA inhibiting melanin synthesis by downregulating MC1R expression. *Int. J. Mol. Sci.* **2023**, *24*, 11017. [CrossRef] [PubMed]
50. Zhang, X.; Li, J.; Li, Y.; Liu, Z.; Lin, Y.; Huang, J.-A. Anti-melanogenic effects of epigallocatechin-3-gallate (EGCG), epicatechin-3-gallate (ECG) and gallocatechin-3-gallate (GCG) via down-regulation of cAMP/CREB/MITF signaling pathway in B16F10 melanoma cells. *Fitoterapia* **2020**, *145*, 104634. [CrossRef]
51. Jiang, T.; Qi, Y.; Wu, Y.; Zhang, J. Application of antioxidant and ultraviolet absorber into HDPE: Enhanced resistance to UV irradiation. *e-Polymers* **2019**, *19*, 499–510. [CrossRef]
52. Laksmiani, N.P.L.; Sanjaya, I.K.N.; Leliqia, N.P.E. The activity of avocado (*Persea americana* Mill.) seed extract containing catechin as a skin lightening agent. *J. Pharm. Pharmacogn. Res.* **2020**, *8*, 449–456.
53. Vale, E.P.; dos Santos Morais, E.; de Souza Tavares, W.; de Sousa, F.F.O. Epigallocatechin-3-gallate loaded-zein nanoparticles: Characterization, stability and associated antioxidant, anti-tyrosinase and sun protection properties. *J. Mol. Liq.* **2022**, *358*, 119107. [CrossRef]
54. Donglikar, M.M.; Deore, S.L. Sunscreens: A review. *Pharmacogn. J.* **2016**, *8*, 171–179. [CrossRef]
55. Sander, M.; Sander, M.; Burbidge, T.; Beecker, J. The efficacy and safety of sunscreen use for the prevention of skin cancer. *CMAJ* **2020**, *192*, E1802–E1808. [CrossRef] [PubMed]
56. Dewi, D.A.R. Sunscreen Protection Against Visible Light: Is It Needed? *Malahayati Nurs. J.* **2022**, *4*, 2527–2536. [CrossRef]
57. Latha, M.; Martis, J.; Shobha, V.; Shinde, R.S.; Bangera, S.; Krishnankutty, B.; Bellary, S.; Varughese, S.; Rao, P.; Kumar, B.N.; et al. Sunscreening agents: A review. *J. Clin. Aesthet. Dermatol.* **2013**, *6*, 16. [PubMed]
58. Portilho, L.; Aiello, L.M.; Vasques, L.I.; Bagatin, E.; Leonardi, G.R. Effectiveness of sunscreens and factors influencing sun protection: A review. *Braz. J. Pharm. Sci.* **2023**, *58*, e20693. [CrossRef]
59. Ebrahimzadeh, M.A.; Enayatifard, R.; Khalili, M.; Ghaffarloo, M.; Saeedi, M.; Charati, J.Y. Correlation between sun protection factor and antioxidant activity, phenol and flavonoid contents of some medicinal plants. *Iran J. Pharm. Res.* **2014**, *13*, 1041.
60. Warnida, H.; Masliyana, A.; Sapri, S.J. Formulasi Ekstrak Etanol Gambir (*Uncaria gambir* Roxb.) dalam Bedak Anti Jerawat. *J. Ilm. Manuntung* **2016**, *2*, 99–106. [CrossRef]
61. Messire, G.; Serreau, R.; Berteina-Raboin, S. Antioxidant Effects of Catechins (EGCG), Andrographolide, and Curcuminoids Compounds for Skin Protection, Cosmetics, and Dermatological Uses: An Update. *Antioxidants* **2023**, *12*, 1317. [CrossRef]
62. Liu, B.; Li, A.; Xu, J.; Cui, Y. Single-Cell Transcriptional Analysis Deciphers the Inflammatory Response of Skin-Resident Stromal Cells. *Front. Surg.* **2022**, *9*, 935107. [CrossRef]
63. Shanbhag, S.; Nayak, A.; Narayan, R.; Nayak, U.Y. Anti-aging and sunscreens: Paradigm shift in cosmetics. *Adv. Pharm. Bull.* **2019**, *9*, 348. [CrossRef]
64. Estikomah, S.; Tussifah, H.; Kusumaningtyas, N.; Sholihatin, B.; Dinta, L. Formulation of Solid Soap Combination of Green Tea Leaf (*Camellia sinnensis* L.) and Corn Kernel (*Zea mays*) Extracts. In Proceedings of the U-Go Healthy International Conference, U-Go Healthy 2020, Pacitan, Indonesia, 29 March 2020.
65. Kamal, S.; Rusdi, M.S. Utilization of catechins in sunscreen lotion formulation. *Borneo J. Pharm.* **2018**, *1*, 68–71. [CrossRef]
66. Winarti, C. Physical Characteristics and UV Protection (In Vitro) of Gambier Leaf Extract Lotion. *IOP Conf. Ser. Earth Environ. Sci.* **2022**, *1024*, 012061. [CrossRef]
67. Brito, I.; Ferreira, S.M.; Santos, L. On the Path to Sustainable Cosmetics: Development of a Value-Added Formulation of Solid Shampoo Incorporating Mango Peel Extract. *Cosmetics* **2023**, *10*, 140. [CrossRef]
68. Supiati, S. The Quality of Cream Formulated From Gambier Leaf Extract. *IOP Conf. Ser. Earth Environ. Sci.* **2022**, *1024*, 012010.
69. Rosaini, H.; Makmur, I.; Lestari, E.A.; Sidoretno, W.M.; Yetti, R.D. Formulation of Gel Peel Off Catechins Mask from Gambir (*Uncaria gambir* (Hunter) Roxb) with the PVP K-30 Concentration Variation. *Int. J. Res. Rev.* **2021**, *8*, 205–211.
70. Zeng, L.; Ma, M.; Li, C.; Luo, L. Stability of tea polyphenols solution with different pH at different temperatures. *Int. J. Food Prop.* **2017**, *20*, 1–18. [CrossRef]
71. Yamamoto, M.; Kirita, M.; Honma, D.; Yokota, T. Novel Methylated Catechin and Composition Containing the Same. U.S. Patent 20100324312A1, 23 December 2010.

72. Figueroa-Robles, A.; Antunes-Ricardo, M.; Guajardo-Flores, D. Encapsulation of phenolic compounds with liposomal improvement in the cosmetic industry. *Int. J. Pharm.* **2021**, *593*, 120125. [CrossRef] [PubMed]
73. Farrar, M.D.; Nicolaou, A.; Clarke, K.A.; Mason, S.; Massey, K.A.; Dew, T.P.; Watson, R.E.; Williamson, G.; Rhodes, L.E. A randomized controlled trial of green tea catechins in protection against ultraviolet radiation–induced cutaneous inflammation. *Am. J. Clin. Nutr.* **2015**, *102*, 608–615. [CrossRef]
74. Hsieh, W.-C.; Fang, C.-W.; Suhail, M.; Vu, Q.L.; Chuang, C.-H.; Wu, P.-C. Improved skin permeability and whitening effect of catechin-loaded transfersomes through topical delivery. *Int. J. Pharm.* **2021**, *607*, 121030. [CrossRef]
75. Chiu, A.E.; Chan, J.L.; Kern, D.G.; Kohler, S.; Rehmus, W.E.; Kimball, A.B. Double-blinded, placebo-controlled trial of green tea extracts in the clinical and histologic appearance of photoaging skin. *Dermatol. Surg.* **2005**, *31*, 855–860. [CrossRef] [PubMed]
76. Morley, N.; Clifford, T.; Salter, L.; Campbell, S.; Gould, D.; Curnow, A. The green tea polyphenol (−)-epigallocatechin gallate and green tea can protect human cellular DNA from ultraviolet and visible radiation-induced damage. *Photodermatol. Photoimmunol. Photomed.* **2005**, *21*, 15–22. [CrossRef] [PubMed]
77. Megow, I.; Darvin, M.E.; Meinke, M.C.; Lademann, J. A randomized controlled trial of green tea beverages on the in vivo radical scavenging activity in human skin. *Skin. Pharmacol. Physiol.* **2017**, *30*, 225–233. [CrossRef] [PubMed]
78. Charoenchon, N.; Rhodes, L.E.; Nicolaou, A.; Williamson, G.; Watson, R.E.; Farrar, M.D. Ultraviolet radiation-induced degradation of dermal extracellular matrix and protection by green tea catechins: A randomized controlled trial. *Clin. Exp. Dermatol.* **2022**, *47*, 1314–1323. [CrossRef] [PubMed]
79. Kim, H.K.; Chang, H.K.; Baek, S.Y.; Chung, J.O.; Rha, C.S.; Kim, S.Y.; Kim, B.J.; Kim, M.N. Treatment of atopic dermatitis associated with *Malassezia sympodialis* by green tea extracts bath therapy: A pilot study. *Mycobiology* **2012**, *40*, 124–128. [CrossRef]
80. Gianeti, M.D.; Mercurio, D.G.; Maia Campos, P.M. The use of green tea extract in cosmetic formulations: Not only an antioxidant active ingredient. *Dermatol. Ther.* **2013**, *26*, 267–271. [CrossRef]

Disclaimer/Publisher's Note: The statements, opinions and data contained in all publications are solely those of the individual author(s) and contributor(s) and not of MDPI and/or the editor(s). MDPI and/or the editor(s) disclaim responsibility for any injury to people or property resulting from any ideas, methods, instructions or products referred to in the content.

cosmetics

MDPI

Article

Effect of Incorporating a Biowax Derived from Hydroprocessing of Crude Palm Oil in a Facial Cream and a Blemish Balm Cream

Laura Aguilar [1,*], Jonathan Hernández [1], Luis Javier López-Giraldo [1] and Ronald Mercado [2]

[1] Grupo de Investigación en Ciencia y Tecnología de Alimentos-CICTA, Escuela de Ingeniería Química-Universidad Industrial de Santander, Bucaramanga 680002, Santander, Colombia; jonathanicolas1@gmail.com (J.H.); ljlopez@uis.edu.co (L.J.L.-G.)

[2] Grupo de Investigación en Fenómenos Interfaciales, Reología y Simulación de Transporte-FIRST, Escuela de Ingeniería Química-Universidad Industrial de Santander, Bucaramanga 680002, Santander, Colombia; ramerca@uis.edu.co

* Correspondence: laura.aguilar.na@gmail.com; Tel.: +57-3163286224

Abstract: Most waxes used as cosmetic ingredients are derived from the petrochemical industry. A modern alternative to this complex synthesis approach is the hydrotreatment of palm oil; thus, the aim of this study is to evaluate the effect of incorporating a biowax derived from hydroprocessing of crude palm oil as a new natural cosmetic ingredient in facial cream and BB cream. Therefore, two water in oil (W/O) emulsions, one including pigments, with five different weight percentages were developed and subjected to further sensory evaluation by a trained panel to estimate the level of acceptance. Moreover, resistance to centrifugation, pH, spreadability, phase separation, viscosity and storage modulus were the parameters evaluated in a preliminary stability study using thermal stress. Sensory analysis showed that the highest level of acceptance was obtained between 3% and 9 wt% biowax. For both prototypes, increasing biowax percentage led to a greater effect on stickiness, the viscosity increased, and extensibility decreased. The formulations were able to maintain their pH. The best stability for BB cream was observed at 9%, since the changes in the properties were slight. For facial cream, the emulsion was more stable at intermediate biowax content. It was observed that biowax exhibits favorable characteristics as an emollient or thickening agent. Finally, the formulations with the best stability and sensory characteristics were obtained at 9 wt% biowax.

Keywords: biowax; preliminary stability; sensory profile; level of acceptance

Citation: Aguilar, L.; Hernández, J.; López-Giraldo, L.J.; Mercado, R. Effect of Incorporating a Biowax Derived from Hydroprocessing of Crude Palm Oil in a Facial Cream and a Blemish Balm Cream. *Cosmetics* **2023**, *10*, 123. https://doi.org/10.3390/cosmetics10050123

Academic Editor: Agnieszka Feliczak-Guzik

Received: 23 June 2023
Revised: 14 August 2023
Accepted: 15 August 2023
Published: 5 September 2023

Copyright: © 2023 by the authors. Licensee MDPI, Basel, Switzerland. This article is an open access article distributed under the terms and conditions of the Creative Commons Attribution (CC BY) license (https://creativecommons.org/licenses/by/4.0/).

1. Introduction

Currently, the global market shows a growing preference for natural cosmetics, with an estimated annual growth rate of 15%, compared to 5% for traditional cosmetics [1]. In fact, the skincare market is expected to contribute significantly to the growth of the industry, followed by hair care and, in third place, color makeup, such as blemish balm cream (BB cream), has great potential due to the protection it provides to the skin [1].

In Colombia, the cosmetics sector has had an annual growth rate of 8.7% [2], higher than Latin America, which has made it one of the most important producers and the one with the best projection. Furthermore, Colombia is one of the countries with the richest soil and a wide variety of climates, which favor the production of natural ingredients. Despite this, Asociación Nacional de Industriales de Colombia (ANDI) shows most raw materials are imported (85%) [3], and many of them are derived from palm oil of the *Elaeis Guineensis* tree. Therefore, ANDI emphasizes that if Colombia's palm oil sector ventures into the production of these raw materials, mainly emulsifiers and emollients, it would increase the country's competitiveness considerably [3].

Palm oil production increased by 8.2% in the last year, from 1.6 to 1.7 million tons [4]. Likewise, 27% of the crude palm oil produced in the country has certified sustainability, which is a differential factor and an added value due to the palm oil industry having a high environmental impact [4].

In addition, the availability of FDA-grade wax for the cosmetics industry is limited to imports from China, Brazil, Argentina and Germany. However, there is an opportunity to produce it from sources available in the national oleochemical industry of Colombia as palm oil is composed of triglycerides, combinations of glycerol and different fatty acids, which makes it a sensorially interesting cosmetic ingredient [5].

Waxes are commonly used as lubricants, adhesives, foodstuffs, pharmaceuticals and cosmetics. Seventy percent of waxes are made from petroleum; nevertheless, the depletion of reserves, fluctuating crude oil prices and sustainability concerns have led to the search for new alternatives. Thus, vegetable oils (VOs) represent a promising source for producing renewable and environmentally friendly wax alternatives [6].

Ecopetrol has been developing a technology that involves FDA-grade wax production from the hydroprocessing of crude palm oil [7,8], known as biowax. This technology has desirable characteristics for the cosmetics industry, especially as an emollient and thickening agent. biowax was obtained following the methodology described in Guzman et al., 2010 and 2013 [9,10], modifying the reaction conditions and catalysts used according to Olarte et al., 2023 [8], as follows: Biowax was obtained in a pilot-scale fixed-bed reactor using a nickel molybdenum catalyst ($NiMo/Al_2O_3$), with temperatures ranging from 240 to 260 °C, pressures between 800 and 1300 psig, and liquid hourly space velocity (LHSV) of 1 to 2 h-1. According to stability studies carried out by Laura Chaparro et al. 2023 [11], this biowax (BPW7) consists of a mixture of fatty esters (24.80%), paraffin (29.60%), triglycerides (TGs, 22.00%), diglycerides (DGs, 1.7%), fatty acids (FFAs, 14.30%) and fatty alcohols (7.60%). In addition, biowax has physicochemical characteristics that make it attractive for use in the cosmetics industry [12].

Especially in cosmetics, waxes have been included in the formulation of many products, one of which is facial creams and BB creams. Both products are emulsified systems, typically in color cosmetics, such as water in silicone (w/si); however, silicones have been rejected due to their environmental impact and have been replaced by oils with similar sensory characteristics [13]. However, switching from silicone to oils brings with it the challenge of achieving stable emulsions with sensory profiles acceptable to the consumer; thus, it is important to include a stability prediction and evaluation of sensory parameters.

Sensory analysis is an important parameter for the cosmetic industries, as sensory evaluation data are used in marketing decisions and have shown to be important in the development of cosmetic products designed to delight the consumer's senses [14]. Another important parameter is the study of accelerated stability studies using thermal stress, which is a good tool for inducing emulsion alterations [15]. Some physicochemical parameters that help to predict stability are viscosity, storage modulus, pH and phase separation.

Despite the above, few studies have been found in which the stability and sensorial performance of facial creams using waxes of natural origin have been evaluated. And, consequently, biowax has not been produced at a pilot scale in Colombia. Therefore, the aim of this study is to evaluate the effect of incorporating a biowax derived from hydroprocessing of crude palm oil, as a new cosmetic ingredient of natural origin, in facial cream and BB cream.

2. Materials and Methods

2.1. Preparation of Formulations

The preparation of the cream began by melting the biowax (3, 6, 9, 12, or 15%) with Plantsil (phase B) at 60 °C, until a single phase is formed; subsequently, the Emulium Illustro (emulsifier), Cetiol 5C, Plantsil (emollients), silicone and, in the case of the BB cream, the pigments, (Phase A) were added to this container; this oily phase was homogenized using an Ultra Turrax (IKA T25 digital) at 3000 rpm for 5 min. Then, the aqueous phase (Phase C) consisting of water, salts and the chelating agent was slowly added to the container and mixed at 3000 rpm for 10 min using an Ultra Turrax and left to cool to 40 °C. Finally, additives such as tocopheryl acetate (antioxidant), Patch H_2O (wetting active), preservative, silica, fragrance and pH stabilizer (citric acid), if necessary, were added below 40 °C (See Table 1). This procedure was carried out for both W/O emulsion formulations (BB cream and facial cream) at 5 different biowax weight percentages (3, 6, 9, 12, and 15%).

Table 1. General formulation, raw materials distributors, and functions.

Phase	Raw Material	Company	% W/W [a]	Function
A	Emulium Illustro	GATTEFOSSÉ	5.239%	Emulsifier
	Natura-Tec Plantsil	NAURA TEC	13.388%	Emollient
	Mirasil C-DML	IMCD	3.492%	Silicone
	Cetiol 5C	BASF Care Creations	5.239%	Emollient
B	Natura-Tec Plantsil	NAURA TEC	5.821%	Emollient
	BIOWAX	--	0% [b]	Biowax
C	Water	Type 1	56.809%	Water
	Sodium chloride	NATIVUS S.A.S.	1.048%	Stabilizer
	Magnesium sulphate	NATIVUS S.A.S.	1.048%	Stabilizer
	Disodium EDTA	NATIVUS S.A.S.	0.117%	Chelating agent
D	Euxyl PE 9010	ASHLAND	1.048%	Preservative
	MSS-5003H	KOBO	3.143%	Powder
	Tocopheryl Acetate	NATIVUS S.A.S.	0.116%	Antioxidant
	Patch H_2O	BASF Care Creations	3.143%	Active ingredient
	Aqua Fragrance	--	0.349%	Fragrance
	TOTAL		100%	

[a]: All the components were normalized in each sample in function of biowax percentage; [b]: the biowax was added at 5 different weight percentages (3, 6, 9, 12, and 15%) for BB cream and facial cream.

2.2. Sensory Analysis

The sensory analysis of the developed prototypes was performed by a sensory panel composed of 19 members, aged between 18 and 25 years. Each member was previously trained according to ASTM E1490-19 [16]. During training, they were introduced to cosmetics and to the general concept of the study, followed by a detailed explanation of the sensory panel and the test that comprises it. Raw materials defined by the standard as sensory extremes for each property to be evaluated were used; these properties are presented in Table 2. The training began with 30 people, and, as indicated in the standard, a test was performed on each of the trainees, which consisted of a sensory evaluation of creams and raw materials. Those who answered correctly were considered suitable for the sensory panel and went on to evaluate the prototypes.

Table 2. Sensory properties evaluated in the prototypes.

Property	Maximum Value
Stickiness	The force required to separate fingertips is minimal
Spreadability	Product spreads very easily on the skin
Absorbency	The number of rubs at which the product loses its wet, moist feel and the resistance perceived is minimal
Gloss	The amount of reflected light from the product is minimal
Softness	The degree to which the skin feels soft is maximum
Coverage	The coverage of color imperfections (freckles, facial spots) is maximum
Uniformity	Maximum product continuity on the skin
Adherence	The product is perfectly adhered to the area of application
Amount of residue	The amount of product remaining on the skin is minimal

The evaluation of each of the prototypes followed the methodology of Renata Moschini Daudt [15] and the standard used during the training. The panelists performed a process of washing the area before each application of the sample, applied an equal amount of product on the forearm for each test and, finally, gave a score from 0 to 10 for each aspect evaluated, with 10 being the score with the highest level of acceptance.

2.3. Stability Tests

When the formulations reached room temperature, they were packaged in transparent glass containers of 10 mL for organoleptic tests and 30 mL for physicochemical tests, which were taken to the stability test by thermal stress. The samples were subjected to 6 cycles of cooling and heating. Each cycle consisted of 24 h in an oven at 40 °C, followed by 24 h in the refrigerator at 4 °C [15]. The parameters analyzed for each sample before and after the cycles were resistance to centrifugation, pH, spreadability, phase separation, viscosity curve and storage modulus.

2.4. Resistance to Centrifugation

The resistance to centrifugation test was performed in a centrifuge (SL8R THERMO SCIENTIFIC) using 10 mL samples (3000 rpm, for 30 min, a 20 °C) [15]. At the end of the test, if there was a separation of the cream at the bottom or the top, this value was measured for subsequent analysis.

2.5. pH

The pH values were measured at room temperature directly in the formulation using a previously calibrated digital pH meter (SI Analytics HandyLab 100).

2.6. Spreadability

The spreadability was evaluated by depositing 0.1 g of sample in the center of a flat glass dish. This container was positioned on a sheet of graph paper with the axis in the center of the sample, then a flat glass dish (45.4 g) was placed on the sample. After 1 min, the surface covered was measured with the diameter at 8 different points. The spreadability was calculated as the area covered by the sample.

2.7. Rheological Characterization

The rheological parameters measured were viscosity curves and storage modulus. The measurements were carried out in a rheometer (Anton Paar MCR72) using parallel plates geometry (50 mm diameter, gap of 1.5 mm). All measurements were performed at 25 °C.

Viscosity curves were obtained using a rotational test, recording shear stress values when shearing the samples at increasing shear rates from 0.01 to 100 s^{-1}. The storage modulus and loss modulus were measured with an oscillatory test at a constant frequency (1 Hz), and the strain was swept from 0.001% to 10%.

2.8. Phase Separation

The observed separation of the creams at the top and bottom of the 10 mL vial was measured before and after the heat stress cycles; the cream was left to stand for the entire duration.

2.9. Statistical Analysis

To determine whether biowax percentage affects the parameters studied above, an analysis of variance (ANOVA) and Tukey HSD test were performed for the result of the sensory panel; for the stability study, a multifactorial analysis of variance was applied. The software used for these studies was Statistica™ V. 14.

2.10. Decision Matrix

For the selection of biowax percentage with the best sensory and physicochemical properties, a decision matrix was employed according to the methodology used by Medina Godoy and Diana Gabriela [17]. For this purpose, with the support of a group of experts, the weights for each evaluated aspect were defined.

The score (St) obtained by each aspect is given by Equation (1).

$$S_t = S * V \quad (1)$$

where S is the weight assigned to the criterion, being negative if its desirability is minimum and positive if it is maximum, and V is the value obtained for the property being weighted.

The total obtained for each biowax percentage is given by Equation (2). The best score was compared with a commercial product (moisturizing cream and BB cream Natura Faces)

$$T_i = \sum S_{ti} \quad (2)$$

3. Results and Discussion

3.1. Sensory Analysis

The skin sensory performance of personal care products is an important factor for the sales potential of any cosmetic product [18]. The p-value statistics of analysis of variance for every sensory characteristic studied in both formulations are shown in Table 3.

Table 3. Table ANOVA, p-value for facial cream and BB cream with respect to the amount of biowax. (p-value < 0.05 is statistically significant).

Property	Facial Cream	BB Cream
Spreadability	0.0001	0.0581
Absorbency	0.172	0.1237
Softness	0.0705	0.0149
Gloss	0.4805	0.1971
Stickiness	0.0018	0.0000
Amount of residue	0.0631	0.0000
Coverage	N.A.*	0.0870
Uniformity	N.A.*	0.1509
Adherence	N.A.*	0.0521

N.A.*: Not applicable.

The p-value presented in Table 3 reveals whether biowax percentage affects the parameters studied above; if it is less than 0.05, there is a statistically significant difference between the means. Thus, regarding facial cream, it is observed that biowax percentage influences stickiness and spreadability, whereas, for BB cream, biowax percentage affects the parameters softness, stickiness and amount of residue. The differences in the statistically significant properties are attributed to the fact that the pigments increase the viscosity of the cream, making it less extensible in all its percentages of biowax; in addition, due to the layer that leaves, the BB cream on the skin increases the amount of residue and the sticky sensation of the cream when absorbed. As a result, biowax percentage is not statistically significant for gloss, absorbency or aspects related to color.

As a verifying factor, Tukey's test was performed for the variables evaluated, and the results are shown in Tables 4 and 5, which confirms that there are statistical differences for both formulations in the aspects mentioned between the means of the biowax percentages of 3% and 15%.

Increasing biowax percentage also increases stickiness and the amount of cream residue on the skin but decreases spreadability and softness (Figure 1), which demonstrates that the biowax significantly influenced in the formulations, agreeing with previous studies [19] that demonstrated that the constituents of the oil phase affected the sensory aspects of emulsions. Figure 1 has a division at 9%; this is because at lower percentages there is a better acceptance, since the cream is perceived as softer and more spreadable and less viscous and sticky and leaves less residue.

Table 4. Tukey's test for stickiness and spreadability for significant different factors for facial cream (if the studied parameter has the same "X" or "XX", its average is similar).

%Biowax	Stickiness		Spreadability	
	Average	Homogeneous Groups	Media	Homogeneous Groups
3%	1.47	X	8.42	X
6%	1.84	XX	7.74	XX
9%	2.26	XX	6.89	XX
12%	2.74	XX	7.26	XX
15%	2.95	X	6.26	X

Table 5. Tukey's test for stickiness and spreadability for significant different factors for BB cream. (if the studied parameter has the same "X" or "XX", its average is similar).

%Biowax	Softness		Stickiness		Amount of residue	
	Average	Homogeneous Groups	Average	Homogeneous Groups	Average	Homogeneous Groups
3%	7.11	X	1.41	X	2.41	X
6%	7.17	X	3.05	X	2.29	X
9%	6.41	XX	3.47	XX	2.23	X
12%	5.35	X	3.58	X	4.35	X
15%	5.88	XX	4.64	X	5.17	X

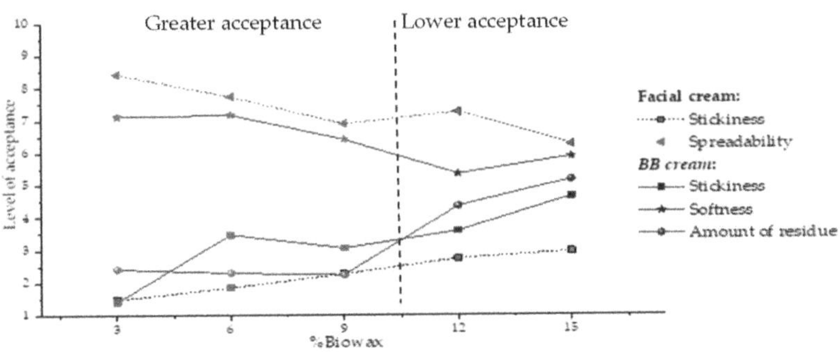

Figure 1. Statistically significant properties with respect to biowax percentage for the facial cream and the BB cream.

Hence, it is possible to affirm that biowax derived from the hydroprocessing of crude palm oil has important effects on the cosmetics formulation developed and could be considered as an emollient, as it is closely related to the sensory properties of emulsions and its performance on the skin [20]. Moreover, emollients have a major impact on physicochemical properties of cosmetic emulsions such as consistency and spreadability, properties that are important to achieve adequate efficacy and user acceptance of the products [21]. Particularly in cosmetics, waxes are formulated in numerous personal care products due to their excellent emollient behavior [22], which is consistent with the results of this study. Regarding the sensory characteristics desired for these cosmetic prototypes, the highest level of pleasure is obtained by incorporating biowax between 3 and 9% (Figure 1), as it is perceived as an easy-to-spread, smooth cream with little residue attributes that are of great importance for its application on areas of the skin, especially on the face.

3.2. Resistance to Centrifugation

After performing the test, the length of each of the differentiated phases in each of the prototypes was measured; the results after stability are shown in Table 6. It is important to note that before thermal stress, prototypes showed no phase separation. As shown in

Table 6 and Figure 2, for facial cream, the highest clarification of the oily phase (top) was presented for concentrations lower than 6%. On the other hand, at concentrations higher than 12%, in addition to the separation of the oily phase, coalescence of water free (bottom) was observed (Figure 2).

In the case of BB cream, a higher phase separation was observed at low biowax values; above 9%, there was no perceptible difference in phase separation. This result can be attributed to the amount of biowax, as mentioned by Norton and Norton [23], because in a W/O emulsion the addition of waxes provides greater connections in the continuous phase of the emulsion, increasing the resistance to flow of the aqueous phase. The difference in the result for the face cream and the BB cream is attributed to the increase in viscosity that occurs when pigments are incorporated into the formulation, and a higher viscosity also increases the resistance to flow of the dispersed phase.

Table 6. Average pH values, centrifugation and phase separation after stability.

% of Biowax	Cream Type	Resistance to Centrifugation (mm)	pH		Phase Separation (mm)
			Initial	Final	
3%	Facial cream BB cream	18 16	5.22 5.11	5.81 4.20	1 0
6%	Facial cream BB cream	14 7	5.29 5.30	5.73 4,35	0 0
9%	Facial cream BB cream	9 48	5.21 5.29	5.88 4.30	3 0
12%	Facial cream BB cream	27 48	5.26 5.21	5.76 4.30	3.5 1
15%	Facial cream BB cream	17 28	5.24 5.34	5.55 4.40	4 3

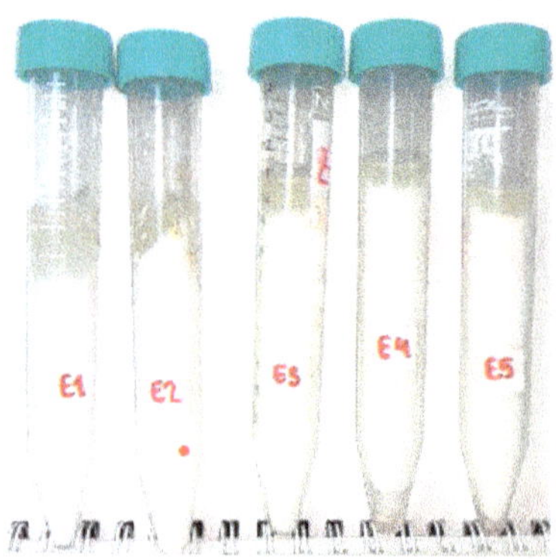

Figure 2. Facial cream subjected to centrifugal test after accelerated stability.

3.3. Phase Separation

The phase separation values were recorded and presented in Table 6. It is observed that at concentrations higher than 12% of biowax, the BB cream presents a mostly white supernatant, which has the texture and consistency of biowax mixed with the oily phase. In the case of the facial cream, this supernatant is opaque white, but sensorially it has the same characteristics as the other supernatant. Figure 3a,b show the results.

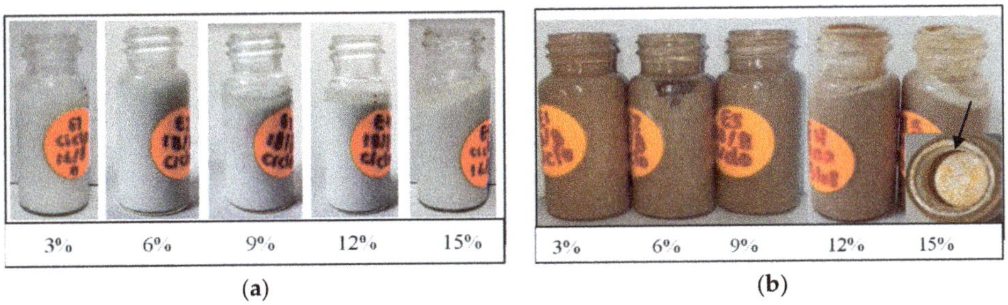

Figure 3. Emulsions after preliminary stability: (**a**) facial cream, (**b**) BB cream.

As has been demonstrated in some studies, a higher viscosity value allows the formulation to remain stable for a longer time because the sedimentation of droplets is slower, and, in some systems, it is also the liquid film drainage; however, the presence of the supernatant indicates that there is a maximum concentration at which the biowax, during the cooling and heating cycles, shows a substantial separation from the emulsion and is located on the surface of the cream. This behavior is due to the crystallization of the biowax as explained by Supratim Ghosh and Dérick Rousseau [24] due to the expansion and compression cycles of the wax crystals, contributing to the coalescence of the dispersed phase.

3.4. pH

The pH of the formulations was measured directly in the creams before and after stability. The results are presented in Table 6, from which it is concluded that the creams are able to maintain their pH because the values obtained for the formulations are located within the range established for a facial cream and a BB cream, 4.2 to 6.5 [25].

The pH for the facial cream after preliminary stability increased as shown in Table 6, which is consistent with the results presented in previous research on stability [25]. On the other hand, the pH of the BB cream decreased (Table 6) after the test. This is a behavior presented in other works [26]. This difference in pH response is attributed to the presence of the pigments either due to the possible oxidation or the coating they present, since it is the only variation between the two formulations developed.

3.5. Spreadability

Spreadability of all the formulations was evaluated, and for both prototypes a similar behavior was obtained, showing a dependence between the amount of biowax and the spreadability value, which means a high concentration makes the formulation less spreadable. This same behavior is shown by commercial waxes whose function in cosmetic creams is to act as emollients and rheological modifiers, such as carnauba wax or castor wax.

This result is congruent with the spreadability determined through the sensory analysis performed by the trained panelists. On the other hand, the facial cream spreads 1.2 times more on the skin than the BB cream; this result indicates a greater distribution of the emulsion per application area, which is desirable for this prototype.

3.6. Rheology

All tests showed a non-Newtonian and viscoplastic behavior, as shown in Figures 2 and 3, a desirable result in this type of formulations because the cream shows shear thinning, for example, when applied on the skin, being a desirable rheological property in cosmetic formulations because it improves the application and dispersion, providing a pleasant sensation [19]. Viscoplasticity is also a desired behavior since provides cosmetic products with consistency and stability.

Additionally, the results obtained in the mentioned studies on the relationship between a high wax concentration and the high viscosity values of the formulation are corroborated, as shown in Figure 4a,b, confirming the action of this raw material as a thickening agent.

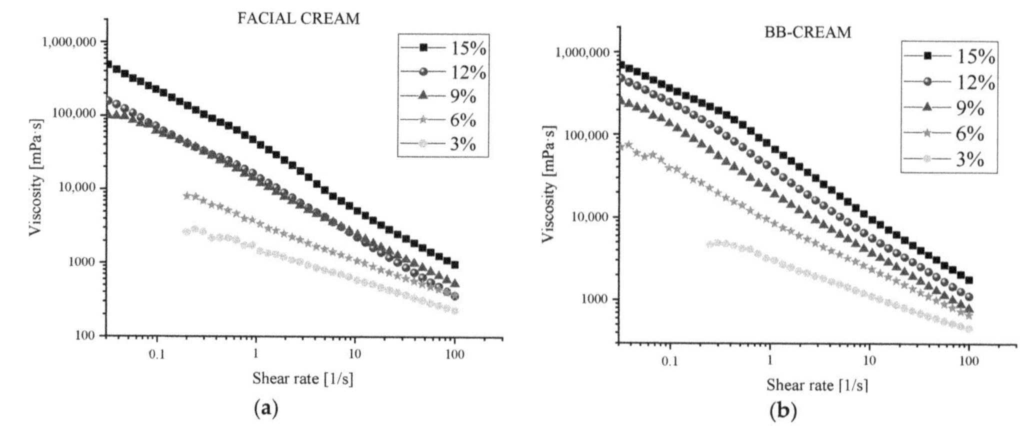

Figure 4. (**a**) Viscosity profile for BB cream after preliminary stability. (**b**) Viscosity profile for the facial cream after preliminary stability.

With respect to the viscoelastic modulus, it is observed that all formulations exhibit an elastic modulus (G′) higher than the viscous modulus (G″), as shown in Figure 5a,b. This is a key parameter of stable emulsions since it can be considered as a viscoplastic fluid. In other words, this fact indicates the viscoelastic character of the emulsions. As G′ > G″, the sample shows a solid-like structure and can be termed as a viscoelastic solid material. Likewise, after subjecting the formulations to thermal stress, the change in the modulus G′ for each prototype is not significant, as in the case of Moschini et al. [15], which means that the prototypes present stability in their rheological behavior.

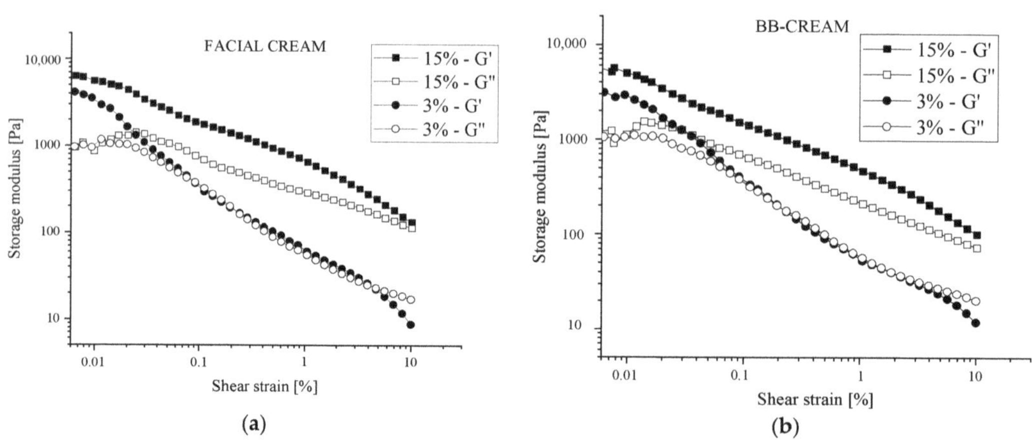

Figure 5. (**a**) Viscoelastic profile for BB cream after preliminary stability. (**b**) Viscoelastic profile for facial cream after preliminary stability.

3.7. Decision Matrix

The result for each prototype (calculated as described in the methodology) is shown in Table 7. Hence, biowax percentage which obtained the best score was selected, 9% for both. This is the suggested dosage for its sensory profile and stability. The formulations selected were compared sensorially with a commercial reference product (moisturizing cream and BB cream of Natura FACES). Figure 6 shows that the property that was the least close was gloss, being higher for the prototypes developed, which decreases the level of user acceptance, as facial creams usually leave a matte finish.

Table 7. Decision matrix results for sensory and stability analysis.

%Biowax	Result	
	Facial Cream	**BB Cream**
3%	201	285
6%	210	272
9%	233	287
12%	171	257
15%	185	252

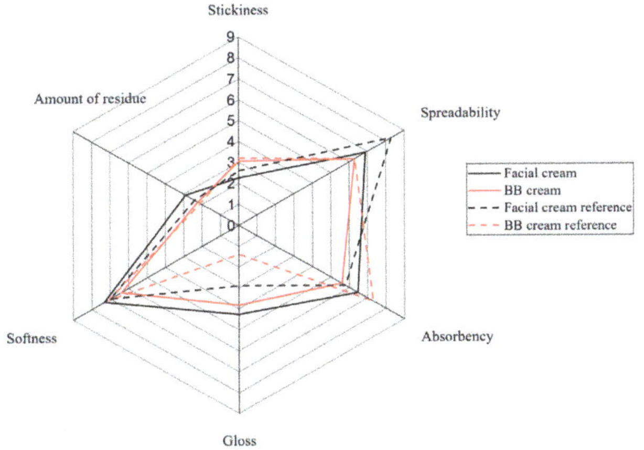

Figure 6. Sensory panel of facial cream, BB cream and two commercial creams.

4. Conclusions

The result of this work reveals that the biowax derived from hydroprocessing of crude palm oil presents high potential as a cosmetic ingredient, firstly as an emollient as it modifies the sensory properties of the formulations developed and secondly as a thickening agent, since the viscosity and the spreadability present a high dependence of biowax percentage used in the formulations studied.

Related to the stability, it is concluded that the formulations are able to maintain their pH. On the other hand, the cream shows phase separation in the prototypes with lower viscosity; however, the optimum value of biowax found where the separation decreases and avoids the formation of a supernatant is 9%. Furthermore, all tests showed a non-Newtonian and shear-thinning behavior, being a desirable rheological property in cosmetic formulation. In addition, the obtained emulsions can be considered as stable solid-like fluids since $G' > G''$.

Finally, the percentage that presented the best stability and sensory profile was 9% of biowax for both creams, which is expected to be easily accepted in the market because, compared to a reference product, its sensory properties are very similar.

Author Contributions: Conceptualization, L.A., J.H., L.J.L.-G. and R.M.; methodology, L.A., J.H., L.J.L.-G. and R.M.; formal analysis, L.A., J.H., L.J.L.-G. and R.M.; Investigation, L.A., J.H., L.J.L.-G. and R.M.; writing—original draft preparation L.A. and J.H.; writing—review and editing, L.A., J.H., L.J.L.-G. and R.M.; supervision, L.J.L.-G. and R.M.; project administration, L.J.L.-G. All authors have read and agreed to the published version of the manuscript.

Funding: This research was funded by Fondo de Ciencia, Tecnología e Innovación Sistema General de Regalías y Gobernación de Santander under the research project "Desarrollo de una tecnología para la producción de bioceras que fomente el biocomercio en el Departamento de Santander código BPIN 2018000100188", Universidad Industrial de Santander (UIS), Servicio Nacional de Aprendizaje (SENA) and Instituto Colombiano de Petróleo (ICP).

Institutional Review Board Statement: The study was conducted in accordance with the Declaration of Helsinki and approved by Ethics Committee of the Industrial University of Santander, Bucaramanga, Colombia, act N.2 from 21 February 2020.

Informed Consent Statement: Informed consent was obtained from all subjects involved in the study.

Data Availability Statement: The data presented in this study are available upon request from the corresponding author. Data are not publicly available due to privacy restrictions.

Acknowledgments: The authors would like to acknowledge Grupo de Investigación En Ciencia y Tecnología de Alimentos—CICTA, Sistema General de Regalías in the project "Desarrollo de una Tecnología para la Producción de Bioceras que Fomente el Biocomercio en el Departamento de Santander-BPIN 2018000100188". We also acknowledge the Instituto Colombiano del Petróleo (ICP), the Universidad Industrial de Santander (UIS) and the Servicio Nacional de Aprendizaje (SENA) and Colciencias in the agreement 907-2021 for their support in this research.

Conflicts of Interest: The authors declare no conflict of interest. The funders had no role in the design of the study; in the collection, analyses, or interpretation of data; in the writing of the manuscript; or in the decision to publish the results.

References

1. "La Investigación En La Cosmética Natural": Aprovechamiento Sostenible de Recursos Naturales. UPS. Available online: https://www.ups.edu.ec/noticias?articleId=5002908 (accessed on 21 January 2023).
2. Colombia: La Industria Cosmética un Sector Que Crece y Promete. Available online: https://www.legiscomex.com/Documentos/colombia-industria-cosmetica-sector-crece-promete (accessed on 21 January 2023).
3. Cluster de Cosméticos, Cámara de Comercio de Bogotá. Available online: https://www.ccb.org.co/Clusters/Cluster-de-Cosmeticos/Noticias/2018/Septiembre-2018/Cosmeticos-un-mercado-que-vale-en-Colombia-3.280-millones-de-dolares (accessed on 20 July 2022).
4. Oportunidades Para la Exportación de Aceite de Palma Colombiano. Available online: https://www.colombiatrade.com.co/noticias/exportacion-de-aceite-de-palma-en-Colombia (accessed on 21 January 2023).
5. Rugeles, L.; Ortiz, J.; Guaitero, B.; Huertas, D. *La Cadena de Valor de los Ingredientes Naturales del Bio-Comercio para las Industrias Farmacéutica, Alimentaria y Cosmética—FAC*; Universidad Jorge Tadeo Lozano: Bogotá, Colombia, 2011; ISBN 978-958-725-095-4.
6. Fei, T.; Wang, T. A Review of Recent Development of Sustainable Waxes Derived from Vegetable Oils. *Curr. Opin. Food Sci.* **2017**, *16*, 7–14. [CrossRef]
7. Isaza, N.; Laureano, M.; Villamizar, P.; Patricia, L. Proceso para la obtención de compuestos parafínicos sólidos por hidrotratamiento de aceites vegetales. PCT WO/2010/067164, 17 June 2010.
8. Olarte, G.; Garzón, L.; Sarmiento, J.; López-Giraldo, L.J.; Vivas-Báez, J.C. Biowax Production from the Hydrotreatment of Refined Palm Oil (RPO). *Processes* **2023**, *11*, 1372. [CrossRef]
9. Guzmán, M.; Kafarov, V.; Guzmán, A.; Garzón, L. Influence of temperature during crude palm oil hydrotreating over NiMo/-Al_2O_3 catalysts. *Rev. ION* **2013**, *26*, 7–14.
10. Guzman, A.; Torres, J.E.; Prada, L.P.; Nuñez, M.L. Hydroprocessing of crude palmoil at pilot plant scale. *Catal. Today* **2010**, *156*, 38–43. [CrossRef]
11. Chaparro, L.M.; Neira, L.F.; Molina, D.; Rivera-Barrera, D.; Castañeda, M.; López-Giraldo, L.J.; Escobar, P. Biowaxes from Palm Oil as Promising Candidates for Cosmetic Matrices and Pharmaceuticals for Human Use. *Materials* **2023**, *16*, 4402. [CrossRef] [PubMed]
12. Murillo-Méndez, C.; López-Giraldo, L.J.; Quintero, A.F.R.; Castañeda-Rodas, M. Planteamiento de un modelo matemático de características macroscópicas de bioceras producidas del aceite de palma con interés comercial. *Rev. ION* **2022**, *35*, 59–69. [CrossRef]
13. Comparatively Speaking: O/W, W/O, Micro, Pickering and Suspo Emulsions. Available online: https://www.cosmeticsandtoiletries.com/research/literature-data/article/21836048/comparatively-speaking-o-w-w-o-micro-pickering-and-suspo-emulsions (accessed on 21 January 2023).
14. Onudi Guía de Estabilidad Final—Estudios de Estabilidad de Productos Cosméticos—Studocu. Available online: https://www.studocu.com/co/document/universidad-icesi/analisis-quimico/onudi-guia-de-estabilidad-final-003/37785513 (accessed on 21 January 2023).
15. Daudt, R.M.; Back, P.I.; Cardozo, N.S.M.; Marczak, L.D.F.; Külkamp-Guerreiro, I.C. Pinhão Starch and Coat Extract as New Natural Cosmetic Ingredients: Topical Formulation Stability and Sensory Analysis. *Carbohydr. Polym.* **2015**, *134*, 573–580. [CrossRef] [PubMed]
16. Standard Guide for Two Sensory Descriptive Analysis Approaches for Skin Creams and Lotions. Available online: https://www.astm.org/e1490-19.html (accessed on 20 January 2023).
17. Medina Godoy, D.G. *Planteamiento de Estrategias para el Aprovechamiento Integral del Fruto de Mangostan a Partir de su Caracterizacion Fisicoquimica*; UIS: Bucaramanga, Colombia, 2017.

18. Wortel, V.A.L.; Wiechers, J.W. Skin sensory performance of individual personal care ingredients and marketed personal care products. *Food Qual. Prefer.* **2000**, *11*, 121–127. [CrossRef]
19. Calixto, L.S.; Infante, V.H.P.; Maia Campos, P.M.B.G. Design and Characterization of Topical Formulations: Correlations Between Instrumental and Sensorial Measurements. *AAPS PharmSciTech* **2018**, *19*, 1512–1519. [CrossRef] [PubMed]
20. Azmi, N.; Mat Radzi, S.; Rehan, M.; Mohd Amin, N.A. A Review on Cosmetic Formulations and Physicochemical Characteristics of Emollient and Day Cream Using Vegetable Based-Wax Ester. *Malays. J. Sci. Health Technol.* **2022**, *8*, 38–45. [CrossRef]
21. Parente, M.E.; Gámbaro, A.; Ares, G. Sensory Characterization of Emollients. *J. Sens. Stud.* **2008**, *23*, 149–161. [CrossRef]
22. Keng, P.S.; Basri, M.; Zakaria, M.R.S.; Rahman, M.B.A.; Ariff, A.B.; Rahman, R.N.Z.A.; Salleh, A.B. Newly Synthesized Palm Esters for Cosmetics Industry. *Ind. Crops Prod.* **2009**, *29*, 37–44. [CrossRef]
23. Beri, A.; Norton, J.E.; Norton, I.T. Effect of emulsifier type and concentration, aqueous phase volume and wax ratio on physical, material and mechanical properties of water in oil lipsticks. *Int. J. Cosmet. Sci.* **2013**, *35*, 613–621. [CrossRef] [PubMed]
24. Ghosh, S.; Rousseau, D. Fat Crystals and Water-in-Oil Emulsion Stability. *Curr. Opin. Colloid Interface Sci.* **2011**, *16*, 421–431. [CrossRef]
25. Kartini, K.; Winarjo, B.; Fitriani, E.; Islamie, R. Formulation and PH-Physical Stability Evaluation of Gel and Cream of Plantago Major Leaves Extract. *Media Pharm. Indones. MPI* **2017**, *1*, 174. [CrossRef]
26. Apriani, E.; Nurleni, N.; Nurfitria, H.; Iskandarsyah, I. Stability Testing of Azelaic Acid Cream Based Ethosome. *Asian J. Pharm. Clin. Res.* **2018**, *11*, 270. [CrossRef]

Disclaimer/Publisher's Note: The statements, opinions and data contained in all publications are solely those of the individual author(s) and contributor(s) and not of MDPI and/or the editor(s). MDPI and/or the editor(s) disclaim responsibility for any injury to people or property resulting from any ideas, methods, instructions or products referred to in the content.

MDPI AG
Grosspeteranlage 5
4052 Basel
Switzerland
Tel.: +41 61 683 77 34

Cosmetics Editorial Office
E-mail: cosmetics@mdpi.com
www.mdpi.com/journal/cosmetics

Disclaimer/Publisher's Note: The statements, opinions and data contained in all publications are solely those of the individual author(s) and contributor(s) and not of MDPI and/or the editor(s). MDPI and/or the editor(s) disclaim responsibility for any injury to people or property resulting from any ideas, methods, instructions or products referred to in the content.

www.ingramcontent.com/pod-product-compliance
Lightning Source LLC
Chambersburg PA
CBHW061618210926
51850CB00009B/2368
* 9 7 8 3 7 2 5 8 1 5 8 5 2 *